Data Mining

Theories, Algorithms, and Examples

Human Factors and Ergonomics Series

PUBLISHED TITLES

Conceptual Foundations of Human Factors Measurement
D. Meister

Content Preparation Guidelines for the Web and Information Appliances:
Cross-Cultural Comparisons
H. Liao, Y. Guo, A. Savoy, and G. Salvendy

Cross-Cultural Design for IT Products and Services
P. Rau, T. Plocher and Y. Choong

Data Mining: Theories, Algorithms, and Examples
Nong Ye

Designing for Accessibility: A Business Guide to Countering Design Exclusion
S. Keates

Handbook of Cognitive Task Design
E. Hollnagel

The Handbook of Data Mining
N. Ye

Handbook of Digital Human Modeling: Research for Applied Ergonomics
and Human Factors Engineering
V. G. Duffy

Handbook of Human Factors and Ergonomics in Health Care and Patient Safety
Second Edition
P. Carayon

Handbook of Human Factors in Web Design, Second Edition
K. Vu and R. Proctor

Handbook of Occupational Safety and Health
D. Koradecka

Handbook of Standards and Guidelines in Ergonomics and Human Factors
W. Karwowski

Handbook of Virtual Environments: Design, Implementation, and Applications
K. Stanney

Handbook of Warnings
M. Wogalter

Human–Computer Interaction: Designing for Diverse Users and Domains
A. Sears and J. A. Jacko

Human–Computer Interaction: Design Issues, Solutions, and Applications
A. Sears and J. A. Jacko

Human–Computer Interaction: Development Process
A. Sears and J. A. Jacko

Human–Computer Interaction: Fundamentals
A. Sears and J. A. Jacko

The Human–Computer Interaction Handbook: Fundamentals
Evolving Technologies, and Emerging Applications, Third Edition
A. Sears and J. A. Jacko

Human Factors in System Design, Development, and Testing
D. Meister and T. Enderwick

PUBLISHED TITLES (CONTINUED)

Introduction to Human Factors and Ergonomics for Engineers, Second Edition
M. R. Lehto

Macroergonomics: Theory, Methods and Applications
H. Hendrick and B. Kleiner

Practical Speech User Interface Design
James R. Lewis

The Science of Footwear
R. S. Goonetilleke

Skill Training in Multimodal Virtual Environments
M. Bergamsco, B. Bardy, and D. Gopher

Smart Clothing: Technology and Applications
Gilsoo Cho

Theories and Practice in Interaction Design
S. Bagnara and G. Crampton-Smith

The Universal Access Handbook
C. Stephanidis

Usability and Internationalization of Information Technology
N. Aykin

User Interfaces for All: Concepts, Methods, and Tools
C. Stephanidis

FORTHCOMING TITLES

Around the Patient Bed: Human Factors and Safety in Health care
Y. Donchin and D. Gopher

Cognitive Neuroscience of Human Systems Work and Everyday Life
C. Forsythe and H. Liao

Computer-Aided Anthropometry for Research and Design
K. M. Robinette

Handbook of Human Factors in Air Transportation Systems
S. Landry

Handbook of Virtual Environments: Design, Implementation and Applications, Second Edition,
K. S. Hale and K M. Stanney

Variability in Human Performance
T. Smith, R. Henning, and M. Wade

Data Mining

Theories, Algorithms, and Examples

NONG YE

CRC Press
Taylor & Francis Group
Boca Raton London New York

CRC Press is an imprint of the
Taylor & Francis Group, an **Informa** business

MATLAB® is a trademark of The MathWorks, Inc. and is used with permission. The MathWorks does not warrant the accuracy of the text or exercises in this book. This book's use or discussion of MATLAB® software or related products does not constitute endorsement or sponsorship by The MathWorks of a particular pedagogical approach or particular use of the MATLAB® software.

CRC Press
Taylor & Francis Group
6000 Broken Sound Parkway NW, Suite 300
Boca Raton, FL 33487-2742

First issued in paperback 2017

© 2014 by Taylor & Francis Group, LLC
CRC Press is an imprint of Taylor & Francis Group, an Informa business

No claim to original U.S. Government works

Version Date: 20140220

ISBN 13: 978-1-4398-0838-2 (hbk)
ISBN 13: 978-1-138-07366-1 (pbk)

This book contains information obtained from authentic and highly regarded sources. Reasonable efforts have been made to publish reliable data and information, but the author and publisher cannot assume responsibility for the validity of all materials or the consequences of their use. The authors and publishers have attempted to trace the copyright holders of all material reproduced in this publication and apologize to copyright holders if permission to publish in this form has not been obtained. If any copyright material has not been acknowledged please write and let us know so we may rectify in any future reprint.

Except as permitted under U.S. Copyright Law, no part of this book may be reprinted, reproduced, transmitted, or utilized in any form by any electronic, mechanical, or other means, now known or hereafter invented, including photocopying, microfilming, and recording, or in any information storage or retrieval system, without written permission from the publishers.

For permission to photocopy or use material electronically from this work, please access www.copyright.com (http://www.copyright.com/) or contact the Copyright Clearance Center, Inc. (CCC), 222 Rosewood Drive, Danvers, MA 01923, 978-750-8400. CCC is a not-for-profit organization that provides licenses and registration for a variety of users. For organizations that have been granted a photocopy license by the CCC, a separate system of payment has been arranged.

Trademark Notice: Product or corporate names may be trademarks or registered trademarks, and are used only for identification and explanation without intent to infringe.

Library of Congress Cataloging-in-Publication Data

Ye, Nong, 1964-
  Data mining : theories, algorithms, and examples / Nong Ye.
    pages cm. -- (Human Factors and Ergonomics Series)
  "A CRC title."
  Includes bibliographical references and index.
  ISBN 978-1-4398-0838-2 (hardcover : alk. paper) 1. Data mining. 2. Data mining--Mathematical models. I. Title.

QA76.9.D343Y4 2014
006.3'12--dc23   2013013338

Visit the Taylor & Francis Web site at
http://www.taylorandfrancis.com

and the CRC Press Web site at
http://www.crcpress.com

Contents

Preface xiii
Acknowledgments xvii
Author xix

Part I An Overview of Data Mining

1. Introduction to Data, Data Patterns, and Data Mining 3
   1.1 Examples of Small Data Sets 3
   1.2 Types of Data Variables 5
      1.2.1 Attribute Variable versus Target Variable 5
      1.2.2 Categorical Variable versus Numeric Variable 8
   1.3 Data Patterns Learned through Data Mining 9
      1.3.1 Classification and Prediction Patterns 9
      1.3.2 Cluster and Association Patterns 12
      1.3.3 Data Reduction Patterns 13
      1.3.4 Outlier and Anomaly Patterns 14
      1.3.5 Sequential and Temporal Patterns 15
   1.4 Training Data and Test Data 17
   Exercises 17

Part II Algorithms for Mining Classification and Prediction Patterns

2. Linear and Nonlinear Regression Models 21
   2.1 Linear Regression Models 21
   2.2 Least-Squares Method and Maximum Likelihood Method
       of Parameter Estimation 23
   2.3 Nonlinear Regression Models and Parameter Estimation 28
   2.4 Software and Applications 29
   Exercises 29

3. Naïve Bayes Classifier 31
   3.1 Bayes Theorem 31
   3.2 Classification Based on the Bayes Theorem and Naïve Bayes
       Classifier 31
   3.3 Software and Applications 35
   Exercises 36

4. Decision and Regression Trees 37
4.1 Learning a Binary Decision Tree and Classifying Data Using a Decision Tree 37
4.1.1 Elements of a Decision Tree 37
4.1.2 Decision Tree with the Minimum Description Length 39
4.1.3 Split Selection Methods 40
4.1.4 Algorithm for the Top-Down Construction of a Decision Tree 44
4.1.5 Classifying Data Using a Decision Tree 49
4.2 Learning a Nonbinary Decision Tree 51
4.3 Handling Numeric and Missing Values of Attribute Variables 56
4.4 Handling a Numeric Target Variable and Constructing a Regression Tree 57
4.5 Advantages and Shortcomings of the Decision Tree Algorithm 59
4.6 Software and Applications 61
Exercises 62

5. Artificial Neural Networks for Classification and Prediction 63
5.1 Processing Units of ANNs 63
5.2 Architectures of ANNs 69
5.3 Methods of Determining Connection Weights for a Perceptron 71
5.3.1 Perceptron 72
5.3.2 Properties of a Processing Unit 72
5.3.3 Graphical Method of Determining Connection Weights and Biases 73
5.3.4 Learning Method of Determining Connection Weights and Biases 76
5.3.5 Limitation of a Perceptron 79
5.4 Back-Propagation Learning Method for a Multilayer Feedforward ANN 80
5.5 Empirical Selection of an ANN Architecture for a Good Fit to Data 86
5.6 Software and Applications 88
Exercises 88

6. Support Vector Machines 91
6.1 Theoretical Foundation for Formulating and Solving an Optimization Problem to Learn a Classification Function 91
6.2 SVM Formulation for a Linear Classifier and a Linearly Separable Problem 93
6.3 Geometric Interpretation of the SVM Formulation for the Linear Classifier 96
6.4 Solution of the Quadratic Programming Problem for a Linear Classifier 98

6.5	SVM Formulation for a Linear Classifier and a Nonlinearly Separable Problem	105
6.6	SVM Formulation for a Nonlinear Classifier and a Nonlinearly Separable Problem	108
6.7	Methods of Using SVM for Multi-Class Classification Problems	113
6.8	Comparison of ANN and SVM	113
6.9	Software and Applications	114
	Exercises	114

7. k-Nearest Neighbor Classifier and Supervised Clustering ... 117
7.1	k-Nearest Neighbor Classifier	117
7.2	Supervised Clustering	122
7.3	Software and Applications	136
	Exercises	136

Part III Algorithms for Mining Cluster and Association Patterns

8. Hierarchical Clustering ... 141
8.1	Procedure of Agglomerative Hierarchical Clustering	141
8.2	Methods of Determining the Distance between Two Clusters	141
8.3	Illustration of the Hierarchical Clustering Procedure	146
8.4	Nonmonotonic Tree of Hierarchical Clustering	150
8.5	Software and Applications	152
	Exercises	152

9. K-Means Clustering and Density-Based Clustering ... 153
9.1	K-Means Clustering	153
9.2	Density-Based Clustering	165
9.3	Software and Applications	165
	Exercises	166

10. Self-Organizing Map ... 167
10.1	Algorithm of Self-Organizing Map	167
10.2	Software and Applications	175
	Exercises	175

11. Probability Distributions of Univariate Data ... 177
11.1	Probability Distribution of Univariate Data and Probability Distribution Characteristics of Various Data Patterns	177
11.2	Method of Distinguishing Four Probability Distributions	182
11.3	Software and Applications	183
	Exercises	184

12. Association Rules 185
12.1 Definition of Association Rules and Measures of Association 185
12.2 Association Rule Discovery 189
12.3 Software and Applications 194
Exercises 194

13. Bayesian Network 197
13.1 Structure of a Bayesian Network and Probability Distributions of Variables 197
13.2 Probabilistic Inference 205
13.3 Learning of a Bayesian Network 210
13.4 Software and Applications 213
Exercises 213

Part IV Algorithms for Mining Data Reduction Patterns

14. Principal Component Analysis 217
14.1 Review of Multivariate Statistics 217
14.2 Review of Matrix Algebra 220
14.3 Principal Component Analysis 228
14.4 Software and Applications 230
Exercises 231

15. Multidimensional Scaling 233
15.1 Algorithm of MDS 233
15.2 Number of Dimensions 246
15.3 INDSCALE for Weighted MDS 247
15.4 Software and Applications 248
Exercises 248

Part V Algorithms for Mining Outlier and Anomaly Patterns

16. Univariate Control Charts 251
16.1 Shewhart Control Charts 251
16.2 CUSUM Control Charts 254
16.3 EWMA Control Charts 257
16.4 Cuscore Control Charts 261
16.5 Receiver Operating Curve (ROC) for Evaluation and Comparison of Control Charts 265
16.6 Software and Applications 267
Exercises 267

17. Multivariate Control Charts 269
17.1 Hotelling's T^2 Control Charts 269
17.2 Multivariate EWMA Control Charts 272
17.3 Chi-Square Control Charts 272
17.4 Applications 274
Exercises 274

Part VI Algorithms for Mining Sequential and Temporal Patterns

18. Autocorrelation and Time Series Analysis 277
18.1 Autocorrelation 277
18.2 Stationarity and Nonstationarity 278
18.3 ARMA Models of Stationary Series Data 279
18.4 ACF and PACF Characteristics of ARMA Models 281
18.5 Transformations of Nonstationary Series Data and ARIMA Models 283
18.6 Software and Applications 284
Exercises 285

19. Markov Chain Models and Hidden Markov Models 287
19.1 Markov Chain Models 287
19.2 Hidden Markov Models 290
19.3 Learning Hidden Markov Models 294
19.4 Software and Applications 305
Exercises 305

20. Wavelet Analysis 307
20.1 Definition of Wavelet 307
20.2 Wavelet Transform of Time Series Data 309
20.3 Reconstruction of Time Series Data from Wavelet Coefficients 316
20.4 Software and Applications 317
Exercises 318

References 319

Index 323

Preface

Technologies have enabled us to collect massive amounts of data in many fields. Our pace of discovering useful information and knowledge from these data falls far behind our pace of collecting the data. Conversion of massive data into useful information and knowledge involves two steps: (1) mining patterns present in the data and (2) interpreting those data patterns in their problem domains to turn them into useful information and knowledge. There exist many data mining algorithms to automate the first step of mining various types of data patterns from massive data. Interpretation of data patterns usually depend on specific domain knowledge and analytical thinking. This book covers data mining algorithms that can be used to mine various types of data patterns. Learning and applying data mining algorithms will enable us to automate and thus speed up the first step of uncovering data patterns from massive data. Understanding how data patterns are uncovered by data mining algorithms is also crucial to carrying out the second step of looking into the meaning of data patterns in problem domains and turning data patterns into useful information and knowledge.

Overview of the Book

The data mining algorithms in this book are organized into five parts for mining five types of data patterns from massive data, as follows:

1. Classification and prediction patterns
2. Cluster and association patterns
3. Data reduction patterns
4. Outlier and anomaly patterns
5. Sequential and temporal patterns

Part I introduces these types of data patterns with examples. Parts II–VI describe algorithms to mine the five types of data patterns, respectively.

Classification and prediction patterns capture relations of attribute variables with target variables and allow us to classify or predict values of target variables from values of attribute variables. Part II describes the following algorithms to mine classification and prediction patterns:

- Linear and nonlinear regression models (Chapter 2)
- Naïve Bayes classifier (Chapter 3)
- Decision and regression trees (Chapter 4)
- Artificial neural networks for classification and prediction (Chapter 5)
- Support vector machines (Chapter 6)
- K-nearest neighbor classifier and supervised clustering (Chapter 7)

Part III describes data mining algorithms to uncover cluster and association patterns. Cluster patterns reveal patterns of similarities and differences among data records. Association patterns are established based on co-occurrences of items in data records. Part III describes the following data mining algorithms to mine cluster and association patterns:

- Hierarchical clustering (Chapter 8)
- K-means clustering and density-based clustering (Chapter 9)
- Self-organizing map (Chapter 10)
- Probability distributions of univariate data (Chapter 11)
- Association rules (Chapter 12)
- Bayesian networks (Chapter 13)

Data reduction patterns look for a small number of variables that can be used to represent a data set with a much larger number of variables. Since one variable gives one dimension of data, data reduction patterns allow a data set in a high-dimensional space to be represented in a low-dimensional space. Part IV describes the following data mining algorithms to mine data reduction patterns:

- Principal component analysis (Chapter 14)
- Multidimensional scaling (Chapter 15)

Outliers and anomalies are data points that differ largely from a normal profile of data, and there are many ways to define and establish a norm profile of data. Part V describes the following data mining algorithms to detect and identify outliers and anomalies:

- Univariate control charts (Chapter 16)
- Multivariate control charts (Chapter 17)

Preface   xv

Sequential and temporal patterns reveal how data change their patterns over time. Part VI describes the following data mining algorithms to mine sequential and temporal patterns:

- Autocorrelation and time series analysis (Chapter 18)
- Markov chain models and hidden Markov models (Chapter 19)
- Wavelet analysis (Chapter 20)

Distinctive Features of the Book

As stated earlier, mining data patterns from massive data is only the first step of turning massive data into useful information and knowledge in problem domains. Data patterns need to be understood and interpreted in their problem domain in order to be useful. To apply a data mining algorithm and acquire the ability of understanding and interpreting data patterns produced by that data mining algorithm, we need to understand two important aspects of the algorithm:

1. Theoretical concepts that establish the rationale of why elements of the data mining algorithm are put together in a specific way to mine a particular type of data pattern
2. Operational steps and details of how the data mining algorithm processes massive data to produce data patterns.

This book aims at providing both theoretical concepts and operational details of data mining algorithms in each chapter in a self-contained, complete manner with small data examples. It will enable readers to understand theoretical and operational aspects of data mining algorithms and to manually execute the algorithms for a thorough understanding of the data patterns produced by them.

This book covers data mining algorithms that are commonly found in the data mining literature (e.g., decision trees artificial neural networks and hierarchical clustering) and data mining algorithms that are usually considered difficult to understand (e.g., hidden Markov models, multidimensional scaling, support vector machines, and wavelet analysis). All the data mining algorithms in this book are described in the self-contained, example-supported, complete manner. Hence, this book will enable readers to achieve the same level of thorough understanding and will provide the same ability of manual execution regardless of the difficulty level of the data mining algorithms.

For the data mining algorithms in each chapter, a list of software packages that support them is provided. Some applications of the data mining algorithms are also given with references.

Teaching Support

The data mining algorithms covered in this book involve different levels of difficulty. The instructor who uses this book as the textbook for a course on data mining may select the book materials to cover in the course based on the level of the course and the level of difficulty of the book materials. The book materials in Chapters 1, 2 (Sections 2.1 and 2.2 only), 3, 4, 7, 8, 9 (Section 9.1 only), 12, 16 (Sections 16.1 through 16.3 only), and 19 (Section 19.1 only), which cover the five types of data patterns, are appropriate for an undergraduate-level course. The remainder is appropriate for a graduate-level course.

Exercises are provided at the end of each chapter. The following additional teaching support materials are available on the book website and can be obtained from the publisher:

- Solutions manual
- Lecture slides, which include the outline of topics, figures, tables, and equations

MATLAB® is a registered trademark of The MathWorks, Inc. For product information, please contact:

The MathWorks, Inc.
3 Apple Hill Drive
Natick, MA 01760-2098 USA
Tel: 508-647-7000
Fax: 508-647-7001
E-mail: info@mathworks.com
Web: www.mathworks.com

Acknowledgments

I would like to thank my family, Baijun and Alice, for their love, understanding, and unconditional support. I appreciate them for always being there for me and making me happy.

I am grateful to Dr. Gavriel Salvendy, who has been my mentor and friend, for guiding me in my academic career. I am also thankful to Dr. Gary Hogg, who supported me in many ways as the department chair at Arizona State University.

I would like to thank Cindy Carelli, senior editor at CRC Press. This book would not have been possible without her responsive, helpful, understanding, and supportive nature. It has been a great pleasure working with her. Thanks also go to Kari Budyk, senior project coordinator at CRC Press, and the staff at CRC Press who helped publish this book.

Author

Nong Ye is a professor at the School of Computing, Informatics, and Decision Systems Engineering, Arizona State University, Tempe, Arizona. She holds a PhD in industrial engineering from Purdue University, West Lafayette, Indiana, an MS in computer science from the Chinese Academy of Sciences, Beijing, People's Republic of China, and a BS in computer science from Peking University, Beijing, People's Republic of China.

Her publications include *The Handbook of Data Mining* and *Secure Computer and Network Systems: Modeling, Analysis and Design*. She has also published over 80 journal papers in the fields of data mining, statistical data analysis and modeling, computer and network security, quality of service optimization, quality control, human–computer interaction, and human factors.

Part I

An Overview of Data Mining

1

Introduction to Data, Data Patterns, and Data Mining

Data mining aims at discovering useful data patterns from massive amounts of data. In this chapter, we give some examples of data sets and use these data sets to illustrate various types of data variables and data patterns that can be discovered from data. Data mining algorithms to discover each type of data patterns are briefly introduced in this chapter. The concepts of training and testing data are also introduced.

1.1 Examples of Small Data Sets

Advanced technologies such as computers and sensors have enabled many activities to be recorded and stored over time, producing massive amounts of data in many fields. In this section, we introduce some examples of small data sets that are used throughout the book to explain data mining concepts and algorithms.

Tables 1.1 through 1.3 give three examples of small data sets from the UCI Machine Learning Repository (Frank and Asuncion, 2010). The balloons data set in Table 1.1 contains data records for 16 instances of balloons. Each balloon has four attributes: Color, Size, Act, and Age. These attributes of the balloon determine whether or not the balloon is inflated. The space shuttle O-ring erosion data set in Table 1.2 contains data records for 23 instances of the *Challenger* space shuttle flights. There are four attributes for each flight: Number of O-rings, Launch Temperature (°F), Leak-Check Pressure (psi), and Temporal Order of Flight, which can be used to determine Number of O-rings with Stress. The lenses data set in Table 1.3 contains data records for 24 instances for the fit of lenses to a patient. There are four attributes of a patient for each instance: Age, Prescription, Astigmatic, and Tear Production Rate, which can be used to determine the type of lenses to be fitted to a patient.

Table 1.4 gives the data set for fault detection and diagnosis of a manufacturing system (Ye et al., 1993). The manufacturing system consists of nine machines, M1, M2, ..., M9, which process parts. Figure 1.1 shows the production flows of parts to go through the nine machines. There are some parts

TABLE 1.1

Balloon Data Set

Instance	Attribute Variables				Target Variable
	Color	Size	Act	Age	Inflated
1	Yellow	Small	Stretch	Adult	T
2	Yellow	Small	Stretch	Child	T
3	Yellow	Small	Dip	Adult	T
4	Yellow	Small	Dip	Child	T
5	Yellow	Large	Stretch	Adult	T
6	Yellow	Large	Stretch	Child	F
7	Yellow	Large	Dip	Adult	F
8	Yellow	Large	Dip	Child	F
9	Purple	Small	Stretch	Adult	T
10	Purple	Small	Stretch	Child	F
11	Purple	Small	Dip	Adult	F
12	Purple	Small	Dip	Child	F
13	Purple	Large	Stretch	Adult	T
14	Purple	Large	Stretch	Child	F
15	Purple	Large	Dip	Adult	F
16	Purple	Large	Dip	Child	F

that go through M1 first, M5 second, and M9 last, some parts that go through M1 first, M5 second, and M7 last, and so on. There are nine variables, x_i, $i = 1, 2, ..., 9$, representing the quality of parts after they go through the nine machines. If parts after machine i pass the quality inspection, x_i takes the value of 0; otherwise, x_i takes the value of 1. There is a variable, y, representing whether or not the system has a fault. The system has a fault if any of the nine machines is faulty. If the system does not have a fault, y takes the value of 0; otherwise, y takes the value of 1. There are nine variables, y_i, $i = 1, 2, ..., 9$, representing whether or not nine machines are faulty, respectively. If machine i does not have a fault, y_i takes the value of 0; otherwise, y_i takes the value of 1. The fault detection problem is to determine whether or not the system has a fault based on the quality information. The fault detection problem involves the nine quality variables, x_i, $i = 1, 2, ..., 9$, and the system fault variable, y. The fault diagnosis problem is to determine which machine has a fault based on the quality information. The fault diagnosis problem involves the nine quality variables, x_i, $i = 1, 2, ..., 9$, and the nine variables of machine fault, y_i, $i = 1, 2, ..., 9$. There may be one or more machines that have a fault at the same time, or no faulty machine. For example, in instance 1 with M1 being faulty (y_1 and y taking the value of 1 and $y_2, y_3, y_4, y_5, y_6, y_7, y_8,$ and y_9 taking the value of 0), parts after M1, M5, M7, M9 fails the quality inspection with $x_1, x_5, x_7,$ and x_9 taking the value of 1 and other quality variables, $x_2, x_3, x_4, x_6,$ and $x_8,$ taking the value of 0.

TABLE 1.2
Space Shuttle O-Ring Data Set

	Attribute Variables				Target Variable
Instance	Number of O-Rings	Launch Temperature	Leak-Check Pressure	Temporal Order of Flight	Number of O-Rings with Stress
1	6	66	50	1	0
2	6	70	50	2	1
3	6	69	50	3	0
4	6	68	50	4	0
5	6	67	50	5	0
6	6	72	50	6	0
7	6	73	100	7	0
8	6	70	100	8	0
9	6	57	200	9	1
10	6	63	200	10	1
11	6	70	200	11	1
12	6	78	200	12	0
13	6	67	200	13	0
14	6	53	200	14	2
15	6	67	200	15	0
16	6	75	200	16	0
17	6	70	200	17	0
18	6	81	200	18	0
19	6	76	200	19	0
20	6	79	200	20	0
21	6	75	200	21	0
22	6	76	200	22	0
23	6	58	200	23	1

1.2 Types of Data Variables

The types of data variables affect what data mining algorithms can be applied to a given data set. This section introduces the different types of data variables.

1.2.1 Attribute Variable versus Target Variable

A data set may have attribute variables and target variable(s). The values of the attribute variables are used to determine the values of the target variable(s). Attribute variables and target variables may also be called as independent variables and dependent variables, respectively, to reflect that the values of

TABLE 1.3

Lenses Data Set

Instance	Attributes				Target
	Age	Spectacle Prescription	Astigmatic	Tear Production Rate	Lenses
1	Young	Myope	No	Reduced	Noncontact
2	Young	Myope	No	Normal	Soft contact
3	Young	Myope	Yes	Reduced	Noncontact
4	Young	Myope	Yes	Normal	Hard contact
5	Young	Hypermetrope	No	Reduced	Noncontact
6	Young	Hypermetrope	No	Normal	Soft contact
7	Young	Hypermetrope	Yes	Reduced	Noncontact
8	Young	Hypermetrope	Yes	Normal	Hard contact
9	Pre-presbyopic	Myope	No	Reduced	Noncontact
10	Pre-presbyopic	Myope	No	Normal	Soft contact
11	Pre-presbyopic	Myope	Yes	Reduced	Noncontact
12	Pre-presbyopic	Myope	Yes	Normal	Hard contact
13	Pre-presbyopic	Hypermetrope	No	Reduced	Noncontact
14	Pre-presbyopic	Hypermetrope	No	Normal	Soft contact
15	Pre-presbyopic	Hypermetrope	Yes	Reduced	Noncontact
16	Pre-presbyopic	Hypermetrope	Yes	Normal	Noncontact
17	Presbyopic	Myope	No	Reduced	Noncontact
18	Presbyopic	Myope	No	Normal	Noncontact
19	Presbyopic	Myope	Yes	Reduced	Noncontact
20	Presbyopic	Myope	Yes	Normal	Hard contact
21	Presbyopic	Hypermetrope	No	Reduced	Noncontact
22	Presbyopic	Hypermetrope	No	Normal	Soft contact
23	Presbyopic	Hypermetrope	Yes	Reduced	Noncontact
24	Presbyopic	Hypermetrope	Yes	Normal	Noncontact

the target variables depend on the values of the attribute variables. In the balloon data set in Table 1.1, the attribute variables are Color, Size, Act, and Age, and the target variable gives the inflation status of the balloon. In the space shuttle data set in Table 1.2, the attribute variables are Number of O-rings, Launch Temperature, Leak-Check Pressure, and Temporal Order of Flight, and the target variable is the Number of O-rings with Stress.

Some data sets may have only attribute variables. For example, customer purchase transaction data may contain the items purchased by each customer at a store. We have attribute variables representing the items purchased. The interest in the customer purchase transaction data is in finding out what items are often purchased together by customers. Such association patterns of items or attribute variables can be used to design the store layout for sale of items and assist customer shopping. Mining such a data set involves only attribute variables.

TABLE 1.4
Data Set for a Manufacturing System to Detect and Diagnose Faults

Instance (Faulty Machine)	Attribute Variables									Target Variables										
	Quality of Parts									System Fault, y	Machine Fault								
	x_1	x_2	x_3	x_4	x_5	x_6	x_7	x_8	x_9		y_1	y_2	y_3	y_4	y_5	y_6	y_7	y_8	y_9
1 (M1)	1	0	0	0	1	0	1	0	1	1	1	0	0	0	0	0	0	0	0
2 (M2)	0	1	0	1	0	0	0	1	0	1	0	1	0	0	0	0	0	0	0
3 (M3)	0	0	1	1	0	1	1	1	0	1	0	0	1	0	0	0	0	0	0
4 (M4)	0	0	0	1	0	0	1	1	0	1	0	0	0	1	0	0	0	0	0
5 (M5)	0	0	0	0	1	0	1	0	1	1	0	0	0	0	1	0	0	0	0
6 (M6)	0	0	0	0	0	1	1	0	0	1	0	0	0	0	0	1	0	0	0
7 (M7)	0	0	0	0	0	0	1	1	0	1	0	0	0	0	0	0	1	0	0
8 (M8)	0	0	0	0	0	0	0	1	0	1	0	0	0	0	0	0	0	1	0
9 (M9)	0	0	0	0	0	0	0	0	1	1	0	0	0	0	0	0	0	0	1
10 (none)	0	0	0	0	0	0	0	0	0	0	0	0	0	0	0	0	0	0	0

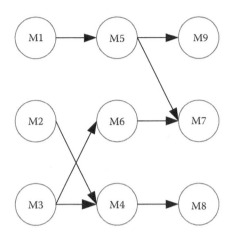

FIGURE 1.1
A manufacturing system with nine machines and production flows of parts.

1.2.2 Categorical Variable versus Numeric Variable

A variable can take categorical or numeric values. All the attribute variables and the target variable in the balloon data set take categorical values. For example, two values of the Color attribute, yellow and purple, give two different categories of Color. All the attribute variables and the target variable in the space shuttle O-ring data set take numeric values. For example, the values of the target variable, 0, 1, and 2, give the quantity of O-rings with Stress. The values of a numeric variable can be used to measure the quantitative magnitude of differences between numeric values. For example, the value of 2 O-rings is 1 unit larger than 1 O-ring and 2 units larger than 0 O-rings. However, the quantitative magnitude of differences cannot be obtained from the values of a categorical variable. For example, although yellow and purple show us a difference in the two colors, it is inappropriate to assign a quantitative measure of the difference. For another example, child and adult are two different categories of Age. Although each person has his/her years of age, we cannot state from child and adult categories in the balloon data set that an instance of child is 20, 30, or 40 years younger than an instance of adult.

Categorical variables have two subtypes: *nominal variables* and *ordinal variables* (Tan et al., 2006). The values of an ordinal variable can be sorted in order, whereas the values of nominal variables can be viewed only as same or different. For example, three values of Age (child, adult, and senior) make Age an ordinal variable since we can sort child, adult, and senior in this order of increasing age. However, we cannot state that the age difference between child and adult is bigger or smaller than the age difference between adult and senior since child, adult, and senior are categorical values instead of numeric values. That is, although the values of an ordinal variable can be sorted, those values are categorical and their quantitative differences are not available. Color is a nominal variable since yellow and purple show two different colors but an order of yellow and purple may be meaningless. Numeric variables have two subtypes: *interval variables* and *ratio variables* (Tan et al., 2006). Quantitative differences between the values of an interval variable (e.g., Launch Temperature in °F) are meaningful, whereas both quantitative differences and ratios between the values of a ratio variable (e.g., Number of O-rings with Stress) are meaningful.

Formally, we denote the attribute variables as $x_1, \ldots, x_p$, and the target variables as $y_1, \ldots, y_q$. We let $x = (x_1, \ldots, x_p)$ and $y = (y_1, \ldots, y_q)$. Instances or data observations of $x_1, \ldots, x_p, y_1, \ldots, y_q$ give data records, $(x_1, \ldots, x_p, y_1, \ldots, y_q)$.

1.3 Data Patterns Learned through Data Mining

The following are the major types of data patterns that are discovered from data sets through data mining algorithms:

- Classification and prediction patterns
- Cluster and association patterns
- Data reduction patterns
- Outlier and anomaly patterns
- Sequential and temporal patterns

Each type of data patterns is described in the following sections.

1.3.1 Classification and Prediction Patterns

Classification and prediction patterns capture relations of attribute variables, $x_1, \ldots, x_p$, with target variables, $y_1, \ldots, y_q$, which are supported by a given set of data records, $(x_1, \ldots, x_p, y_1, \ldots, y_q)$. Classification and prediction patterns allow us to classify or predict values of target variables from values of attribute variables.

For example, all the 16 data records of the balloon data set in Table 1.1 support the following relation of the attribute variables, Color, Size, Age, and Act, with the target variable, Inflated (taking the value of T for true or F for false):

IF (Color = Yellow AND Size = Small) OR (Age = Adult AND Act = Stretch), THEN Inflated = T; OTHERWISE, Inflated = F.

This relation allows us to classify a given balloon into a categorical value of the target variable using a specific value of its Color, Size, Age, and Act attributes. Hence, the relation gives us data patterns that allow us to perform the classification of a balloon. Although we can extract this relation pattern by examining the 16 data records in the balloon data set, learning such a pattern manually from a much larger set of data with noise can be a difficult task. A data mining algorithm allows us to learn from a large data set automatically.

For another example, the following linear model fits the 23 data records of the attribute variable, Launch Temperature, and the target variable, Number of O-rings with Stress, in the space shuttle O-ring data set in Table 1.2:

$$y = -0.05746x + 4.301587 \tag{1.1}$$

where
  y denotes the target variable, Number of O-rings with Stress
  x denotes the attribute variable, Launch Temperature

Figure 1.2 illustrates the values of Launch Temperature and Number of O-rings with Stress in the 23 data records and the fitted line given by Equation 1.1. Table 1.5 shows the value of O-rings with Stress for each data record that is predicted from the value of Launch Temperature using the linear relation model of Launch Temperature with Number of O-rings with Stress in Equation 1.1. Except two data records for instances 2 and 11, the linear model in Equation 1.1 captures the relation of Launch Temperature with Number of O-rings with Stress well in that a lower value of Launch Temperature increases the value of O-rings with Stress. The highest predicted value of O-rings with Stress is produced for the data record of instance 14 with 2 O-rings experiencing thermal stress. Two predicted values in the

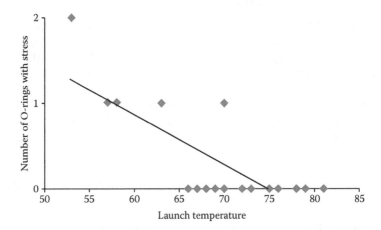

FIGURE 1.2
The fitted linear relation model of Launch Temperature with Number of O-rings with Stress in the space shuttle O-ring data set.

TABLE 1.5

Predicted Value of O-Rings with Stress

Instance	Attribute Variable Launch Temperature	Target Variable	
		Number of O-Rings with Stress	Predicted Value of O-Rings with Stress
1	66	0	0.509227
2	70	1	0.279387
3	69	0	0.336847
4	68	0	0.394307
5	67	0	0.451767
6	72	0	0.164467
7	73	0	0.107007
8	70	0	0.279387
9	57	1	1.026367
10	63	1	0.681607
11	70	1	0.279387
12	78	0	−0.180293
13	67	0	0.451767
14	53	2	1.256207
15	67	0	0.451767
16	75	0	−0.007913
17	70	0	0.279387
18	81	0	−0.352673
19	76	0	−0.065373
20	79	0	−0.237753
21	75	0	−0.007913
22	76	0	−0.065373
23	58	1	0.968907

middle range, 1.026367 and 0.681607, are produced for two data records of instances 9 and 10 with 1 O-rings with Stress. The predicted values in the low range from −0.352673 to 0.509227 are produced for all the data records with 0 O-rings with Stress. The negative coefficient of x, −0.05746, in Equation 1.1, also reveals this relation. Hence, the linear relation in Equation 1.1 gives data patterns that allow us to predict the target variable, Number of O-rings with Stress, from the attribute variable, Launch Temperature, in the space shuttle O-ring data set.

Classification and prediction patterns, which capture the relation of attribute variables, $x_1, \ldots, x_p$, with target variables, $y_1, \ldots, y_q$, can be represented in the general form of $y = F(x)$. For the balloon data set, classification patterns for F take the form of decision rules. For the space shuttle O-ring data set, prediction patterns for F take the form of a linear model. Generally, the term, "classification patterns," is used if the target variable is a categorical variable, and the term, "prediction patterns," is used if the target variable is a numeric variable.

Part II of the book introduces the following data mining algorithms that are used to discover classification and prediction patterns from data:

- Regression models in Chapter 2
- Naïve Bayes classifier in Chapter 3
- Decision and regression trees in Chapter 4
- Artificial neural networks for classification and prediction in Chapter 5
- Support vector machines in Chapter 6
- K-nearest neighbor classifier and supervised clustering in Chapter 7

Chapters 20, 21, and 23 in *The Handbook of Data Mining* (Ye, 2003) and Chapters 12 and 13 in *Secure Computer and Network Systems: Modeling, Analysis and Design* (Ye, 2008) give applications of classification and prediction algorithms to human performance data, text data, science and engineering data, and computer and network data.

1.3.2 Cluster and Association Patterns

Cluster and association patterns usually involve only attribute variables, $x_1, \ldots, x_p$. Cluster patterns give groups of similar data records such that data records in one group are similar but have larger differences from data records in another group. In other words, cluster patterns reveal patterns of similarities and differences among data records. Association patterns are established based on co-occurrences of items in data records. Sometimes target variables, $y_1, \ldots, y_q$, are also used in clustering but are treated in the same way as attribute variables.

For example, 10 data records in the data set of a manufacturing system in Table 1.4 can be clustered into seven groups, as shown in Figure 1.3. The horizontal axis of each chart in Figure 1.3 lists the nine quality variables, and the vertical axis gives the value of these nine quality variables. There are three groups that consist of more than one data record: group 1, group 2, and group 3. Within each of these groups, the data records are similar with different values in only one of the nine quality variables. Adding any other data record to each of these three groups makes the group having at least two data records with different values in more than one quality variable.

For the same data set of a manufacturing system, the quality variables, x_4 and x_8, are highly associated because they have the same value in all the data records except that of instance 8. There are other pairs of variables, e.g., x_5 and x_9, that are highly associated for the same reason. These are some association patterns that exist in the data set of a manufacturing system in Table 1.4.

Introduction to Data, Data Patterns, and Data Mining

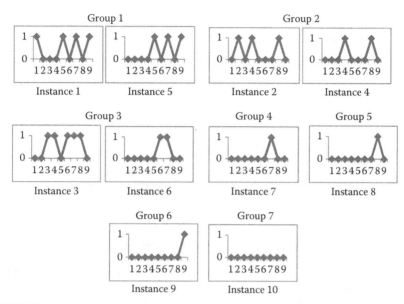

FIGURE 1.3
Clustering of 10 data records in the data set of a manufacturing system.

Part III of the book introduces the following data mining algorithms that are used to discover cluster and association patterns from data:

- Hierarchical clustering in Chapter 8
- K-means clustering and density-based clustering in Chapter 9
- Self-organizing map in Chapter 10
- Probability distribution of univariate data in Chapter 11
- Association rules in Chapter 12
- Bayesian networks in Chapter 13

Chapters 10, 21, 22, and 27 in *The Handbook of Data Mining* (Ye, 2003) give applications of cluster algorithms to market basket data, web log data, text data, geospatial data, and image data. Chapter 24 in *The Handbook of Data Mining* (Ye, 2003) gives an application of the association rule algorithm to protein structure data.

1.3.3 Data Reduction Patterns

Data reduction patterns look for a small number of variables that can be used to represent a data set with a much larger number of variables. Since one variable gives one dimension of data, data reduction patterns allow a data set in a high-dimensional space to be represented in a low-dimensional space. For example, Figure 1.4 gives 10 data points in a two-dimensional

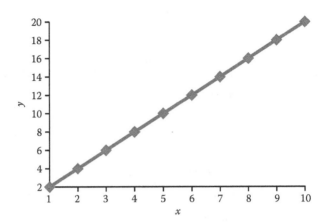

FIGURE 1.4
Reduction of a two-dimensional data set to a one-dimensional data set.

space, (x, y), with $y = 2x$ and $x = 1, 2, \ldots, 10$. This two-dimensional data set can be represented as the one-dimensional data set with z as the axis, and z is related to the original variables, x and y, as follows:

$$z = x * \sqrt{1^2 + 1 * \left(\frac{y}{x}\right)^2}. \tag{1.2}$$

The 10 data points of z are 2.236, 4.472, 6.708, 8.944, 11.180, 13.416, 15.652, 17.889, 20.125, and 22.361.

Part IV of the book introduces the following data mining algorithms that are used to discover data reduction patterns from data:

- Principal component analysis in Chapter 14
- Multidimensional scaling in Chapter 15

Chapters 23 and 8 in *The Handbook of Data Mining* (Ye, 2003) give applications of principal component analysis to volcano data and science and engineering data.

1.3.4 Outlier and Anomaly Patterns

Outliers and anomalies are data points that differ largely from the norm of data. The norm can be defined in many ways. For example, the norm can be defined by the range of values that a majority of data points take, and a data point with a value outside this range can be considered as an outlier. Figure 1.5 gives the frequency histogram of Launch Temperature values for the data points in the space shuttle data set in Table 1.2. There are 3 values of Launch Temperature in the range of [50, 59], 7 values in the range of [60, 69], 12 values in the range of [70, 79], and only 1 value in the range of [80, 89]. Hence, the majority of values in Launch Temperature are in the range of [50, 79]. The value of 81 in instance 18 can be considered as an outlier or anomaly.

Introduction to Data, Data Patterns, and Data Mining

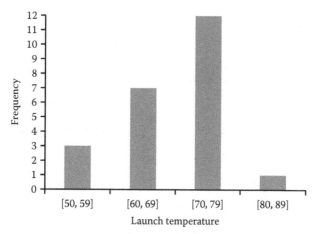

FIGURE 1.5
Frequency histogram of Launch Temperature in the space shuttle data set.

Part V of the book introduces the following data mining algorithms that are used to define some statistical norms of data and detect outliers and anomalies according to these statistical norms:

- Univariate control charts in Chapter 16
- Multivariate control charts in Chapter 17

Chapters 26 and 28 in *The Handbook of Data Mining* (Ye, 2003) and Chapter 14 in *Secure Computer and Network Systems: Modeling, Analysis and Design* (Ye, 2008) give applications of outlier and anomaly detection algorithms to manufacturing data and computer and network data.

1.3.5 Sequential and Temporal Patterns

Sequential and temporal patterns reveal patterns in a sequence of data points. If the sequence is defined by the time over which data points are observed, we call the sequence of data points as a time series. Figure 1.6 shows a time

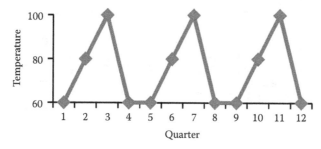

FIGURE 1.6
Temperature in each quarter of a 3-year period.

TABLE 1.6
Test Data Set for a Manufacturing System to Detect and Diagnose Faults

Instance (Faulty Machine)	Attribute Variables									System Fault, y	Target Variables								
	Quality of Parts										Machine Fault								
	x_1	x_2	x_3	x_4	x_5	x_6	x_7	x_8	x_9		y_1	y_2	y_3	y_4	y_5	y_6	y_7	y_8	y_9
1 (M1, M2)	1	1	0	1	1	0	1	1	1	1	1	1	0	0	0	0	0	0	0
2 (M2, M3)	0	1	1	1	0	1	1	1	0	1	0	1	1	0	0	0	0	0	0
3 (M1, M3)	1	0	1	1	1	1	1	1	1	1	1	0	1	0	0	0	0	0	0
4 (M1, M4)	1	0	0	1	1	0	1	1	1	1	1	0	0	1	0	0	0	0	0
5 (M1, M6)	1	0	0	0	1	1	1	0	1	1	1	0	0	0	0	1	0	0	0
6 (M2, M6)	0	1	0	1	0	1	1	1	0	1	0	1	0	0	0	1	0	0	0
7 (M2, M5)	0	1	0	1	1	0	1	1	0	1	0	1	0	0	1	0	0	0	0
8 (M3, M5)	0	0	1	1	1	1	1	1	1	1	0	0	1	0	1	0	0	0	0
9 (M4, M7)	0	0	0	1	0	0	1	1	0	1	0	0	0	1	0	0	1	0	0
10 (M5, M8)	0	0	0	0	1	0	1	1	1	1	0	0	0	0	1	0	0	1	0
11 (M3, M9)	0	0	1	1	0	1	1	1	1	1	0	0	1	0	0	0	0	0	1
12 (M1, M8)	1	0	0	0	1	0	1	1	0	1	1	0	0	0	0	0	0	1	0
13 (M1, M2, M3)	1	1	1	1	1	1	1	1	1	1	1	1	1	0	0	0	0	0	0
14 (M2, M3, M5)	0	1	1	1	1	1	1	1	1	1	0	1	1	0	1	0	0	0	0
15 (M2, M3, M9)	0	1	1	1	0	1	1	1	1	1	0	1	1	0	0	0	0	0	1
16 (M1, M6, M8)	1	0	0	0	1	1	1	1	1	1	1	0	0	0	0	1	0	1	0

series of temperature values for a city over quarters of a 3-year period. There is a cyclic pattern of 60, 80, 100, and 60, which repeats every year. A variety of sequential and temporal patterns can be discovered using the data mining algorithms covered in Part VI of the book, including

- Autocorrelation and time series analysis in Chapter 18
- Markov chain models and hidden Markov models in Chapter 19
- Wavelet analysis in Chapter 20

Chapters 10, 11, and 16 in *Secure Computer and Network Systems: Modeling, Analysis and Design* (Ye, 2008) give applications of sequential and temporal pattern mining algorithms to computer and network data for cyber attack detection.

1.4 Training Data and Test Data

The training data set is a set of data records that is used to learn and discover data patterns. After data patterns are discovered, they should be tested to see how well they can generalize to a wide range of data records, including those that are different from the training data records. A test data set is used for this purpose and includes new, different data records. For example, Table 1.6 shows a test data set for a manufacturing system and its fault detection and diagnosis. The training data set for this manufacturing system in Table 1.4 has data records for nine single-machine faults and a case where there is no machine fault. The test data set in Table 1.6 has data records for some two-machine and three-machine faults.

Exercises

1.1 Find and describe a data set of at least 20 data records that has been used in a data mining application for discovering classification patterns. The data set contains multiple categorical attribute variables and one categorical target variable.

1.2 Find and describe a data set of at least 20 data records that has been used in a data mining application for discovering prediction patterns. The data set contains multiple numeric attribute variables and one numeric target variable.

1.3 Find and describe a data set of at least 20 data records that has been used in a data mining application for discovering cluster patterns. The data set contains multiple numeric attribute variables.

1.4 Find and describe a data set of at least 20 data records that has been used in a data mining application for discovering association patterns. The data set contains multiple categorical variables.

1.5 Find and describe a data set of at least 20 data records that has been used in a data mining application for discovering data reduction patterns, and identify the type(s) of data variables in this data set.

1.6 Find and describe a data set of at least 20 data records that has been used in a data mining application for discovering outlier and anomaly patterns, and identify the type(s) of data variables in this data set.

1.7 Find and describe a data set of at least 20 data records that has been used in a data mining application for discovering sequential and temporal patterns, and identify the type(s) of data variables in this data set.

Part II

Algorithms for Mining Classification and Prediction Patterns

2

Linear and Nonlinear Regression Models

Regression models capture how one or more target variables vary with one or more attribute variables. They can be used to predict the values of the target variables using the values of the attribute variables. In this chapter, we introduce linear and nonlinear regression models. This chapter also describes the least-squares method and the maximum likelihood method of estimating parameters in regression models. A list of software packages that support building regression models is provided.

2.1 Linear Regression Models

A simple linear regression model, as shown next, has one target variable y and one attribute variable x:

$$y_i = \beta_0 + \beta_1 x_i + \varepsilon_i \qquad (2.1)$$

where
  (x_i, y_i) denotes the ith observation of x and y
  ε_i represents random noise (e.g., measurement error) contributing to the ith observation of y

For a given value of x_i, both y_i and ε_i are random variables whose values may follow a probability distribution as illustrated in Figure 2.1. In other words, for the same value of x, different values of y and ε may be observed at different times. There are three assumptions about ε_i:

1. $E(\varepsilon_i) = 0$, that is, the mean of ε_i is zero
2. $\text{var}(\varepsilon_i) = \sigma^2$, that is, the variance of ε_i is σ^2
3. $\text{cov}(\varepsilon_i, \varepsilon_j) = 0$ for $i \neq j$, that is, the covariance of ε_i and ε_j for any two different data observations, the ith observation and the jth observation, is zero

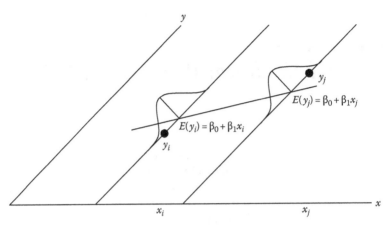

FIGURE 2.1
Illustration of a simple regression model.

These assumptions imply

1. $E(y_i) = \beta_0 + \beta_1 x_i$
2. $\text{var}(y_i) = \sigma^2$
3. $\text{cov}(y_i, y_j) = 0$ for any two different data observations of y, the ith observation and the jth observation

The simple linear regression model in Equation 2.1 can be extended to include multiple attribute variables:

$$y_i = \beta_0 + \beta_1 x_{i,1} + \cdots + \beta_p x_{i,p} + \varepsilon_i, \tag{2.2}$$

where

p is an integer greater than 1
$x_{i,j}$ denotes the ith observation of jth attribute variable

The linear regression models in Equations 2.1 and 2.2 are linear in the parameters $\beta_0, \ldots, \beta_p$ and the attribute variables $x_{i,1}, \ldots, x_{i,p}$. In general, linear regression models are linear in the parameters but are not necessarily linear in the attribute variables. The following regression model with polynomial terms of x_1 is also a linear regression model:

$$y_i = \beta_0 + \beta_1 x_{i,1} + \cdots + \beta_k x_{i,1}^k + \varepsilon_i, \tag{2.3}$$

where k is an integer greater than 1. The general form of a linear regression model is

$$y_i = \beta_0 + \beta_1 \Phi_1(x_{i,1}, \ldots, x_{i,p}) + \cdots + \beta_k \Phi_k(x_{i,1}, \ldots, x_{i,p}) + \varepsilon_i, \tag{2.4}$$

where Φ_l, $l = 1, \ldots, k$, is a linear or nonlinear function involving one or more of the variables $x_1, \ldots, x_p$. The following is another example of a linear regression model that is linear in the parameters:

$$y_i = \beta_0 + \beta_1 x_{i,1} + \beta_2 x_{i,2} + \beta_3 \log x_{i,1} x_{i,2} + \varepsilon_i. \qquad (2.5)$$

2.2 Least-Squares Method and Maximum Likelihood Method of Parameter Estimation

To fit a linear regression model to a set of training data (x_i, y_i), $x_i = (x_{i,1}, \ldots, x_{i,p})$, $i = 1, \ldots, n$, the parameters βs need to be estimated. The least-squares method and the maximum likelihood method are usually used to estimate the parameters βs. We illustrate both methods using the simple linear regression model in Equation 2.1.

The least-squares method looks for the values of the parameters β_0 and β_1 that minimize the sum of squared errors (SSE) between observed target values (y_i, $i = 1, \ldots, n$) and the estimated target values ($\hat{y}_i$, $i = 1, \ldots, n$) using the estimated parameters $\hat{\beta}_0$ and $\hat{\beta}_1$. SSE is a function of $\hat{\beta}_0$ and $\hat{\beta}_1$:

$$SSE = \sum_{i=1}^{n}(y_i - \hat{y}_i)^2 = \sum_{i=1}^{n}\left(y_i - \hat{\beta}_0 - \hat{\beta}_1 x_i\right)^2. \qquad (2.6)$$

The partial derivatives of SSE with respect to $\hat{\beta}_0$ and $\hat{\beta}_1$ should be zero at the point where SSE is minimized. Hence, the values of $\hat{\beta}_0$ and $\hat{\beta}_1$ that minimize SSE are obtained by differentiating SSE with respect to $\hat{\beta}_0$ and $\hat{\beta}_1$ and setting these partial derivatives equal to zero:

$$\frac{\partial SSE}{\partial \hat{\beta}_0} = -2\sum_{i=1}^{n}\left(y_i - \hat{\beta}_0 - \hat{\beta}_1 x_i\right) = 0 \qquad (2.7)$$

$$\frac{\partial SSE}{\partial \hat{\beta}_1} = -2\sum_{i=1}^{n}x_i\left(y_i - \hat{\beta}_0 - \hat{\beta}_1 x_i\right) = 0. \qquad (2.8)$$

Equations 2.7 and 2.8 are simplified to

$$\sum_{i=1}^{n}\left(y_i - \hat{\beta}_0 - \hat{\beta}_1 x_i\right) = \sum_{i=1}^{n}y_i - n\hat{\beta}_0 - \hat{\beta}_1\sum_{i=1}^{n}x_i = 0 \qquad (2.9)$$

$$\sum_{i=1}^{n}x_i\left(y_i - \hat{\beta}_0 - \hat{\beta}_1 x_i\right) = \sum_{i=1}^{n}x_i y_i - \hat{\beta}_0\sum_{i=1}^{n}x_i - \hat{\beta}_1\sum_{i=1}^{n}x_i^2 = 0. \qquad (2.10)$$

Solving Equations 2.9 and 2.10 for $\hat{\beta}_0$ and $\hat{\beta}_1$, we obtain:

$$\hat{\beta}_1 = \frac{\sum_{i=1}^{n}(x_i - \bar{x})(y_i - \bar{y})}{\sum_{i=1}^{n}(x_i - \bar{x})^2} = \frac{n\sum_{i=1}^{n}x_i y_i - \left(\sum_{i=1}^{n}x_i\right)\left(\sum_{i=1}^{n}y_i\right)}{n\sum_{i=1}^{n}x_i^2 - \left(\sum_{i=1}^{n}x_i\right)^2} \quad (2.11)$$

$$\hat{\beta}_0 = \frac{1}{n}\left(\sum_{i=1}^{n}y_i - \hat{\beta}_1\sum_{i=1}^{n}x_i\right) = \bar{y} - \hat{\beta}_1\bar{x}. \quad (2.12)$$

The estimation of the parameters in the simple linear regression model based on the least-squares method does not require that the random error ε_i has a specific form of the probability distribution. If we add to the simple linear regression model in Equation 2.1 an assumption that ε_i is normally distributed with the mean of zero and the constant, unknown variance of σ^2, denoted by $N(0, \sigma^2)$, the maximum likelihood method can also be used to estimate the parameters in the simple linear regression model. The assumption that ε_is are independent $N(0, \sigma^2)$ gives the normal distribution of y_i with

$$E(y_i) = \beta_0 + \beta_1 x_i \quad (2.13)$$

$$\text{var}(y_i) = \sigma^2 \quad (2.14)$$

and the density function of the normal probability distribution:

$$f(y_i) = \frac{1}{\sqrt{2\pi}\sigma}e^{-\frac{1}{2}\left(\frac{y_i - E(y_i)}{\sigma}\right)^2} = \frac{1}{\sqrt{2\pi}\sigma}e^{-\frac{1}{2}\left(\frac{y_i - \beta_0 - \beta_1 x_i}{\sigma}\right)^2}. \quad (2.15)$$

Because y_is are independent, the likelihood of observing $y_1, \ldots, y_n$, L, is the product of individual densities $f(y_i)$s and is the function of β_0, β_1, and σ^2:

$$L(\beta_0, \beta_1, \sigma) = \prod_{i=1}^{n}\frac{1}{(2\pi\sigma^2)^{1/2}}e^{-\frac{1}{2}\left(\frac{y_i - \beta_0 - \beta_1 x_i}{\sigma}\right)^2}. \quad (2.16)$$

The estimated values of the parameters, $\hat{\beta}_0$, $\hat{\beta}_1$, and $\hat{\sigma}^2$, which maximize the likelihood function in Equation 2.16 are the maximum likelihood estimators and can be obtained by differentiating this likelihood function with respect to β_0, β_1, and σ^2 and setting these partial derivatives to zero. To ease the computation, we use the natural logarithm transformation (ln) of the likelihood function to obtain

$$\frac{\partial lnL(\hat{\beta}_0, \hat{\beta}_1, \hat{\sigma}^2)}{\partial \hat{\beta}_0} = \frac{1}{\hat{\sigma}^2}\sum_{i=1}^{n}(y_i - \hat{\beta}_0 - \hat{\beta}_1 x_i) = 0 \quad (2.17)$$

$$\frac{\partial lnL(\hat{\beta}_0,\hat{\beta}_1,\hat{\sigma}^2)}{\partial \hat{\beta}_1} = \frac{1}{\hat{\sigma}^2}\sum_{i=1}^{n}x_i\left(y_i-\hat{\beta}_0-\hat{\beta}_1 x_i\right)=0 \quad (2.18)$$

$$\frac{\partial lnL(\hat{\beta}_0,\hat{\beta}_1,\hat{\sigma}^2)}{\partial \hat{\sigma}^2} = -\frac{n}{2\hat{\sigma}^2}+\frac{1}{2\hat{\sigma}^4}\sum_{i=1}^{n}\left(y_i-\hat{\beta}_0-\hat{\beta}_1 x_i\right)^2=0. \quad (2.19)$$

Equations 2.17 through 2.19 are simplified to

$$\sum_{i=1}^{n}\left(y_i-\hat{\beta}_0-\hat{\beta}_1 x_i\right)=0 \quad (2.20)$$

$$\sum_{i=1}^{n}x_i\left(y_i-\hat{\beta}_0-\hat{\beta}_1 x_i\right)=0 \quad (2.21)$$

$$\hat{\sigma}^2 = \frac{\sum_{i=1}^{n}\left(y_i-\hat{\beta}_0-\hat{\beta}_1 x_i\right)^2}{n}. \quad (2.22)$$

Equations 2.20 and 2.21 are the same as Equations 2.9 and 2.10. Hence, the maximum likelihood estimators of β_0 and β_1 are the same as the least-squares estimators of β_0 and β_1 that are given in Equations 2.11 and 2.12.

For the linear regression model in Equation 2.2 with multiple attribute variables, we define $x_0 = 1$ and rewrite Equation 2.2 to

$$y_i = \beta_0 x_{i,0} + \beta_1 x_{i,1} + \cdots + \beta_1 x_{i,p} + \varepsilon_i. \quad (2.23)$$

Defining the following matrices

$$y = \begin{bmatrix} y_1 \\ \vdots \\ y_n \end{bmatrix} \quad x = \begin{bmatrix} 1 & x_{1,1} & \cdots & x_{1,p} \\ \vdots & \vdots & \ddots & \vdots \\ 1 & x_{n,1} & \cdots & x_{n,p} \end{bmatrix} \quad \beta = \begin{bmatrix} \beta_0 \\ \vdots \\ \beta_p \end{bmatrix} \quad \varepsilon = \begin{bmatrix} \varepsilon_1 \\ \vdots \\ \varepsilon_n \end{bmatrix},$$

we rewrite Equation 2.23 in the matrix form

$$y = x\beta + \varepsilon. \quad (2.24)$$

The least-squares and maximum likelihood estimators of the parameters are

$$\hat{\beta} = (x'x)^{-1}(x'y), \quad (2.25)$$

where $(x'x)^{-1}$ represents the inverse of the matrix $x'x$.

Example 2.1

Use the least-squares method to fit a linear regression model to the space shuttle O-rings data set in Table 1.5, which is also given in Table 2.1, and determine the predicted target value for each observation using the linear regression model.

This data has one attribute variable x representing Launch Temperature and one target variable y representing Number of O-rings with Stress. The linear regression model for this data set is

$$y_i = \beta_0 + \beta_1 x_i + \varepsilon_i.$$

Table 2.2 shows the calculation for estimating β_1 using Equation 2.11. Using Equation 2.11, we obtain:

$$\hat{\beta}_1 = \frac{\sum_{i=1}^{n}(x_i - \bar{x})(y_i - \bar{y})}{\sum_{i=1}^{n}(x_i - \bar{x})^2} = \frac{-65.91}{1382.82} = -0.05.$$

TABLE 2.1

Data Set of O-Rings with Stress along with the Predicted Target Value from the Linear Regression

Instance	Launch Temperature	Number of O-Rings with Stress
1	66	0
2	70	1
3	69	0
4	68	0
5	67	0
6	72	0
7	73	0
8	70	0
9	57	1
10	63	1
11	70	1
12	78	0
13	67	0
14	53	2
15	67	0
16	75	0
17	70	0
18	81	0
19	76	0
20	79	0
21	75	0
22	76	0
23	58	1

TABLE 2.2
Calculation for Estimating the Parameters of the Linear Model in Example 2.1

Instance	Launch Temperature	Number of O-Rings	$x_i - \bar{x}$	$y_i - \bar{y}$	$(x_i - \bar{x})(y_i - \bar{y})$	$(x_i - \bar{x})^2$
1	66	0	−3.57	−0.30	1.07	12.74
2	70	1	0.43	0.70	0.30	0.18
3	69	0	−0.57	−0.30	0.17	0.32
4	68	0	−1.57	−0.30	0.47	2.46
5	67	0	−2.57	−0.30	0.77	6.60
6	72	0	2.43	−0.30	−0.73	5.90
7	73	0	3.43	−0.30	−1.03	11.76
8	70	0	0.43	−0.30	−0.13	0.18
9	57	1	−12.57	0.70	−8.80	158.00
10	63	1	−6.57	0.70	−4.60	43.16
11	70	1	0.43	0.70	0.30	0.18
12	78	0	8.43	−0.30	−2.53	71.06
13	67	0	−2.57	−0.30	0.77	6.60
14	53	2	−16.53	1.70	−28.10	273.24
15	67	0	−2.57	−0.30	0.77	6.60
16	75	0	5.43	−0.30	−1.63	29.48
17	70	0	0.43	−0.30	−0.13	0.18
18	81	0	11.43	−0.30	−3.43	130.64
19	76	0	6.43	−0.30	−1.93	41.34
20	79	0	19.43	−0.30	−5.83	377.52
21	75	0	5.43	−0.30	−1.63	29.48
22	76	0	6.43	−0.30	−1.93	41.34
23	58	1	−11.57	0.70	−8.10	133.86
Sum	1600	7			−65.91	1382.82
Average	$\bar{x} = 69.57$	$\bar{y} = 0.30$				

Using Equation 2.12, we obtain:

$$\hat{\beta}_0 = \bar{y} - \hat{\beta}_1 \bar{x} = 0.30 - (-.05)(69.57) = 3.78.$$

Hence, the linear regression model is

$$y_i = 3.78 - 0.05 x_i + \varepsilon_i.$$

The parameters in this linear regression model are similar to the parameters $\hat{\beta}_0 = 4.301587$ and $\hat{\beta}_1 = -0.05746$ in Equation 1.1, which are obtained from Excel for the same data set. The differences in the parameters are caused by rounding in the calculation.

2.3 Nonlinear Regression Models and Parameter Estimation

Nonlinear regression models are nonlinear in model parameters and take the following general form:

$$y_i = f(x_i, \beta) + \varepsilon_i, \qquad (2.26)$$

where

$$x_i = \begin{bmatrix} 1 \\ x_{i,1} \\ \vdots \\ x_{i,p} \end{bmatrix} \quad \beta = \begin{bmatrix} \beta_0 \\ \beta_1 \\ \vdots \\ \beta_p \end{bmatrix}$$

and f is nonlinear in β. The exponential regression model given next is an example of nonlinear regression models:

$$y_i = \beta_0 + \beta_1 e^{\beta_2 x_i} + \varepsilon_i. \qquad (2.27)$$

The logistic regression model given next is another example of nonlinear regression models:

$$y_i = \frac{\beta_0}{1 + \beta_1 e^{\beta_2 x_i}} + \varepsilon_i. \qquad (2.28)$$

The least-squares method and the maximum likelihood method are used to estimate the parameters of a nonlinear regression model. Unlike Equations 2.9, 2.10, 2.20, and 2.21 for a linear regression model, the equations for a nonlinear regression model generally do not have analytical solutions because a nonlinear regression model is nonlinear in the parameters. Numerical search methods using an iterative search procedure such as the Gauss–Newton method and the gradient decent search method are used to determine the solution for the values of the estimated parameters. A detailed description of the Gauss–Newton method is given in Neter et al. (1996). Computer software programs in many statistical software packages are usually used to estimate the parameters of a nonlinear regression model because intensive computation is involved in a numerical search procedure.

2.4 Software and Applications

Many statistical software packages, including the following, support building a linear or nonlinear regression model:

- Statistica (http://www.statsoft.com)
- SAS (http://www.sas.com)
- SPSS (http://www.ibm/com/software/analytics/spss/)

Applications of linear and nonlinear regression models are common in many fields.

Exercises

2.1 Given the space shuttle data set in Table 2.1, use Equation 2.25 to estimate the parameters of the following linear regression model:

$$y_i = \beta_0 + \beta_1 \sqrt{x_i} + \varepsilon_i,$$

where
x_i is Launch Temperature
y_i is Number of O-rings with Stress

Compute the sum of squared errors that are produced by the predicted y values from the regression model.

2.2 Given the space shuttle data set in Table 2.1, use Equations 2.11 and 2.12 to estimate the parameters of the following linear regression model:

$$y_i = \beta_0 + \beta_1 \sqrt{x_i} + \varepsilon_i,$$

where
x_i is Launch Temperature
y_i is Number of O-rings with Stress

Compute the sum of squared errors that are produced by the predicted y values from the regression model.

2.3 Use the data set found in Exercise 1.2 to build a linear regression model and compute the sum of squared errors that are produced by the predicted y values from the regression model.

3
Naïve Bayes Classifier

A naïve Bayes classifier is based on the Bayes theorem. Hence, this chapter first reviews the Bayes theorem and then describes naïve Bayes classifier. A list of data mining software packages that support the learning of a naïve Bayes classifier is provided. Some applications of naïve Bayes classifiers are given with references.

3.1 Bayes Theorem

Given two events A and B, the conjunction ($\wedge$) of the two events represents the occurrence of both A and B. The probability, $P(A \wedge B)$ is computed using the probability of A and B, $P(A)$ and $P(B)$, and the conditional probability of A given B, $P(A|B)$, or B given A, $P(B|A)$:

$$P(A \wedge B) = P(A|B)P(B) = P(B|A)P(A). \tag{3.1}$$

The Bayes theorem is derived from Equation 3.1:

$$P(A|B) = \frac{P(B|A)P(A)}{P(B)}. \tag{3.2}$$

3.2 Classification Based on the Bayes Theorem and Naïve Bayes Classifier

For a data vector x whose target class y needs to be determined, the maximum a posterior (MAP) classification y of x is

$$y_{MAP} = \arg\max_{y \in Y} P(y|x) = \arg\max_{y \in Y} \frac{p(y)P(x|y)}{P(x)} \approx \arg\max_{y \in Y} p(y)P(x|y), \tag{3.3}$$

where Y is the set of all target classes. The sign $\approx$ in Equation 3.3 is used because $P(x)$ is the same for all y values and thus can be ignored when we compare $p(y)P(x|y)/P(x)$ for all y values. $P(x)$ is the prior probability that we observe x without any knowledge about what the target class of x is. $P(y)$ is the prior probability that we expect y, reflecting our prior knowledge about the data set of x and the likelihood of the target class y in the data set without referring to any specific x. $P(y|x)$ is the posterior probability of y given the observation of x. $\arg\max_{y \in Y} P(y|x)$ compares the posterior probabilities of all target classes given x and chooses the target class y with the maximum posterior probability.

$P(x|y)$ is the probability that we observe x if the target class is y. A classification y that maximizes $P(x|y)$ among all target classes is the maximum likelihood (ML) classification:

$$y_{ML} = \arg\max_{y \in Y} P(x|y). \tag{3.4}$$

If $P(y) = P(y')$ for any $y \neq y'$, $y \in Y$, $y' \in Y$, then

$$y_{MAP} \approx \arg\max_{y \in Y} p(y)P(x|y) \approx \arg\max_{y \in Y} P(x|y),$$

and thus

$$y_{MAP} = y_{ML}.$$

A naïve Bayes classifier is based on a MAP classification with the additional assumption about the attribute variables $x = (x_1, \ldots, x_p)$ that these attribute variables x_is are independent of each other. With this assumption, we have

$$y_{MAP} \approx \arg\max_{y \in Y} p(y)P(x|y) = \arg\max_{y \in Y} p(y) \prod_{i=1}^{p} P(x_i|y). \tag{3.5}$$

The naïve Bayes classifier estimates the probability terms in Equation 3.5 in the following way:

$$P(y) = \frac{n_y}{n} \tag{3.6}$$

$$P(x_i|y) = \frac{n_{y \& x_i}}{n_y}, \tag{3.7}$$

where
  n is the total number of data points in the training data set
  n_y is the number of data points with the target class y
  $n_{y \& x_i}$ is the number of data points with the target class y the ith attribute variable taking the value of x_i

Naïve Bayes Classifier

An application of the naïve Bayes classifier is given in Example 3.1.

Example 3.1

Learn and use a naïve Bayes classifier for classifying whether or not a manufacturing system is faulty using the values of the nine quality variables. The training data set in Table 3.1 gives a part of the data set in Table 1.4 and includes nine single-fault cases and the nonfault case in a manufacturing system. There are nine attribute variables for the quality of parts, $(x_1, ..., x_9)$, and one target variable y for the system fault. Table 3.2 gives the test cases for some multiple-fault cases.

Using the training data set in Table 3.1, we compute the following:

$n = 10$

$n_{y=1} = 9$ $\quad n_{y=0} = 1$

$n_{y=1 \& x_1=1} = 1 \quad n_{y=1 \& x_1=0} = 8 \quad n_{y=0 \& x_1=1} = 0 \quad n_{y=0 \& x_1=0} = 1$

$n_{y=1 \& x_2=1} = 1 \quad n_{y=1 \& x_2=0} = 8 \quad n_{y=0 \& x_2=1} = 0 \quad n_{y=0 \& x_2=0} = 1$

$n_{y=1 \& x_3=1} = 1 \quad n_{y=1 \& x_3=0} = 8 \quad n_{y=0 \& x_3=1} = 0 \quad n_{y=0 \& x_3=0} = 1$

$n_{y=1 \& x_4=1} = 3 \quad n_{y=1 \& x_4=0} = 6 \quad n_{y=0 \& x_4=1} = 0 \quad n_{y=0 \& x_4=0} = 1$

$n_{y=1 \& x_5=1} = 2 \quad n_{y=1 \& x_5=0} = 7 \quad n_{y=0 \& x_5=1} = 0 \quad n_{y=0 \& x_5=0} = 1$

$n_{y=1 \& x_6=1} = 2 \quad n_{y=1 \& x_6=0} = 7 \quad n_{y=0 \& x_6=1} = 0 \quad n_{y=0 \& x_6=0} = 1$

$n_{y=1 \& x_7=1} = 5 \quad n_{y=1 \& x_7=0} = 4 \quad n_{y=0 \& x_7=1} = 0 \quad n_{y=0 \& x_7=0} = 1$

$n_{y=1 \& x_8=1} = 4 \quad n_{y=1 \& x_8=0} = 5 \quad n_{y=0 \& x_8=1} = 0 \quad n_{y=0 \& x_8=0} = 1$

$n_{y=1 \& x_9=1} = 3 \quad n_{y=1 \& x_9=0} = 6 \quad n_{y=0 \& x_9=1} = 0 \quad n_{y=0 \& x_9=0} = 1$

TABLE 3.1

Training Data Set for System Fault Detection

Instance (Faulty Machine)	Attribute Variables									Target Variable
	Quality of Parts									
	x_1	x_2	x_3	x_4	x_5	x_6	x_7	x_8	x_9	System Fault y
1 (M1)	1	0	0	0	1	0	1	0	1	1
2 (M2)	0	1	0	1	0	0	0	1	0	1
3 (M3)	0	0	1	1	0	1	1	1	0	1
4 (M4)	0	0	0	1	0	0	0	1	0	1
5 (M5)	0	0	0	0	1	0	1	0	1	1
6 (M6)	0	0	0	0	0	1	1	0	0	1
7 (M7)	0	0	0	0	0	0	1	0	0	1
8 (M8)	0	0	0	0	0	0	0	1	0	1
9 (M9)	0	0	0	0	0	0	0	0	1	1
10 (none)	0	0	0	0	0	0	0	0	0	0

TABLE 3.2

Classification of Data Records in the Testing Data Set for System Fault Detection

Instance (Faulty Machine)	Attribute Variables (Quality of Parts)									Target Variable (System Fault y)	
	x_1	x_2	x_3	x_4	x_5	x_6	x_7	x_8	x_9	True Value	Classified Value
1 (M1, M2)	1	1	0	1	1	0	1	1	1	1	1
2 (M2, M3)	0	1	1	1	0	1	1	1	0	1	1
3 (M1, M3)	1	0	1	1	1	1	1	1	1	1	1
4 (M1, M4)	1	0	0	1	1	0	1	1	1	1	1
5 (M1, M6)	1	0	0	0	1	1	1	0	1	1	1
6 (M2, M6)	0	1	0	1	0	1	1	1	0	1	1
7 (M2, M5)	0	1	0	1	1	0	1	1	0	1	1
8 (M3, M5)	0	0	1	1	1	1	1	1	1	1	1
9 (M4, M7)	0	0	0	1	0	0	1	1	0	1	1
10 (M5, M8)	0	0	0	0	1	0	1	1	0	1	1
11 (M3, M9)	0	0	1	1	0	1	1	1	1	1	1
12 (M1, M8)	1	0	0	0	1	0	1	1	1	1	1
13 (M1, M2, M3)	1	1	1	1	1	1	1	1	1	1	1
14 (M2, M3, M5)	0	1	1	1	1	1	1	1	1	1	1
15 (M2, M3, M9)	0	1	1	1	0	1	1	1	1	1	1
16 (M1, M6, M8)	1	0	0	0	1	1	1	1	1	1	1

Instance #1 in Table 3.1 with $x = (1, 0, 0, 0, 1, 0, 1, 0, 1)$ is classified as follows:

$$p(y=1) \prod_{i=1}^{9} P(x_i|y=1) = \frac{n_{y=1}}{n} \prod_{i=1}^{9} \frac{n_{y=1 \& x_i}}{n_{y=1}}$$

$$= \frac{n_{y=1}}{n} \left(\frac{n_{y=1 \& x_1=1}}{n_{y=1}} \times \frac{n_{y=1 \& x_2=0}}{n_{y=1}} \times \frac{n_{y=1 \& x_3=0}}{n_{y=1}} \times \frac{n_{y=1 \& x_4=0}}{n_{y=1}} \right.$$

$$\times \frac{n_{y=1 \& x_5=1}}{n_{y=1}} \times \frac{n_{y=1 \& x_6=0}}{n_{y=1}} \times \frac{n_{y=1 \& x_7=1}}{n_{y=1}}$$

$$\left. \times \frac{n_{y=1 \& x_8=0}}{n_{y=1}} \times \frac{n_{y=1 \& x_9=1}}{n_{y=1}} \right)$$

$$= \frac{9}{10} \left(\frac{1}{9} \times \frac{8}{9} \times \frac{8}{9} \times \frac{6}{9} \times \frac{2}{9} \times \frac{7}{9} \times \frac{5}{9} \times \frac{5}{9} \times \frac{3}{9} \right) > 0$$

$$p(y=0)\prod_{i=1}^{9}P(x_i|y=0) = \frac{n_{y=0}}{n}\prod_{i=1}^{9}\frac{n_{y=0\&x_i}}{n_{y=0}}$$

$$= \frac{n_{y=0}}{n}\left(\frac{n_{y=0\&x_1=1}}{n_{y=0}} \times \frac{n_{y=0\&x_2=0}}{n_{y=0}} \times \frac{n_{y=0\&x_3=0}}{n_{y=0}}\right.$$

$$\times \frac{n_{y=0\&x_4=0}}{n_{y=0}} \times \frac{n_{y=0\&x_5=1}}{n_{y=0}}$$

$$\times \frac{n_{y=0\&x_6=0}}{n_{y=0}} \times \frac{n_{y=0\&x_7=1}}{n_{y=0}}$$

$$\left.\times \frac{n_{y=0\&x_8=0}}{n_{y=0}} \times \frac{n_{y=0\&x_9=1}}{n_{y=0}}\right)$$

$$= \frac{1}{10}\left(\frac{0}{1}\times\frac{1}{1}\times\frac{1}{1}\times\frac{1}{1}\times\frac{0}{1}\times\frac{1}{1}\times\frac{0}{1}\times\frac{1}{1}\times\frac{0}{1}\right) = 0$$

$$y_{MAP} \approx \arg\max_{y\in Y} p(y)\prod_{i=1}^{9}P(x_i|y) = 1 \quad \text{(system is faulty)}.$$

Instance #2 to Case #9 in Table 3.1 and all the cases in Table 3.2 can be classified similarly to produce $y_{MAP}=1$ since there exist $x_i=1$ and $n_{y=0\&x_i=1}/n_{y=0} = 0/1$, which make $p(y=0)P(x|y=0)=0$. Instance #10 in Table 3.1 with $x = (0, 0, 0, 0, 0, 0, 0, 0, 0)$ is classified as follows:

$$y_{MAP} \approx \arg\max_{y\in Y} p(y)\prod_{i=1}^{9}P(x_i|y) = 0 \quad \text{(system is not faulty)}.$$

Hence, all the instances in Tables 3.1 and 3.2 are correctly classified by the naïve Bayes classifier.

3.3 Software and Applications

The following software packages support the learning of a naïve Bayes classifier:

- Weka (http://www.cs.waikato.ac.nz/ml/weka/)
- MATLAB® (http://www.mathworks.com), statistics toolbox

The naïve Bayes classifier has been successfully applied in many fields, including text and document classification (http://www.cs.waikato.ac.nz/~eibe/pubs/FrankAndBouckaertPKDD06new.pdf).

Exercises

3.1 Build a naïve Bayes classifier to classify the target variable from the attribute variable in the balloon data set in Table 1.1 and evaluate the classification performance of the naïve Bayes classifier by computing what percentage of the date records in the data set are classified correctly by the naïve Bayes classifier.

3.2 In the space shuttle O-ring data set in Table 1.2, consider the Leak-Check Pressure as a categorical attribute with three categorical values and the Number of O-rings with Stress as a categorical target variable with three categorical values. Build a naïve Bayes classifier to classify the Number of O-rings with Stress from the Leak-Check Pressure and evaluate the classification performance of the naïve Bayes classifier by computing what percentage of the date records in the data set are classified correctly by the naïve Bayes classifier.

3.3 Build a naïve Bayes classifier to classify the target variable from the attribute variables in the lenses data set in Table 1.3 and evaluate the classification performance of the naïve Bayes classifier by computing what percentage of the date records in the data set are classified correctly by the naïve Bayes classifier.

3.4 Build a naïve Bayes classifier to classify the target variable from the attribute variables in the data set found in Exercise 1.1 and evaluate the classification performance of the naïve Bayes classifier by computing what percentage of the date records in the data set are classified correctly by the naïve Bayes classifier.

4

Decision and Regression Trees

Decision and regression tress are used to learn classification and prediction patterns from data and express the relation of attribute variables x with a target variable y, $y = F(x)$, in the form of a tree. A decision tree classifies the categorical target value of a data record using its attribute values. A regression tree predicts the numeric target value of a data record using its attribute values. In this chapter, we first define a binary decision tree and give the algorithm to learn a binary decision tree from a data set with categorical attribute variables and a categorical target variable. Then the method of learning a nonbinary decision tree is described. Additional concepts are introduced to handle numeric attribute variables and missing values of attribute variables, and to handle a numeric target variable for constructing a regression tree. A list of data mining software packages that support the learning of decision and regression trees is provided. Some applications of decision and regression trees are given with references.

4.1 Learning a Binary Decision Tree and Classifying Data Using a Decision Tree

In this section, we introduce the elements of a decision tree. The rationale of seeking a decision tree with the minimum description length is provided and followed by the split selection methods. Finally, the top-down construction of a decision tree is illustrated.

4.1.1 Elements of a Decision Tree

Table 4.1 gives a part of the data set for a manufacturing system shown in Table 1.4. The data set in Table 4.1 includes nine attribute variables for the quality of parts and one target variable for system fault. This data set is used as the training data set to learn a binary decision tree for classifying whether or not the system is faulty using the values of the nine quality variables. Figure 4.1 shows the resulting binary decision tree to illustrate the elements of the decision tree. How this decision tree is learned is explained later.

As shown in Figure 4.1, a binary decision tree is a graph with nodes. The *root node* at the top of the tree consists of all data records in the training

TABLE 4.1

Data Set for System Fault Detection

Instance (Faulty Machine)	Attribute Variables									Target Variable
	Quality of Parts									System Fault y
	x_1	x_2	x_3	x_4	x_5	x_6	x_7	x_8	x_9	
1 (M1)	1	0	0	0	1	0	1	0	1	1
2 (M2)	0	1	0	1	0	0	0	1	0	1
3 (M3)	0	0	1	1	0	1	1	1	0	1
4 (M4)	0	0	0	1	0	0	0	1	0	1
5 (M5)	0	0	0	0	1	0	1	0	1	1
6 (M6)	0	0	0	0	0	1	1	0	0	1
7 (M7)	0	0	0	0	0	0	1	0	0	1
8 (M8)	0	0	0	0	0	0	0	1	0	1
9 (M9)	0	0	0	0	0	0	0	0	1	1
10 (none)	0	0	0	0	0	0	0	0	0	0

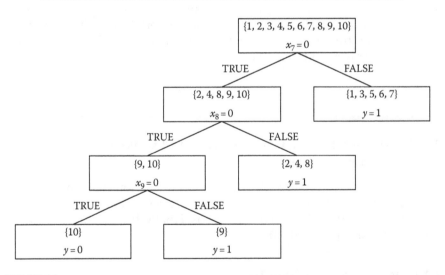

FIGURE 4.1
Decision tree for system fault detection.

data set. For the data set of system fault detection, the root node contains a set with all the 10 data records in the training data set, {1, 2, ..., 10}. Note that the numbers in the data set are the instance numbers. The root node is split into two subsets, {2, 4, 8, 9, 10} and {1, 2, 5, 6, 7}, using the attribute variable, x_7, and its two categorical values, $x_7 = 0$ and $x_7 = 1$. All the instances in the subset, {2, 4, 8, 9, 10}, have $x_7 = 0$. All the instances in the subset, {1, 2, 5, 6, 7}, have $x_7 = 1$. Each subset is represented as a node in the decision tree. A Boolean expression is used in the decision tree to express $x_7 = 0$ by $x_7 = 0$ is

TRUE, and $x_7 = 1$ by $x_7 = 0$ is FALSE. $x_7 = 0$ is called a split condition or a *split criterion*, and its TRUE and FALSE values allow a binary split of the set at the root node into two branches with a node at the end of each branch. Each of the two new nodes can be further divided using one of the remaining attribute variables in the split criterion. A node cannot be further divided if the data records in the data set at this node have the same value of the target variable. Such a node becomes a *leaf node* in the decision tree. Except the root node and leaf nodes, all other nodes in the decision trees are called *internal nodes*.

The decision tree can classify a data record by passing the data record through the decision tree using the attribute values in the data record. For example, the data record for instance 10 is first checked with the first split condition at the root node. With $x_7 = 0$, the data record is passed down to the left branch. With $x_8 = 0$ and then $x_9 = 0$, the data record is passed down to the left-most leaf node. The data record takes the target value for this leaf node, $y = 0$, which classifies the data record as not faulty.

4.1.2 Decision Tree with the Minimum Description Length

Starting with the root node with all the data records in the training data set, there are nine possible ways to split the root node using the nine attribute variables individually in the split condition. For each node at the end of a branch from the split of the root node, there are eight possible ways to split the node using each of the remaining eight attribute variables individually. This process continues and can result in many possible decision trees. All the possible decision trees differ in their size and complexity. A decision tree can be large to have as many leaf nodes as data records in the training data set with each leaf node containing each data record. Which one of all the possible decision trees should be used to represent F, the relation of the attribute variables with the target variable? A decision tree algorithm aims at obtaining the smallest decision tree that can capture F, that is, the decision tree that requires the minimum description length. Given both the smallest decision tree and a larger decision tree that classify all the data records in the training data set correctly, it is expected that the smallest decision tree generalizes classification patterns better than the larger decision tree and better-generalized classification patterns allow the better classification of more data points including those not in the training data set. Consider a large decision tree that has as many leaf nodes as data records in the training data set with each leaf node containing each data record. Although this large decision tree classifies all the training data records correctly, it may perform poorly in classifying new data records not in the training data set. Those new data records have different sets of attribute values from those of data records in the training data set and thus do not follow the same paths of the data records to leaf nodes in the decision tree. We need a decision tree that captures generalized classification patterns for the F relation. The more generalized the *F* relation, the smaller description length it has because it eliminates specific differences among individual data records. Hence, the smaller a decision tree is, the more generalization capacity the decision tree is expected to have.

4.1.3 Split Selection Methods

With the goal of seeking a decision tree with the minimum description length, we need to know how to split a node so that we can achieve the goal of obtaining the decision tree with the minimum description length. Take an example of learning a decision tree from the data set in Table 4.1. There are nine possible ways to split the root node using the nine attribute variables individually, as shown in Table 4.2.

Which one of the nine split criteria should we use so we will obtain the smallest decision tree? A common approach of split selection is to select the split that produces the most homogeneous subsets. A homogenous data set is a data set whose data records have the same target value. There are various measures of data homogeneity: information entropy, gini-index, etc. (Breiman et al., 1984; Quinlan, 1986; Ye, 2003).

Information entropy is originally introduced to measure the number of bits of information needed to encode data. Information entropy is defined as follows:

$$\text{entropy}(D) = \sum_{i=1}^{c} -P_i \log_2 P_i \quad (4.1)$$

$$-0\log_2 0 = 0 \quad (4.2)$$

$$\sum_{i=1}^{c} P_i = 1, \quad (4.3)$$

where
  D denotes a given data set
  c denotes the number of different target values
  P_i denotes the probability that a data record in the data set takes the *i*th target value

An entropy value falls in the range, $[0, \log_2 c]$. For example, given the data set in Table 4.1, we have $c = 2$ (for two target values, $y = 0$ and $y = 1$), $P_1 = 9/10$ (9 of the 10 records with $y = 0$) = 0.9, $P_2 = 1/10$ (1 of the 10 records with $y = 1$) = 0.1, and

$$\text{entropy}(D) = \sum_{i=1}^{2} -P_i \log_2 P_i = -0.9\log_2 0.9 - 0.1\log_2 0.1 = 0.47.$$

TABLE 4.2

Binary Split of the Root Node and Calculation of Information Entropy for the Data Set of System Fault Detection

Split Criterion	Resulting Subsets and Average Information Entropy of Split
$x_1 = 0$: TRUE or FALSE	$\{2, 3, 4, 5, 6, 7, 8, 9, 10\}, \{1\}$ $\text{entropy}(S) = \dfrac{9}{10}\text{entropy}(D_{true}) + \dfrac{1}{10}\text{entropy}(D_{false})$ $= \dfrac{9}{10} \times \left(-\dfrac{8}{9}\log_2 \dfrac{8}{9} - \dfrac{1}{9}\log_2 \dfrac{1}{9}\right) + \dfrac{1}{10} \times 0 = 0.45$
$x_2 = 0$: TRUE or FALSE	$\{1, 3, 4, 5, 6, 7, 8, 9, 10\}, \{2\}$ $\text{entropy}(S) = \dfrac{9}{10}\text{entropy}(D_{true}) + \dfrac{1}{10}\text{entropy}(D_{false})$ $= \dfrac{9}{10} \times \left(-\dfrac{8}{9}\log_2 \dfrac{8}{9} - \dfrac{1}{9}\log_2 \dfrac{1}{9}\right) + \dfrac{1}{10} \times 0 = 0.45$
$x_3 = 0$: TRUE or FALSE	$\{1, 2, 4, 5, 6, 7, 8, 9, 10\}, \{3\}$ $\text{entropy}(S) = \dfrac{9}{10}\text{entropy}(D_{true}) + \dfrac{1}{10}\text{entropy}(D_{false})$ $= \dfrac{9}{10} \times \left(-\dfrac{8}{9}\log_2 \dfrac{8}{9} - \dfrac{1}{9}\log_2 \dfrac{1}{9}\right) + \dfrac{1}{10} \times 0 = 0.45$
$x_4 = 0$: TRUE or FALSE	$\{1, 5, 6, 7, 8, 9, 10\}, \{2, 3, 4\}$ $\text{entropy}(S) = \dfrac{7}{10}\text{entropy}(D_{true}) + \dfrac{3}{10}\text{entropy}(D_{false})$ $= \dfrac{7}{10} \times \left(-\dfrac{6}{7}\log_2 \dfrac{6}{7} - \dfrac{1}{7}\log_2 \dfrac{1}{7}\right) + \dfrac{3}{10} \times 0 = 0.41$
$x_5 = 0$: TRUE or FALSE	$\{2, 3, 4, 6, 7, 8, 9, 10\}, \{1, 5\}$ $\text{entropy}(S) = \dfrac{8}{10}\text{entropy}(D_{true}) + \dfrac{2}{10}\text{entropy}(D_{false})$ $= \dfrac{8}{10} \times \left(-\dfrac{7}{8}\log_2 \dfrac{7}{8} - \dfrac{1}{8}\log_2 \dfrac{1}{8}\right) + \dfrac{2}{10} \times 0 = 0.43$
$x_6 = 0$: TRUE or FALSE	$\{1, 2, 4, 5, 7, 8, 9, 10\}, \{3, 6\}$ $\text{entropy}(S) = \dfrac{8}{10}\text{entropy}(D_{true}) + \dfrac{2}{10}\text{entropy}(D_{false})$ $= \dfrac{8}{10} \times \left(-\dfrac{7}{8}\log_2 \dfrac{7}{8} - \dfrac{1}{8}\log_2 \dfrac{1}{8}\right) + \dfrac{2}{10} \times 0 = 0.43$
$x_7 = 0$: TRUE or FALSE	$\{2, 4, 8, 9, 10\}, \{1, 3, 5, 6, 7\}$ $\text{entropy}(S) = \dfrac{5}{10}\text{entropy}(D_{true}) + \dfrac{5}{10}\text{entropy}(D_{false})$ $= \dfrac{5}{10} \times \left(-\dfrac{4}{5}\log_2 \dfrac{4}{5} - \dfrac{1}{5}\log_2 \dfrac{1}{5}\right) + \dfrac{5}{10} \times 0 = 0.36$

(continued)

TABLE 4.2 (continued)
Binary Split of the Root Node and Calculation of Information Entropy for the Data Set of System Fault Detection

Split Criterion	Resulting Subsets and Average Information Entropy of Split
$x_8 = 0$: TRUE or FALSE	$\{1, 5, 6, 7, 9, 10\}, \{2, 3, 4, 8\}$ $\text{entropy}(S) = \dfrac{6}{10}\text{entropy}(D_{true}) + \dfrac{4}{10}\text{entropy}(D_{false})$ $= \dfrac{6}{10} \times \left(-\dfrac{5}{6}\log_2\dfrac{5}{6} - \dfrac{1}{6}\log_2\dfrac{1}{6}\right) + \dfrac{4}{10} \times 0 = 0.39$
$x_9 = 0$: TRUE or FALSE	$\{2, 3, 4, 6, 7, 8, 10\}, \{1, 5, 9\}$ $\text{entropy}(S) = \dfrac{7}{10}\text{entropy}(D_{true}) + \dfrac{3}{10}\text{entropy}(D_{false})$ $= \dfrac{7}{10} \times \left(-\dfrac{6}{7}\log_2\dfrac{6}{7} - \dfrac{1}{7}\log_2\dfrac{1}{7}\right) + \dfrac{3}{10} \times 0 = 0.41$

Figure 4.2 shows how the entropy value changes with P_1 ($P_2 = 1 - P_1$) when $c = 2$. Especially, we have

- $P_1 = 0.5, P_2 = 0.5, \text{entropy}(D) = 1$
- $P_1 = 0, P_2 = 1, \text{entropy}(D) = 0$
- $P_1 = 1, P_2 = 0, \text{entropy}(D) = 0$

If all the data records in a data set take one target value, we have $P_1 = 0, P_2 = 1$ or $P_1 = 1, P_2 = 0$, and the value of information entropy is 0, that is, we need 0 bit of information because we already know the target value that all the data records take. Hence, the entropy value of 0 indicates that the data set is homogenous

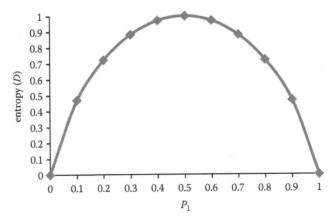

FIGURE 4.2
Information entropy.

with regard to the target value. If one half set of the data records in a data set takes one target value and the other half set takes another target value, we have $P_1 = 0.5$, $P_2 = 0.5$, and the value of information entropy is 1, meaning that we need 1 bit of information to convey what target value is. Hence, the entropy value of 1 indicates that the data set is inhomogeneous. When we use the information entropy to measure data homogeneity, the lower the entropy value is, the more homogenous the data set is with regard to the target value.

After a split of a data set into several subsets, the following formula is used to compute the average information entropy of subsets:

$$\text{entropy}(S) = \sum_{v \in Values(S)} \frac{|D_v|}{|D|} \text{entropy}(D_v), \quad (4.4)$$

where
S denotes the split
$Values(S)$ denotes a set of values that are used in the split
v denotes a value in $Values(S)$
D denotes the data set being split
$|D|$ denotes the number of data records in the data set D
D_v denotes the subset resulting from the split using the split value v
$|D_v|$ denotes the number of data records in the data set D_v

For example, the root node of a decision tree for the data set in Table 4.1 has the data set, $D = \{1, 2, ..., 10\}$, whose entropy value is 0.47 as shown previously. Using the split criterion, $x_1 = 0$: TRUE or FALSE, the root node is split into two subsets: $D_{false} = \{1\}$, which is homogenous, and $D_{true} = \{2, 3, 4, 5, 6, 7, 8, 9, 10\}$, which is inhomogeneous with eight data records taking the target value of 1 and one data record taking the target value of 0. The average entropy of the two subsets after the split is

$$\text{entropy}(S) = \frac{9}{10} \text{entropy}(D_{true}) + \frac{1}{10} \text{entropy}(D_{false})$$

$$= \frac{9}{10} \times \left(-\frac{8}{9} \log_2 \frac{8}{9} - \frac{1}{9} \log_2 \frac{1}{9}\right) + \frac{1}{10} \times 0 = 0.45.$$

Since this average entropy of subsets after the split is better than entropy $(D) = 0.47$, the split improves data homogeneity. Table 4.2 gives the average entropy of subsets after each of the other eight splits of the root node. Among the nine possible splits, the split using the criterion of $x_7 = 0$: TRUE or FALSE produces the smallest average information entropy, which indicates the most homogeneous subsets. Hence, the split criterion of $x_7 = 0$: TRUE or FALSE is selected to split the root node, resulting in two internal nodes as shown in Figure 4.1. The internal node with the subset, $\{2, 4, 8, 9, 10\}$, is not homogenous. Hence, the decision tree is further expanded with more splits until all leaf nodes are homogenous.

The gini-index, another measure of data homogeneity, is defined as follows:

$$\text{gini}(D) = 1 - \sum_{i=1}^{c} P_i^2. \tag{4.5}$$

For example, given the data set in Table 4.1, we have $c = 2$, $P_1 = 0.9$, $P_2 = 0.1$, and

$$\text{gini}(D) = 1 - \sum_{i=1}^{c} P_i^2 = 1 - 0.9^2 - 0.1^2 = 0.18.$$

The gini-index values are computed for $c = 2$ and the following values of P_i:

- $P_1 = 0.5$, $P_2 = 0.5$, gini(D) = $1 - 0.5^2 - 0.5^2 = 0.5$
- $P_1 = 0$, $P_2 = 1$, gini(D) = $1 - 0^2 - 1^2 = 0$
- $P_1 = 1$, $P_2 = 0$, gini(D) = $1 - 1^2 - 0^2 = 0$

Hence, the smaller the gini-index value is, the more homogeneous the data set is. The average gini-index value of data subsets after a split is calculated as follows:

$$\text{gini}(S) = \sum_{v \in Values(S)} \frac{|D_v|}{|D|} \text{gini}(D_v). \tag{4.6}$$

Table 4.3 gives the average gini-index value of subsets after each of the nine splits of the root node for the training data set of system fault detection. Among the nine possible splits, the split criterion of $x_7 = 0$: TRUE or FALSE produces the smallest average gini-index value, which indicates the most homogeneous subsets. The split criterion of $x_7 = 0$: TRUE or FALSE is selected to split the root node. Hence, using the gini-index produces the same split as using the information entropy.

4.1.4 Algorithm for the Top-Down Construction of a Decision Tree

This section describes and illustrates the algorithm of constructing a complete decision tree. The algorithm for the top-down construction of a binary decision tree has the following steps:

1. Start with the root node that includes all the data records in the training data set and select this node to split.
2. Apply a split selection method to the selected node to determine the best split along with the split criterion and partition the set of the training data records at the selected node into two nodes with two subsets of data records, respectively.
3. Check if the stopping criterion is satisfied. If so, the tree construction is completed; otherwise, go back to Step 2 to continue by selecting a node to split.

TABLE 4.3

Binary Split of the Root Node and Calculation of the Gini-Index for the Data Set of System Fault Detection

Split Criterion	Resulting Subsets and Average Gini-Index Value of Split
$x_1 = 0$: TRUE or FALSE	$\{2, 3, 4, 5, 6, 7, 8, 9, 10\}, \{1\}$
	$\text{gini}(S) = \dfrac{9}{10}\text{gini}(D_{true}) + \dfrac{1}{10}\text{gini}(D_{false})$
	$= \dfrac{9}{10} \times \left[1 - \left(\dfrac{1}{9}\right)^2 - \left(\dfrac{8}{9}\right)^2\right] + \dfrac{1}{10} \times 0 = 0.18$
$x_2 = 0$: TRUE or FALSE	$\{1, 3, 4, 5, 6, 7, 8, 9, 10\}, \{2\}$
	$\text{gini}(S) = \dfrac{9}{10}\text{gini}(D_{true}) + \dfrac{1}{10}\text{gini}(D_{false})$
	$= \dfrac{9}{10} \times \left[1 - \left(\dfrac{1}{9}\right)^2 - \left(\dfrac{8}{9}\right)^2\right] + \dfrac{1}{10} \times 0 = 0.18$
$x_3 = 0$: TRUE or FALSE	$\{1, 2, 4, 5, 6, 7, 8, 9, 10\}, \{3\}$
	$\text{gini}(S) = \dfrac{9}{10}\text{gini}(D_{true}) + \dfrac{1}{10}\text{gini}(D_{false})$
	$= \dfrac{9}{10} \times \left[1 - \left(\dfrac{1}{9}\right)^2 - \left(\dfrac{8}{9}\right)^2\right] + \dfrac{1}{10} \times 0 = 0.18$
$x_4 = 0$: TRUE or FALSE	$\{1, 5, 6, 7, 8, 9, 10\}, \{2, 3, 4\}$
	$\text{gini}(S) = \dfrac{7}{10}\text{gini}(D_{true}) + \dfrac{3}{10}\text{gini}(D_{false})$
	$= \dfrac{7}{10} \times \left[1 - \left(\dfrac{6}{7}\right)^2 - \left(\dfrac{1}{7}\right)^2\right] + \dfrac{3}{10} \times 0 = 0.17$
$x_5 = 0$: TRUE or FALSE	$\{2, 3, 4, 6, 7, 8, 9, 10\}, \{1, 5\}$
	$\text{gini}(S) = \dfrac{8}{10}\text{gini}(D_{true}) + \dfrac{2}{10}\text{gini}(D_{false})$
	$= \dfrac{8}{10} \times \left[1 - \left(\dfrac{7}{8}\right)^2 - \left(\dfrac{1}{8}\right)^2\right] + \dfrac{2}{10} \times 0 = 0.175$
$x_6 = 0$: TRUE or FALSE	$\{1, 2, 4, 5, 7, 8, 9, 10\}, \{3, 6\}$
	$\text{gini}(S) = \dfrac{8}{10}\text{gini}(D_{true}) + \dfrac{2}{10}\text{gini}(D_{false})$
	$= \dfrac{8}{10} \times \left[1 - \left(\dfrac{7}{8}\right)^2 - \left(\dfrac{1}{8}\right)^2\right] + \dfrac{2}{10} \times 0 = 0.175$

(continued)

TABLE 4.3 (continued)

Binary Split of the Root Node and Calculation of the Gini-Index for the Data Set of System Fault Detection

Split Criterion	Resulting Subsets and Average Gini-Index Value of Split
$x_7 = 0$: TRUE or FALSE	$\{2, 4, 8, 9, 10\}, \{1, 3, 5, 6, 7\}$ $$\text{gini}(S) = \frac{5}{10}\text{gini}(D_{true}) + \frac{5}{10}\text{gini}(D_{false})$$ $$= \frac{5}{10} \times \left(1 - \left(\frac{4}{5}\right)^2 - \left(\frac{1}{5}\right)^2\right) + \frac{5}{10} \times 0 = 0.16$$
$x_8 = 0$: TRUE or FALSE	$\{1, 5, 6, 7, 9, 10\}, \{2, 3, 4, 8\}$ $$\text{gini}(S) = \frac{6}{10}\text{gini}(D_{true}) + \frac{4}{10}\text{gini}(D_{false})$$ $$= \frac{6}{10} \times \left(1 - \left(\frac{5}{6}\right)^2 - \left(\frac{1}{6}\right)^2\right) + \frac{4}{10} \times 0 = 0.167$$
$x_9 = 0$: TRUE or FALSE	$\{2, 3, 4, 6, 7, 8, 10\}, \{1, 5, 9\}$ $$\text{gini}(S) = \frac{7}{10}\text{gini}(D_{true}) + \frac{3}{10}\text{gini}(D_{false})$$ $$= \frac{7}{10} \times \left(1 - \left(\frac{6}{7}\right)^2 - \left(\frac{1}{7}\right)^2\right) + \frac{3}{10} \times 0 = 0.17$$

The stopping criterion based on data homogeneity is to stop when each leaf node has homogeneous data, that is, a set of data records with the same target value. Many large sets of real-world data are noisy, making it difficult to obtain homogeneous data sets at leaf nodes. Hence, the stopping criterion is often set to have the measure of data homogeneity to be smaller than a threshold value, e.g., entropy(D) < 0.1.

We show the construction of the complete binary decision tree for the data set of system fault detection next.

Example 4.1

Construct a binary decision tree for the data set of system fault detection in Table 4.1.

We first use the information entropy as the measure of data homogeneity. As shown in Figure 4.1, the data set at the root node is partitioned into two subsets, $\{2, 4, 8, 9, 10\}$, and $\{1, 3, 5, 6, 7\}$, which are already homogeneous with the target value, $y = 1$, and do not need a split. For the subset, $D = \{2, 4, 8, 9, 10\}$,

$$\text{entropy}(D) = \sum_{i=1}^{2} -P_i \log_2 P_i = -\frac{1}{5}\log_2\frac{1}{5} - \frac{4}{5}\log_2\frac{4}{5} = 0.72.$$

Decision and Regression Trees

Except x_7, which has been used to split the root node, the other eight attribute variables, $x_1, x_2, x_3, x_4, x_5, x_6, x_8,$ and $x_9,$ can be used to split D. The split criteria using $x_1 = 0, x_3 = 0, x_5 = 0,$ and $x_6 = 0$ do not produce a split of D. Table 4.4 gives the calculation of information entropy for the splits using $x_2, x_4, x_7, x_8,$ and x_9. Since the split criterion, $x_8 = 0$: TRUE or FALSE, produces the smallest average entropy of the split, this split criterion is selected to split $D = \{2, 4, 8, 9, 10\}$ into $\{9, 10\}$ and $\{2, 4, 8\}$, which are already homogeneous with the target value, $y = 1$, and do not need a split. Figure 4.1 shows this split.

For the subset, $D = \{9, 10\}$,

$$\text{entropy}(D) = \sum_{i=1}^{2} -P_i \log_2 P_i = -\frac{1}{2}\log_2 \frac{1}{2} - \frac{1}{2}\log_2 \frac{1}{2} = 1.$$

Except x_7 and x_8, which have been used to split the root node, the other seven attribute variables, $x_1, x_2, x_3, x_4, x_5, x_6,$ and $x_9,$ can be used to split D. The split criteria using $x_1 = 0, x_2 = 0, x_3 = 0, x_4 = 0, x_5 = 0,$ and $x_6 = 0$ do not produce a split of D. The split criterion of $x_9 = 0$: TRUE or FALSE,

TABLE 4.4

Binary Split of an Internal Node with $D = \{2, 4, 5, 9, 10\}$ and Calculation of Information Entropy for the Data Set of System Fault Detection

Split Criterion	Resulting Subsets and Average Information Entropy of Split
$x_2 = 0$: TRUE or FALSE	$\{4, 8, 9, 10\}, \{2\}$
	$\text{entropy}(S) = \frac{4}{5}\text{entropy}(D_{true}) + \frac{1}{5}\text{entropy}(D_{false})$
	$= \frac{4}{5} \times \left(-\frac{3}{4}\log_2 \frac{8}{9} - \frac{1}{4}\log_2 \frac{1}{4}\right) + \frac{1}{5} \times 0 = 0.64$
$x_4 = 0$: TRUE or FALSE	$\{8, 9, 10\}, \{2, 4\}$
	$\text{entropy}(S) = \frac{3}{5}\text{entropy}(D_{true}) + \frac{2}{5}\text{entropy}(D_{false})$
	$= \frac{3}{5} \times \left(-\frac{2}{3}\log_2 \frac{2}{3} - \frac{1}{3}\log_2 \frac{1}{3}\right) + \frac{2}{5} \times 0 = 0.55$
$x_8 = 0$: TRUE or FALSE	$\{9, 10\}, \{2, 4, 8\}$
	$\text{entropy}(S) = \frac{2}{5}\text{entropy}(D_{true}) + \frac{3}{5}\text{entropy}(D_{false})$
	$= \frac{2}{5} \times \left(-\frac{1}{2}\log_2 \frac{1}{2} - \frac{1}{2}\log_2 \frac{1}{2}\right) + \frac{3}{5} \times 0 = 0.4$
$x_9 = 0$: TRUE or FALSE	$\{2, 4, 8, 10\}, \{9\}$
	$\text{entropy}(S) = \frac{4}{5}\text{entropy}(D_{true}) + \frac{1}{5}\text{entropy}(D_{false})$
	$= \frac{4}{5} \times \left(-\frac{3}{4}\log_2 \frac{3}{4} - \frac{1}{4}\log_2 \frac{1}{4}\right) + \frac{1}{5} \times 0 = 0.64$

produces two subsets, {9} with the target value of $y = 1$, and {10} with the target value of $y = 0$, which are homogeneous and do not need a split. Figure 4.1 shows this split. Since all leaf nodes of the decision tree are homogeneous, the construction of the decision tree is stopped with the complete decision tree shown in Figure 4.1.

We now show the construction of the decision tree using the gini-index as the measure of data homogeneity. As described previously, the data set at the root node is partitioned into two subsets, {2, 4, 8, 9, 10} and {1, 3, 5, 6, 7}, which are already homogeneous with the target value, $y = 1$, and do not need a split. For the subset, $D = \{2, 4, 8, 9, 10\}$,

$$\text{gini}(D) = 1 - \sum_{i=1}^{c} P_i^2 = 1 - \left(\frac{4}{5}\right)^2 - \left(\frac{1}{5}\right)^2 = 0.32.$$

The split criteria using $x_1 = 0$, $x_3 = 0$, $x_5 = 0$, and $x_6 = 0$ do not produce a split of D. Table 4.5 gives the calculation of the gini-index values for the splits

TABLE 4.5

Binary Split of an Internal Node with $D = \{2, 4, 5, 9, 10\}$ and Calculation of the Gini-Index Values for the Data Set of System Fault Detection

Split Criterion	Resulting Subsets and Average Gini-Index Value of Split
$x_2 = 0$: TRUE or FALSE	{4, 8, 9, 10}, {2} $$\text{gini}(S) = \frac{4}{5}\text{gini}(D_{true}) + \frac{1}{5}\text{gini}(D_{false})$$ $$= \frac{4}{5} \times \left(1 - \left(\frac{3}{4}\right)^2 - \left(\frac{1}{4}\right)^2\right) + \frac{1}{5} \times 0 = 0.3$$
$x_4 = 0$: TRUE or FALSE	{8, 9, 10}, {2, 4} $$\text{gini}(S) = \frac{3}{5}\text{gini}(D_{true}) + \frac{2}{5}\text{gini}(D_{false})$$ $$= \frac{3}{5} \times \left(1 - \left(\frac{3}{4}\right)^2 - \left(\frac{1}{4}\right)^2\right) + \frac{2}{5} \times 0 = 0.27$$
$x_8 = 0$: TRUE or FALSE	{9, 10}, {2, 4, 8} $$\text{gini}(S) = \frac{2}{5}\text{gini}(D_{true}) + \frac{3}{5}\text{gini}(D_{false})$$ $$= \frac{2}{5} \times \left(1 - \left(\frac{1}{2}\right)^2 - \left(\frac{1}{2}\right)^2\right) + \frac{3}{5} \times 0 = 0.2$$
$x_9 = 0$: TRUE or FALSE	{2, 4, 8, 10}, {9} $$\text{gini}(S) = \frac{4}{5}\text{gini}(D_{true}) + \frac{1}{5}\text{gini}(D_{false})$$ $$= \frac{4}{5} \times \left(1 - \left(\frac{3}{4}\right)^2 - \left(\frac{1}{4}\right)^2\right) + \frac{1}{5} \times 0 = 0.3$$

using x_2, x_4, x_7, x_8, and x_9. Since the split criterion, $x_8 = 0$: TRUE or FALSE, produces the smallest average gini-index value of the split, this split criterion is selected to split $D = \{2, 4, 8, 9, 10\}$ into $\{9, 10\}$ and $\{2, 4, 8\}$, which are already homogeneous with the target value, $y = 1$, and do not need a split. For the subset, $D = \{9, 10\}$,

$$\text{gini}(D) = 1 - \sum_{i=1}^{c} P_i^2 = 1 - \left(\frac{1}{2}\right)^2 - \left(\frac{1}{2}\right)^2 = 0.5.$$

Except x_7 and x_8, which have been used to split the root node, the other seven attribute variables, x_1, x_2, x_3, x_4, x_5, x_6, and x_9, can be used to split D. The split criteria using $x_1 = 0$, $x_2 = 0$, $x_3 = 0$, $x_4 = 0$, $x_5 = 0$, and $x_6 = 0$ do not produce a split of D. The split criterion of $x_9 = 0$: TRUE or FALSE, produces two subsets, $\{9\}$ with the target value of $y = 1$, and $\{10\}$ with the target value of $y = 0$, which are homogeneous and do not need a split. Since all leaf nodes of the decision tree are homogeneous, the construction of the decision tree is stopped with the complete decision tree, which is the same as the decision tree from using the information entropy as the measure of data homogeneity.

4.1.5 Classifying Data Using a Decision Tree

A decision tree is used to classify a data record by passing the data record into a leaf node of the decision tree using the values of the attribute variables and assigning the target value of the leaf node to the data record. Figure 4.3 highlights in bold the path of passing the training data record, for instance 10 in Table 4.1, from the root node to a leaf node with the target value, $y = 0$. Hence, the data record is classified to have no system fault. For the data records in the testing data set of system fault detection in Table 4.6,

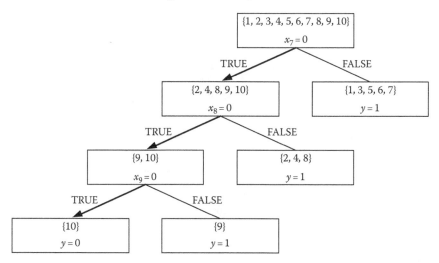

FIGURE 4.3
Classifying a data record for no system fault using the decision tree for system fault detection.

50 Data Mining

TABLE 4.6

Classification of Data Records in the Testing Data Set for System Fault Detection

Instance (Faulty Machine)	Attribute Variables (Quality of Parts)									Target Variable y (System Faults)	
	x_1	x_2	x_3	x_4	x_5	x_6	x_7	x_8	x_9	True Value	Classified Value
1 (M1, M2)	1	1	0	1	1	0	1	1	1	1	1
2 (M2, M3)	0	1	1	1	0	1	1	1	0	1	1
3 (M1, M3)	1	0	1	1	1	1	1	1	1	1	1
4 (M1, M4)	1	0	0	1	1	0	1	1	1	1	1
5 (M1, M6)	1	0	0	0	1	1	1	0	1	1	1
6 (M2, M6)	0	1	0	1	0	1	1	1	0	1	1
7 (M2, M5)	0	1	0	1	1	0	1	1	0	1	1
8 (M3, M5)	0	0	1	1	1	1	1	1	1	1	1
9 (M4, M7)	0	0	0	1	0	0	1	1	0	1	1
10 (M5, M8)	0	0	0	0	1	0	1	1	0	1	1
11 (M3, M9)	0	0	1	1	0	1	1	1	1	1	1
12 (M1, M8)	1	0	0	0	1	0	1	1	1	1	1
13 (M1, M2, M3)	1	1	1	1	1	1	1	1	1	1	1
14 (M2, M3, M5)	0	1	1	1	1	1	1	1	1	1	1
15 (M2, M3, M9)	0	1	1	1	0	1	1	1	1	1	1
16 (M1, M6, M8)	1	0	0	0	1	1	1	1	1	1	1

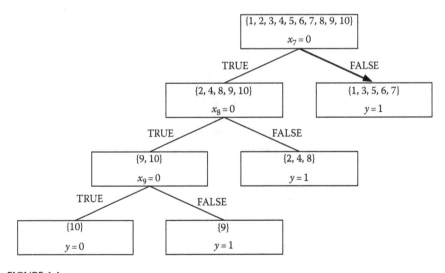

FIGURE 4.4

Classifying a data record for multiple machine faults using the decision tree for system fault detection.

Decision and Regression Trees 51

their target values are obtained using the decision tree in Figure 4.1 and are shown in Table 4.6. Figure 4.4 highlights the path of passing a testing data record for instance 1 in Table 4.6 from the root node to a leaf node with the target value, $y = 1$. Hence, the data record is classified to have a system fault.

4.2 Learning a Nonbinary Decision Tree

In the lenses data set in Table 1.4, the attribute variable, Age, has three categorical values: Young, Pre-presbyopic, and Presbyopic. If we want to construct a binary decision tree for this data set, we need to convert the three categorical values of the attribute variable, Age, into two categorical values if Age is used to split the root node. We may have Young and Pre-presbyopic together as one category, Presbyopic as another category, and Age = Presbyopic: TRUE or FALSE as the split criterion. We may also have Young as one category, Pre-presbyopic and Presbyopic together as another category, and Age = Young: TRUE or FALSE as the split criterion. However, we can construct a nonbinary decision tree to allow partitioning a data set at a node into more than two subsets by using each of multiple categorical values for each branch of the split. Example 4.2 shows the construction of a nonbinary decision tree for the lenses data set.

Example 4.2

Construct a nonbinary decision tree for the lenses data set in Table 1.3.

If the attribute variable, Age, is used to split the root node for the lenses data set, all three categorical values of Age can be used to partition the set of 24 data records at the root node using the split criterion, Age = Young, Pre-presbyopic, or Presbyopic, as shown in Figure 4.5. We use the data set of 24 data records in Table 1.3 as the training data set, D, at the root node of the nonbinary decision tree. In the lenses data set, the target variable has three categorical values, Non-Contact in 15 data records, Soft-Contact in 5 data records, and Hard-Contact in 4 data records. Using the information entropy as the measure of data homogeneity, we have

$$\text{entropy}(D) = \sum_{i=1}^{3} -P_i \log_2 P_i = -\frac{15}{24}\log_2\frac{15}{24} - \frac{5}{24}\log_2\frac{5}{24} - \frac{4}{24}\log_2\frac{4}{24} = 1.3261.$$

Table 4.7 shows the calculation of information entropy to split the root node using the split criterion, Tear Production Rate = Reduced or Normal, which produces a homogenous subset of {1, 3, 5, 7, 9, 11, 13, 15, 17, 19, 21, 23} and an inhomogeneous subset of {2, 4, 6, 8, 10, 12, 14, 16, 18, 20, 22, 24}. Table 4.8 shows the calculation of information

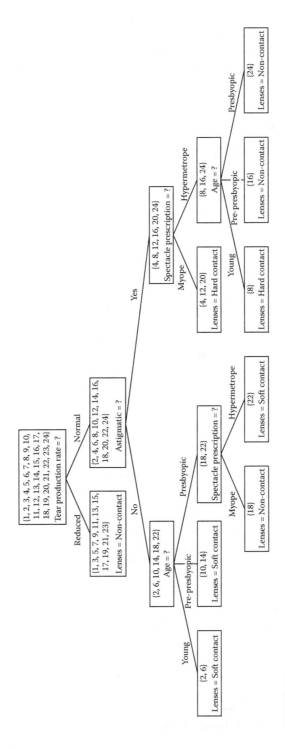

FIGURE 4.5
Decision tree for the lenses data set.

TABLE 4.7

Nonbinary Split of the Root Node and Calculation of Information Entropy for the Lenses Data Set

Split Criterion	Resulting Subsets and Average Information Entropy of Split
Age = Young, Pre-presbyopic, or Presbyopic	$\{1, 2, 3, 4, 5, 6, 7, 8\}, \{9, 10, 11, 12, 13, 14, 15, 16\}, \{17, 18, 19, 20, 21, 22, 23, 24\}$ $\text{entropy}(S) = \frac{8}{24} \text{entropy}(D_{Young}) + \frac{8}{24} \text{entropy}(D_{Pre-presbyopic})$ $\quad + \frac{8}{24} \text{entropy}(D_{Presbyopic})$ $= \frac{8}{24} \times \left(-\frac{4}{8}\log_2\frac{4}{8} - \frac{2}{8}\log_2\frac{2}{8} - \frac{2}{8}\log_2\frac{2}{8}\right)$ $\quad + \frac{8}{24} \times \left(-\frac{5}{8}\log_2\frac{5}{8} - \frac{2}{8}\log_2\frac{2}{8} - \frac{1}{8}\log_2\frac{1}{8}\right)$ $\quad + \frac{8}{24} \times \left(-\frac{6}{8}\log_2\frac{6}{8} - \frac{1}{8}\log_2\frac{1}{8} - \frac{1}{8}\log_2\frac{1}{8}\right) = 1.2867$
Spectacle Prescription = Myope or Hypermetrope	$\{1, 2, 3, 4, 9, 10, 11, 12, 17, 18, 19, 20\}, \{5, 6, 7, 8, 13, 14, 15, 16, 21, 22, 23, 24\}$ $\text{entropy}(S) = \frac{12}{24} \text{entropy}(D_{Myope}) + \frac{12}{24} \text{entropy}(D_{Hypermetrope})$ $= \frac{12}{24} \times \left(-\frac{7}{12}\log_2\frac{7}{12} - \frac{2}{12}\log_2\frac{2}{12} - \frac{3}{12}\log_2\frac{3}{12}\right)$ $\quad + \frac{12}{24} \times \left(-\frac{8}{12}\log_2\frac{8}{12} - \frac{3}{12}\log_2\frac{3}{12} - \frac{1}{12}\log_2\frac{1}{12}\right)$ $= 1.2866$
Astigmatic = No or Yes	$\{1, 2, 5, 6, 9, 10, 13, 14, 17, 18, 21, 22\}, \{3, 4, 7, 8, 11, 12, 15, 16, 19, 20, 23, 24\}$ $\text{entropy}(S) = \frac{12}{24} \text{entropy}(D_{No}) + \frac{12}{24} \text{entropy}(D_{Yes})$ $= \frac{12}{24} \times \left(-\frac{7}{12}\log_2\frac{7}{12} - \frac{5}{12}\log_2\frac{5}{12} - \frac{0}{12}\log_2\frac{0}{12}\right)$ $\quad + \frac{12}{24} \times \left(-\frac{8}{12}\log_2\frac{8}{12} - \frac{4}{12}\log_2\frac{4}{12} - \frac{0}{12}\log_2\frac{0}{12}\right)$ $= 0.9491$
Tear Production Rate = Reduced or Normal	$\{1, 3, 5, 7, 9, 11, 13, 15, 17, 19, 21, 23\}, \{2, 4, 6, 8, 10, 12, 14, 16, 18, 20, 22, 24\}$ $\text{entropy}(S) = \frac{12}{24} \text{entropy}(D_{Reduced}) + \frac{12}{24} \text{entropy}(D_{Normal})$ $= \frac{12}{24} \times \left(-\frac{12}{12}\log_2\frac{12}{12} - \frac{0}{12}\log_2\frac{0}{12} - \frac{0}{12}\log_2\frac{0}{12}\right)$ $\quad + \frac{12}{24} \times \left(-\frac{3}{12}\log_2\frac{3}{12} - \frac{5}{12}\log_2\frac{5}{12} - \frac{4}{12}\log_2\frac{4}{12}\right)$ $= 0.7773$

TABLE 4.8

Nonbinary Split of an Internal Node, {2, 4, 6, 8, 10, 12, 14, 16, 18, 20, 22, 24}, and Calculation of Information Entropy for the Lenses Data Set

Split Criterion	Resulting Subsets and Average Information Entropy of Split
Age = Young, Pre-presbyopic, or Presbyopic	{2, 4, 6, 8}, {10, 12, 14, 16}, {18, 20, 22, 24} $$\text{entropy}(S) = \frac{4}{12}\text{entropy}(D_{Young})$$ $$+ \frac{4}{12}\text{entropy}(D_{Pre-presbyopic})$$ $$+ \frac{4}{12}\text{entropy}(D_{Presbyopic})$$ $$= \frac{4}{12} \times \left(-\frac{0}{4}\log_2\frac{0}{4} - \frac{2}{4}\log_2\frac{2}{4} - \frac{2}{4}\log_2\frac{2}{4}\right)$$ $$+ \frac{4}{12} \times \left(-\frac{1}{4}\log_2\frac{1}{4} - \frac{2}{4}\log_2\frac{2}{4} - \frac{1}{4}\log_2\frac{1}{4}\right)$$ $$+ \frac{4}{12} \times \left(-\frac{2}{4}\log_2\frac{2}{4} - \frac{1}{4}\log_2\frac{1}{4} - \frac{1}{4}\log_2\frac{1}{4}\right)$$ $$= 1.3333$$
Spectacle Prescription = Myope or Hypermetrope	{2, 4, 10, 12, 18, 20}, {6, 7, 14, 16, 22, 24} $$\text{entropy}(S) = \frac{6}{12}\text{entropy}(D_{Myope})$$ $$+ \frac{6}{12}\text{entropy}(D_{Hypermetrope})$$ $$= \frac{6}{12} \times \left(-\frac{1}{6}\log_2\frac{1}{6} - \frac{2}{6}\log_2\frac{2}{6} - \frac{3}{6}\log_2\frac{3}{6}\right)$$ $$+ \frac{6}{12} \times \left(-\frac{2}{6}\log_2\frac{2}{6} - \frac{3}{6}\log_2\frac{3}{6} - \frac{1}{6}\log_2\frac{1}{6}\right)$$ $$= 1.4591$$
Astigmatic = No or Yes	{2, 6, 10, 14, 18, 22}, {4, 8, 12, 16, 20, 24} $$\text{entropy}(S) = \frac{6}{12}\text{entropy}(D_{No})$$ $$+ \frac{6}{12}\text{entropy}(D_{Yes})$$ $$= \frac{6}{12} \times \left(-\frac{1}{6}\log_2\frac{1}{6} - \frac{5}{6}\log_2\frac{5}{6} - \frac{0}{6}\log_2\frac{0}{6}\right)$$ $$+ \frac{6}{12} \times \left(-\frac{2}{6}\log_2\frac{2}{6} - \frac{0}{6}\log_2\frac{0}{6} - \frac{4}{6}\log_2\frac{4}{6}\right)$$ $$= 0.7842$$

TABLE 4.9
Nonbinary Split of an Internal Node, {2, 6, 10, 14, 18, 22}, and Calculation of Information Entropy for the Lenses Data Set

Split Criterion	Resulting Subsets and Average Information Entropy of Split
Age = Young, Pre-presbyopic, or Presbyopic	{2, 6}, {10, 14}, {18, 22} $\text{entropy}(S) = \frac{2}{6}\text{entropy}(D_{Young}) + \frac{2}{6}\text{entropy}(D_{Pre-presbyopic})$ $+ \frac{2}{6}\text{entropy}(D_{Presbyopic})$ $= \frac{2}{6} \times \left(-\frac{0}{2}\log_2\frac{0}{2} - \frac{2}{2}\log_2\frac{2}{2} - \frac{0}{2}\log_2\frac{0}{2} \right)$ $+ \frac{2}{6} \times \left(-\frac{0}{2}\log_2\frac{0}{2} - \frac{2}{2}\log_2\frac{2}{2} - \frac{0}{2}\log_2\frac{0}{2} \right)$ $+ \frac{2}{6} \times \left(-\frac{1}{2}\log_2\frac{1}{2} - \frac{1}{2}\log_2\frac{1}{2} - \frac{0}{2}\log_2\frac{0}{2} \right)$ $= 0.3333$
Spectacle Prescription = Myope or Hypermetrope	{2, 10, 18}, {6, 14, 22} $\text{entropy}(S) = \frac{3}{6}\text{entropy}(D_{Myope}) + \frac{3}{6}\text{entropy}(D_{Hypermetrope})$ $= \frac{3}{6} \times \left(-\frac{1}{3}\log_2\frac{1}{3} - \frac{2}{3}\log_2\frac{2}{3} - \frac{0}{3}\log_2\frac{0}{3} \right)$ $+ \frac{3}{6} \times \left(-\frac{0}{3}\log_2\frac{0}{3} - \frac{3}{3}\log_2\frac{3}{3} - \frac{0}{3}\log_2\frac{0}{3} \right)$ $= 0.4591$

entropy to split the node with the data set of {2, 4, 6, 8, 10, 12, 14, 16, 18, 20, 22, 24} using the split criterion, Astigmatic = No or Yes, which produces two subsets of {2, 6, 10, 14, 18, 22} and {4, 8, 12, 16, 20, 24}. Table 4.9 shows the calculation of information entropy to split the node with the data set of {2, 6, 10, 14, 18, 22} using the split criterion, Age = Young, Pre-presbyopic, or Presbyopic, which produces three subsets of {2, 6}, {10, 14}, and {18, 22}. These subsets are further partitioned using the split criterion, Spectacle Prescription = Myope or Hypermetrope, to produce leaf nodes with homogeneous data sets. Table 4.10 shows the calculation of information entropy to split the node with the data set of {4, 8, 12, 16, 20, 24} using the split criterion, Spectacle Prescription = Myope or Hypermetrope, which produces two subsets of {4, 12, 20} and {8, 16, 24}. These subsets are further partitioned using the split criterion, Age = Young, Pre-presbyopic, or Presbyopic, to produce leaf nodes with homogeneous data sets. Figure 4.5 shows the complete nonbinary decision tree for the lenses data set.

TABLE 4.10

Nonbinary Split of an Internal Node, {4, 8, 12, 16, 20, 24}, and Calculation of Information Entropy for the Lenses Data Set

Split Criterion	Resulting Subsets and Average Information Entropy of Split
Age = Young, Pre-presbyopic, or Presbyopic	{4, 8}, {12, 16}, {20, 24} $$\text{entropy}(S) = \frac{2}{6}\text{entropy}(D_{Young}) + \frac{2}{6}\text{entropy}(D_{Pre-presbyopic})$$ $$+ \frac{2}{6}\text{entropy}(D_{Presbyopic})$$ $$= \frac{2}{6} \times \left(-\frac{0}{2}\log_2\frac{0}{2} - \frac{0}{2}\log_2\frac{0}{2} - \frac{2}{2}\log_2\frac{2}{2}\right)$$ $$+ \frac{2}{6} \times \left(-\frac{1}{2}\log_2\frac{1}{2} - \frac{0}{2}\log_2\frac{0}{2} - \frac{1}{2}\log_2\frac{1}{2}\right)$$ $$+ \frac{2}{6} \times \left(-\frac{1}{2}\log_2\frac{1}{2} - \frac{0}{2}\log_2\frac{0}{2} - \frac{1}{2}\log_2\frac{1}{2}\right)$$ $$= 0.6667$$
Spectacle Prescription = Myope or Hypermetrope	{4, 12, 20}, {8, 16, 24} $$\text{entropy}(S) = \frac{3}{6}\text{entropy}(D_{Myope}) + \frac{3}{6}\text{entropy}(D_{Hypermetrope})$$ $$= \frac{3}{6} \times \left(-\frac{0}{3}\log_2\frac{0}{3} - \frac{0}{3}\log_2\frac{0}{3} - \frac{3}{3}\log_2\frac{3}{3}\right)$$ $$+ \frac{3}{6} \times \left(-\frac{2}{3}\log_2\frac{2}{3} - \frac{0}{3}\log_2\frac{0}{3} - \frac{1}{3}\log_2\frac{1}{3}\right)$$ $$= 0.4591$$

4.3 Handling Numeric and Missing Values of Attribute Variables

If a data set has a numeric attribute variable, the variable needs to be transformed into a categorical variable before being used to construct a decision tree. We present a common method to perform the transformation. Suppose that a numeric attribute variable, x, has the following numeric values in the training data set, $a_1, a_2, ..., a_k$, which are sorted in an increasing order of values. The middle point of two adjacent numeric values, a_i and a_j, is computed as follows:

$$c_i = \frac{a_i + a_j}{2}. \tag{4.7}$$

Using c_i for $i = 1, \ldots, k-1$, we can create the following $k+1$ categorical values of x:

$$\text{Category 1:} \quad x \leq c_1$$
$$\text{Category 2:} \quad c_1 < x \leq c_2$$
$$\vdots$$
$$\text{Category } k: \quad c_{k-1} < x \leq c_k$$
$$\text{Category } k+1: \quad c_k < x.$$

A numeric value of x is transformed into a categorical value according to the aforementioned definition of the categorical values. For example, if $c_1 < x \leq c_2$, the categorical value of x is Category 2.

In many real-world data sets, we may find an attribute variable that does not have a value in a data record. For example, if there are attribute variables of name, address, and email address for customers in a database for a store, we may not have the email address for a particular customer. That is, we may have missing email addresses for some customers. One way to treat a data record with a missing value is to discard the data record. However, when the training data set is small, we may need all the data records in the training data set to construct a decision tree. To use a data record with a missing value, we may estimate the missing value and use the estimated value to fill in the missing value. For a categorical attribute variable, its missing value can be estimated to be the value that is taken by the majority of data records in the training data set that have the same target value as that of the data record with a missing value of the attribute variable. For a numeric attribute variable, its missing value can be estimated to be the average of values that are taken by data records in the training data set that have the same target value as that of the data record with a missing value of the attribute variable. Other methods of estimating a missing value are given in Ye (2003).

4.4 Handling a Numeric Target Variable and Constructing a Regression Tree

If we have a numeric target variable, measures of data homogeneity such as information entropy and gini-index cannot be applied. Formula 4.7 is introduced (Breiman et al., 1984) to compute the average difference of values from their average value, R, and use it to measure data homogeneity for constructing a regression tree when the target variable takes numeric values.

The average difference of values in a data set from their average value indicates how values are similar or homogenous. The smaller the R value is, the more homogenous the data set is. Formula 4.9 shows the computation of the average R value after a split:

$$R(D) = \sum_{y \in D}(y - \bar{y})^2 \qquad (4.8)$$

$$\bar{y} = \frac{\sum_{y \in D} y}{n} \qquad (4.9)$$

$$R(S) = \sum_{v \in Values(S)} \frac{|D_v|}{|D|} R(D_v) \qquad (4.10)$$

The space shuttle data set D in Table 1.2 has one numeric target variable and four numeric attribute variables. The R value of the data set D with the 23 data records at the root node of the regression tree is computed as

$$\bar{y} = \frac{\sum_{y \in D} y}{n}$$

$$= \frac{0+1+0+0+0+0+0+0+1+1+1+0+0+2+0+0+0+0+0+0+0+0+1}{23}$$

$$= 0.3043$$

$$R(D) = \sum_{y \in D}(y-\bar{y})^2 = (0-0.3043)^2 + (1-0.3043)^2 + (0-0.3043)^2 + (0-0.3043)^2$$

$$+(0-0.3043)^2 + (0-0.3043)^2 + (0-0.3043)^2 + (0-0.3043)^2 + (1-0.3043)^2$$

$$+(1-0.3043)^2 + (1-0.3043)^2 + (0-0.3043)^2 + (0-0.3043)^2 + (2-0.3043)^2$$

$$+(0-0.3043)^2 + (0-0.3043)^2 + (0-0.3043)^2 + (0-0.3043)^2 + (0-0.3043)^2$$

$$+(0-0.3043)^2 + (0-0.3043)^2 + (0-0.3043)^2 + (1-0.3043)^2$$

$$= 6.8696$$

The average of target values in data records at a leaf node of a decision tree with a numeric target variable is often taken as the target value for the leaf node. When passing a data record along the decision tree to determine the target value of the data record, the target value of the leaf node where the data record arrives is assigned as the target value of the data record. The decision tree for a numeric target variable is called a regression tree.

4.5 Advantages and Shortcomings of the Decision Tree Algorithm

An advantage of using the decision tree algorithm to learn classification and prediction patterns is the explicit expression of classification and prediction patterns in the decision and regression tree. The decision tree in Figure 4.1 uncovers the following three patterns of part quality leading to three leaf nodes with the classification of system fault, respectively,

- $x_7 = 1$
- $x_7 = 0$ & $x_8 = 1$
- $x_7 = 0$ & $x_8 = 0$ & $x_9 = 1$

and the following pattern of part quality to one leaf node with the classification of no system fault:

- $x_7 = 0$ & $x_8 = 0$ & $x_9 = 0$.

The aforementioned explicit classification patterns reveal the following key knowledge for detecting the fault of this manufacturing system:

- Among the nine quality variables, only the three quality variables, x_7, x_8, and x_9, matter for system fault detection. This knowledge allows us to reduce the cost of part quality inspection by inspecting the part quality after M7, M8, and M9 only rather than all the nine machines.
- If one of these three variables, x_7, x_8, and x_9, shows a quality failure, the system has a fault; otherwise, the system has no fault.

A decision tree has its shortcoming in expressing classification and prediction patterns because it uses only one attribute variable in a split criterion. This may result in a large decision tree. From a large decision tree, it is difficult to see clear patterns for classification and prediction. For example, in Chapter 1, we presented the following classification pattern for the balloon data set in Table 1.1:

IF (Color = Yellow AND Size = Small) OR (Age = Adult AND Act = Stretch), THEN Inflated = T; OTHERWISE, Inflated = F.

This classification pattern for the target value of Inflated = T, (Color = Yellow AND Size = Small) OR (Age = Adult AND Act = Stretch), involves all the four attribute variables of Color, Size, Age, and Act. It is difficult to express this simple pattern in a decision tree. We cannot use all the four attribute variables to partition the root node. Instead, we have to select only one attribute variable. The average information entropy of a split to partition the root node using each of the four attribute variables is the same with the computation shown next:

$$\text{entropy}(S) = \frac{8}{16}\text{entropy}(D_{Yellow}) + \frac{8}{16}\text{entropy}(D_{Purple})$$

$$= \frac{8}{12} \times \left(-\frac{5}{8}\log_2\frac{5}{8} - \frac{3}{8}\log_2\frac{3}{8}\right)$$

$$+ \frac{8}{12} \times \left(-\frac{2}{8}\log_2\frac{2}{8} - \frac{6}{8}\log_2\frac{6}{8}\right)$$

$$= 0.8829.$$

We arbitrarily select Color = Yellow or Purple as the split criterion to partition the root node. Figure 4.6 gives the complete decision tree for the balloon data set. The decision tree is large with the following seven classification patterns leading to seven leaf nodes, respectively:

- Color = Yellow AND Size = Small, with Inflated = T
- Color = Yellow AND Size = Large AND Age = Adult AND Act = Stretch, with Inflated = T

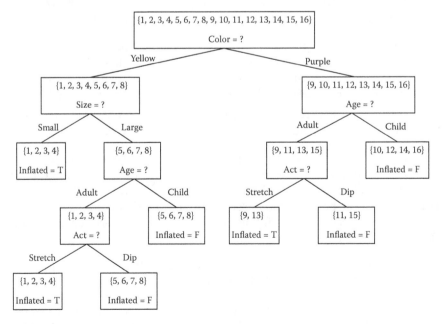

FIGURE 4.6
Decision tree for the balloon data set.

Decision and Regression Trees 61

- Color = Yellow AND Size = Large AND Age = Adult AND Act = Dip, with Inflated = F
- Color = Yellow AND Size = Large AND Age = Child, with Inflated = F
- Color = Purple AND Age = Adult AND Act = Stretch, with Inflated = T
- Color = Purple AND Age = Adult AND Act = Dip, with Inflated = F
- Color = Purple AND Age = Child, with Inflated = F

From these seven classification patterns, it is difficult to see the simple classification pattern:

> IF (Color = Yellow AND Size = Small) OR (Age = Adult AND Act = Stretch), THEN Inflated = T; OTHERWISE, Inflated = F.

Moreover, selecting the best split criterion with only one attribute variable without looking ahead the combination of this split criterion with the following-up criteria to the leaf node is like making a locally optimal decision. There is no guarantee that making locally optimal decisions at separate times leads to the smallest decision tree or a globally optimal decision.

However, considering all the attribute variables and their combinations of conditions for each split would correspond to an exhaustive search of all combination values of all the attribute variables. This is computationally costly or sometimes impossible for a large data set with a large number of attribute variables.

4.6 Software and Applications

The website http://www.knuggets.com has information about various data mining tools. The following software packages support the learning of decision and regression trees:

- Weka (http://www.cs.waikato.ac.nz/ml/weka/)
- SPSS AnswerTree (http://www.spss.com/answertree/)
- SAS Enterprise Miner (http://sas.com/products/miner/)
- IBM Inteligent Miner (http://www.ibm.com/software/data/iminer/)
- CART (http://www.salford-systems.com/)
- C4.5 (http://www.cse.unsw.edu.au/quinlan)

Some applications of decision trees can be found in (Ye, 2003, Chapter 1) and (Li and Ye, 2001; Ye et al., 2001).

Exercises

4.1 Construct a binary decision tree for the balloon data set in Table 1.1 using the information entropy as the measure of data homogeneity.

4.2 Construct a binary decision tree for the lenses data set in Table 1.3 using the information entropy as the measure of the data homogeneity.

4.3 Construct a non-binary regression tree for the space shuttle data set in Table 1.2 using only Launch Temperature and Leak-Check Pressure as the attribute variables and considering two categorical values of Launch Temperature (low for Temperature <60, normal for other temperatures) and three categorical values of Leak-Check Pressure (50, 100, and 200).

4.4 Construct a binary decision tree or a nonbinary decision tree for the data set found in Exercise 1.1.

4.5 Construct a binary decision tree or a nonbinary decision tree for the data set found in Exercise 1.2.

4.6 Construct a dataset for which using the decision tree algorithm based on the best split for data homogeneity does not produce the smallest decision tree.

5

Artificial Neural Networks for Classification and Prediction

Artificial neural networks (ANNs) are designed to mimic the architecture of the human brain in order to create artificial intelligence like human intelligence. Hence, ANNs use the basic architecture of the human brain, which consists of neurons and connections among neurons. ANNs have processing units like neurons and connections among processing units. This chapter introduces two types of ANNs for classification and prediction: perceptron and multilayer feedforward ANN. In this chapter, we first describe the processing units and how these units can be used to construct various types of ANN architectures. We then present the perceptron, which is a single-layer feedforward ANN, and the learning of classification and prediction patterns by a perceptron. Finally, multilayer feedforward ANNs with the back-propagation learning algorithm are described. A list of software packages that support ANNs is provided. Some applications of ANNs are given with references.

5.1 Processing Units of ANNs

Figure 5.1 illustrates a processing unit in an ANN, unit j. The unit takes p inputs, $x_1, x_2, ..., x_p$, another special input, $x_0 = 1$, and produces an output, o. The inputs $x_1, x_2, ..., x_p$ and the output o are used to represent the inputs and the output of a given problem. Take an example of the space shuttle data set in Table 1.2. We may have x_1, x_2, and x_3 to represent Launch Temperature, Leak-Check Pressure, and Temporal Order of Flight, respectively, and have o represent Number of O-rings with Stress. The input x_0 is an inherent part of every processing unit and always takes the value of 1.

Each input, x_i, is connected to the unit j with a connection weight, $w_{j,i}$. The connection weight, $w_{j,0}$, is called the bias or threshold for a reason that is explained later. The unit j processes the inputs by first obtaining the net sum, which is the weighted sum of the inputs, as follows:

$$net_j = \sum_{i=0}^{p} w_{j,i} x_i. \tag{5.1}$$

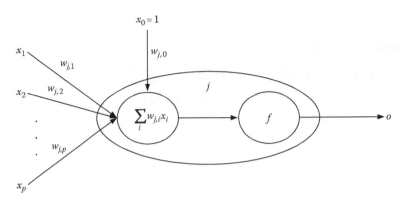

FIGURE 5.1
Processing unit of ANN.

Let the vectors of x and w be defined as follows:

$$x = \begin{bmatrix} x_0 \\ \vdots \\ x_p \end{bmatrix} \quad w' = \begin{bmatrix} w_{j,0} & \cdots & w_{j,p} \end{bmatrix}$$

Equation 5.1 can be represented as follows:

$$net_j = w'x. \tag{5.2}$$

The unit then applies a transfer function, f, to the net sum and obtains the output, o, as follows:

$$o = f(net_j). \tag{5.3}$$

Five of the common transfer functions are given next and illustrated in Figure 5.2.

1. Sign function:

$$o = sgn(net) = \begin{cases} 1 & \text{if } net > 0 \\ -1 & \text{if } net \leq 0 \end{cases} \tag{5.4}$$

2. Hard limit function:

$$o = hardlim(net) = \begin{cases} 1 & \text{if } net > 0 \\ 0 & \text{if } net \leq 0 \end{cases} \tag{5.5}$$

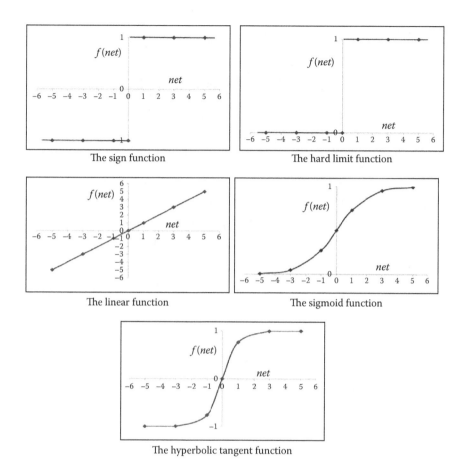

FIGURE 5.2
Examples of transfer functions.

3. Linear function:

$$o = lin(net) = net \tag{5.6}$$

4. Sigmoid function:

$$o = sig(net) = \frac{1}{1+e^{-net}} \tag{5.7}$$

5. Hyperbolic tangent function:

$$O = tanh(net) = \frac{e^{net} - e^{-net}}{e^{net} + e^{-net}}. \tag{5.8}$$

Given the following input vector and connection weight vector

$$x = \begin{bmatrix} 1 \\ 5 \\ -6 \end{bmatrix} \quad w' = \begin{bmatrix} -1.2 & 3 & 2 \end{bmatrix},$$

the output of the unit with each of the five transfer functions is computed as follows:

$$net = w'x = \begin{bmatrix} -1.2 & 3 & 2 \end{bmatrix} \begin{bmatrix} 1 \\ 5 \\ -6 \end{bmatrix} = 1.8$$

$$o = sgn(net) = 1$$

$$o = hardlim(net) = 1$$

$$o = lin(net) = 1.8$$

$$o = sig(net) = 0.8581$$

$$o = tahn(net) = 0.9468.$$

One processing unit is sufficient to implement a logical AND function. Table 5.1 gives the inputs and the output of the AND function and four data records of this function. The AND function has the output values of −1 and 1. Figure 5.3 gives the implementation of the AND function using one processing unit. Among the five transfer functions in Figure 5.2, the sign function and the hyperbolic tangent function can produce the range of

TABLE 5.1
AND Function

Inputs		Output
x_1	x_2	o
−1	−1	−1
−1	1	−1
1	−1	−1
1	1	1

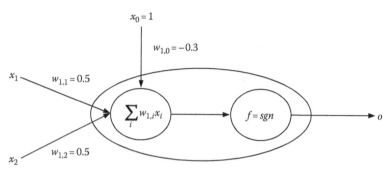

FIGURE 5.3
Implementation of the AND function using one processing unit.

output values from −1 to 1. The sign function is used as the transfer function for the processing unit to implement the AND function. The first three data records require the output value of −1. The weighted sum of the inputs for the first three data records, $w_{1,0}x_0 + w_{1,1}x_1 + w_{1,2}x_2$, should be in the range of [−1, 0]. The last data record requires the output value of 1, and the weighted sum of the inputs should be in the range of (0, 1]. The connection weight $w_{1,0}$ must be a negative value to make *net* for the first three data records less than zero and also make *net* for the last data record greater than zero. Hence, the connection weight $w_{1,0}$ acts as a threshold against the weighted sum of the inputs to drive *net* greater than or less than zero. This is why the connection weight for $x_0 = 1$ is called the threshold or bias. In Figure 5.3, $w_{1,0}$ is set to −0.3. Equation 5.1 can be represented as follows to show the role of the threshold or bias, *b*:

$$net = w'x + b, \quad (5.9)$$

where

$$x = \begin{bmatrix} x_1 \\ \vdots \\ x_p \end{bmatrix} \quad w' = \begin{bmatrix} w_{j,1} & \cdots & w_{j,p} \end{bmatrix}.$$

The computation of the output value for each input is illustrated next.

$$o = sgn(net) = sgn\left(\sum_{i=0}^{2} w_{1,i}x_i\right) = sgn[-0.3 \times 1 + 0.5 \times (-1) + 0.5 \times (-1)]$$

$$= sgn(-0.3 - 1) = sgn(-1.3) = -1$$

$$o = sgn(net) = sgn\left(\sum_{i=0}^{2} w_{1,i} x_i\right) = sgn\left[-0.3 \times 1 + 0.5 \times (-1) + 0.5 \times (1)\right]$$

$$= sgn(-0.3 + 0) = sgn(-0.3) = -1$$

$$o = sgn(net) = sgn\left(\sum_{i=0}^{2} w_{1,i} x_i\right) = sgn\left[-0.3 \times 1 + 0.5 \times (1) + 0.5 \times (-1)\right]$$

$$= sgn(-0.3 + 0) = sgn(-0.3) = -1$$

$$o = sgn(net) = sgn\left(\sum_{i=0}^{2} w_{1,i} x_i\right) = sgn\left[-0.3 \times 1 + 0.5 \times (1) + 0.5 \times (1)\right]$$

$$= sgn(-0.3 + 1) = sgn(0.7) = 1$$

Table 5.2 gives the inputs and the output of the logical OR function. Figure 5.4 shows the implementation of the OR function using one processing unit. Only the first data record requires the output value of −1, and the

TABLE 5.2

OR Function

Inputs		Output
x_1	x_2	o
−1	−1	−1
−1	1	1
1	−1	1
1	1	1

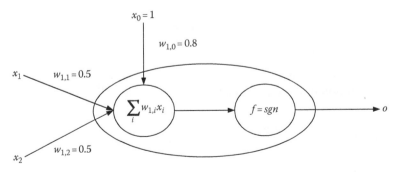

FIGURE 5.4
Implementation of the OR function using one processing unit.

other three data records require the output value of 1. Only the first data record produces the weighted sum −1 from the inputs, and the other three data records produce the weighted sum of the inputs in the range [−0.5, 1]. Hence, any threshold value $w_{1,0}$ in the range (0.5, 1) will make *net* for the first data record less than zero and make *net* for the last three data records greater than zero.

5.2 Architectures of ANNs

Processing units of ANNs can be used to construct various types of ANN architectures. We present two ANN architectures: feedforward ANNs and recurrent ANNs. Feedforward ANNs are widely used. Figure 5.5 shows a one-layer, fully connected feedforward ANN in which each input is connected to each processing unit. Figure 5.6 shows a two-layer, fully connected

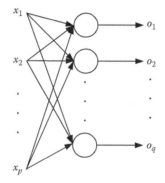

FIGURE 5.5
Architecture of a one-layer feedforward ANN.

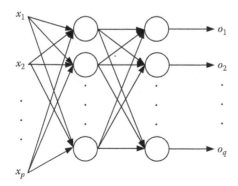

FIGURE 5.6
Architecture of a two-layer feedforward ANN.

feedforward ANN. Note that the input x_0 for each processing unit is not explicitly shown in the ANN architectures in Figures 5.5 and 5.6. The two-layer feedforward ANN in Figure 5.6 contains the output layer of processing units to produce the outputs and a hidden layer of processing units whose outputs are the inputs to the processing units at the output layer. Each input is connected to each processing unit at the hidden layer, and each processing unit at the hidden layer is connected to each processing unit at the output layer. In a feedforward ANN, there are no backward connections between processing units in that the output of a processing unit is not used as a part of inputs to that processing unit directly or indirectly. An ANN is not necessarily fully connected as those in Figures 5.5 and 5.6. Processing units may use the same transfer function or different transfer functions.

The ANNs in Figures 5.3 and 5.4, respectively, are examples of one-layer feedforward ANNs. Figure 5.7 shows a two-layer, fully connected feedforward ANN with one hidden layer of two processing units and the output layer of one processing unit to implement the logical exclusive-OR (XOR) function. Table 5.3 gives the inputs and output of the XOR function.

The number of inputs and the number of outputs in an ANN depend on the function that the ANN is set to capture. For example, the XOR function

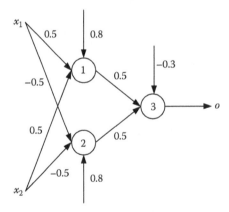

FIGURE 5.7
A two-layer feedforward ANN to implement the XOR function.

TABLE 5.3

XOR Function

Inputs		Output
x_1	x_2	o
−1	−1	−1
−1	1	1
1	−1	1
1	1	−1

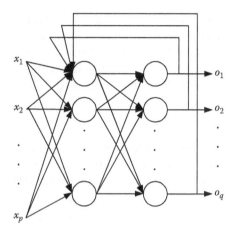

FIGURE 5.8
Architecture of a recurrent ANN.

has two inputs and one output that can be represented by two inputs and one output of an ANN, respectively. The number of processing units at the hidden layer, called hidden units, is often determined empirically to account for the complexity of the function that an ANN implements. In general, more complex the function is, more hidden units are needed. A two-layer feedforward ANN with a sigmoid or hyperbolic tangent function has the capability of implementing a given function (Witten et al., 2011).

Figure 5.8 shows the architecture of a recurrent ANN with backward connections that feed the outputs back as the inputs to the first hidden unit (shown) and other hidden units (not shown). The backward connections allow the ANN to capture the temporal behavior in that the outputs at time $t + 1$ depend on the outputs or state of the ANN at time t. Hence, recurrent ANNs such as that in Figure 5.8 have backward connections to capture temporal behaviors.

5.3 Methods of Determining Connection Weights for a Perceptron

To use an ANN for implementing a function, we first determine the architecture of the ANN, including the number of inputs, the number of outputs, the number of layers, the number of processing units in each layer, and the transfer function for each processing unit. Then we need to determine connection weights. In this section, we describe a graphical method and a learning method to determine connection weights for a perceptron, which is a one-layer feedforward ANN with a sign or hard limit transfer function. Although concepts and methods in this section are explained using the sign transfer function for each processing unit in a perceptron, these concepts and methods are also applicable to a perceptron with a hard limit transfer function for each processing unit. In Section 5.4, we present the back-propagation learning method to determine connection weights for multiple-layer feedforward ANNs.

5.3.1 Perceptron

The following notations are used to represent a fully connected perceptron with p inputs, q processing units at the output layer to produce q outputs, and the sign transfer function for each processing unit, as shown in Figure 5.5:

$$x = \begin{bmatrix} x_1 \\ \vdots \\ x_p \end{bmatrix} \quad o = \begin{bmatrix} o_1 \\ \vdots \\ o_q \end{bmatrix} \quad w' = \begin{bmatrix} w_{1,1} & \cdots & w_{1,p} \\ \vdots & \ddots & \vdots \\ w_{q,1} & \cdots & w_{q,p} \end{bmatrix} = \begin{bmatrix} w'_1 \\ \vdots \\ w'_q \end{bmatrix} \quad w_j = \begin{bmatrix} w_{j,1} \\ \vdots \\ w_{j,p} \end{bmatrix} \quad b = \begin{bmatrix} b_1 \\ \vdots \\ b_q \end{bmatrix}$$

$$o = sgn(w'x + b). \tag{5.10}$$

5.3.2 Properties of a Processing Unit

For a processing unit j, $o = sgn(net) = sgn(w'_j x + b_j)$ separates input vectors, xs, into two regions: one with $net > 0$ and $o = 1$, and another with $net \leq 0$ and $o = -1$. The equation, $net = w'_j x + b_j = 0$, is the decision boundary in the input space that separates the two regions. For example, given x in a two-dimensional space and the following weight and bias values:

$$x = \begin{bmatrix} x_1 \\ x_2 \end{bmatrix} \quad w'_j = \begin{bmatrix} -1 & 1 \end{bmatrix} \quad b_j = -1,$$

the decision boundary is

$$w'_j x + b_j = 0$$

$$-x_1 + x_2 - 1 = 0$$

$$x_2 = x_1 + 1.$$

Figure 5.9 illustrates the decision boundary and the separation of the input space into two regions by the decision boundary. The slope and the intercept of the line representing the decision boundary in Figure 5.9 are

$$\text{slope} = \frac{-w_{j,1}}{w_{j,2}} = \frac{1}{1} = 1$$

Artificial Neural Networks for Classification and Prediction 73

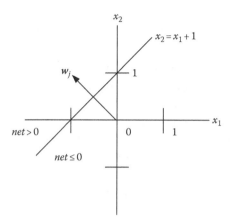

FIGURE 5.9
Example of the decision boundary and the separation of the input space into two regions by a processing unit.

$$\text{intercept} = \frac{-b_j}{w_{j,2}} = \frac{1}{1} = 1.$$

As illustrated in Figure 5.9, a processing unit has the following properties:

- The weight vector is orthogonal to the decision boundary.
- The weight vector points to the positive side ($net > 0$) of the decision boundary.
- The position of the decision boundary can be shifted by changing b. If $b = 0$, the decision boundary passes through the origin, e.g., (0, 0) in the two-dimensional space.
- Because the decision boundary is a linear equation, a processing unit can implement a linearly separable function only.

Those properties of a processing unit are used in the graphical method of determining connection weights in Section 5.3.3 and the learning method of determining connection weights in Section 5.3.4.

5.3.3 Graphical Method of Determining Connection Weights and Biases

The following steps are taken in the graphical method to determine connection weights of a perceptron with p inputs, one output, one processing unit to produce the output, and the sign transfer function for the processing unit:

1. Plot the data points for the data records in the training data set for the function.
2. Draw the decision boundary to separate the data points with $o = 1$ from the data points with $o = -1$.

3. Draw the weight vector to make it be orthogonal to the decision boundary and point to the positive side of the decision boundary. The coordinates of the weight vector define the connection weights.
4. Use one of the two methods to determine the bias:
   a. Use the intercept of the decision boundary and connection weights to determine the bias.
   b. Select one data point on the positive side and one data point on the negative side that are closest to the decision boundary on each side and use these data points and the connection weights to determine the bias.

These steps are illustrated in Example 5.1.

Example 5.1

Use the graphical method to determine the connection weights of a perceptron with one processing unit for the AND function in Table 5.1.

In Step 1, we plot the four circles in Figure 5.10 to represent the four data points of the AND function. The output value of each data point is noted inside the circle for the data point. In Step 2, we use the decision boundary, $x_2 = -x_1 + 1$, to separate the three data points with $o = -1$ from the data point with $o = 1$. The intercept of the line for the decision boundary is 1 with $x_2 = 1$ when x_1 is set to 0. In Step 3, we draw the weight vector, $w_1 = (0.5, 0.5)$, which is orthogonal to the decision boundary and points to the positive side of the decision boundary. Hence, we have $w_{1,1} = 0.5$, $w_{1,2} = 0.5$. In Step 4, we use the following equation to determine the bias:

$$w_{1,1}x_1 + w_{1,2}x_2 + b_1 = 0$$

$$w_{1,2}x_2 = -w_{1,1}x_1 - b_1$$

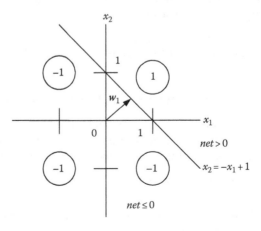

FIGURE 5.10
Illustration of the graphical method to determine connection weights.

$$intercept = -\frac{b_1}{w_{1,2}}$$

$$1 = -\frac{b_1}{0.5}$$

$$b_1 = -0.5.$$

If we move the decision boundary so that it has the intercept of 0.6, we obtain $b_1 = -0.3$ and exactly the same ANN for the AND function as shown in Figure 5.3.

Using another method in Step 4, we select the data point (1, 1) on the positive side of the decision boundary, the data points (−1, 1) on the negative side of the decision boundary, and the connection weights $w_{1,1} = 0.5$, $w_{1,2} = 0.5$ to determine the bias b_1 as follows:

$$net = w_{1,1}x_1 + w_{1,2}x_2 + b_1$$

$$net = 0.5 \times 1 + 0.5 \times 1 + b_1 > 0$$

$$b_1 > -1$$

and

$$net = w_{1,1}x_1 + w_{1,2}x_2 + b_1$$

$$net = 0.5 \times (-1) + 0.5 \times 1 + b_1 \leq 0$$

$$b_1 \leq 0.$$

Hence, we have

$$-1 < b_1 \leq 0.$$

By letting $b_1 = -0.3$, we obtain the same ANN for the AND function as shown in Figure 5.3.

The ANN with the weights, bias, and decision boundary as those in Figure 5.10 produces the correct output for the inputs in each data record in Table 5.1. The ANN also has the generalization capability of classifying any input vector on the negative side of the decision boundary into $o = -1$ and any input vector on the positive side of the decision boundary into $o = 1$.

For a perceptron with multiple output units, the graphical method is applied to determine connection weights and bias for each output unit.

5.3.4 Learning Method of Determining Connection Weights and Biases

We use the following two of the four data records for the AND function in the training data set to illustrate the learning method of determining connection weights for a perceptron with one processing unit without a bias:

1. $x_1 = -1 \quad x_2 = -1 \quad t_1 = -1$
2. $x_1 = 1 \quad x_2 = 1 \quad t_1 = 1$,

where t_1 denotes the target output of processing unit 1 that needs to be produced for each data record. The two data records are plotted in Figure 5.11. We initialize the connection weights using random values, $w_{1,1}(k) = -1$ and $w_{1,2}(k) = 0.8$, with k denoting the iteration number when the weights are assigned or updated. Initially, we have $k = 0$. We present the inputs of the first data record to the perceptron of one processing unit:

$$net = w_{1,1}(0)x_1 + w_{1,2}(0)x_2 = (-1) \times (-1) + 0.8 \times (-1) = -1.8.$$

Since $net < 0$, we have $o_1 = -1$. Hence, the perceptron with the weight vector $(-1, 0.8)$ produces the target output for the inputs of the first data record, $t_1 = -1$. There is no need to change the connection weights. Next, we present the inputs of the second data record to the perceptron:

$$net = w_{1,1}(0)x_1 + w_{1,2}(0)x_2 = (-1) \times 1 + 0.8 \times 1 = -0.2.$$

Since $net < 0$, we have $o_1 = -1$, which is different from the target output for this data record $t_1 = 1$. Hence, the connection weights must be changed in order

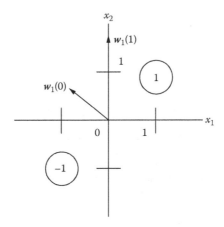

FIGURE 5.11
Illustration of the learning method to change connection weights.

to produce the target output. The following equations are used to change the connection weights for processing unit j:

$$\Delta w_j = \frac{1}{2}(t_j - o_j)x \qquad (5.11)$$

$$w_j(k+1) = w_j(k) + \Delta w_j. \qquad (5.12)$$

In Equation 5.11, if $(t-o)$ is zero, that is, $t=o$, then there is no change of weights. If $t = 1$ and $o = -1$,

$$\Delta w_j = \frac{1}{2}(t_j - o_j)x = \frac{1}{2}(1-(-1))x = x.$$

By adding x to $w_j(k)$, that is, performing $w_j(k)+x$ in Equation 5.12, we move the weight vector closer to x and make the weight vector point more to the direction of x because we want the weight vector to point to the positive side of the decision boundary and x lies on the positive side of the decision boundary. If $t_1 = -1$ and $o_1 = 1$,

$$\Delta w_j = \frac{1}{2}(t_j - o_j)x = \frac{1}{2}(-1-1)x = -x.$$

By subtracting x from $w_j(k)$, that is, performing $w_j(k)-x$ in Equation 5.12, we move the weight vector away from x and make the weight vector point more to the opposite direction of x because x lies on the negative side of the decision boundary with $t = -1$ and we want the weight vector to eventually point to the positive side of the decision boundary.

Using Equations 5.11 and 5.12, we update the connection weights based on the inputs and the target and actual outputs for the second data record as follows:

$$\Delta w_1 = \frac{1}{2}(t_1 - o_1)x = \frac{1}{2}(1-(-1))\begin{bmatrix}1\\1\end{bmatrix} = \begin{bmatrix}1\\1\end{bmatrix}$$

$$w_1(1) = w_1(0) + \Delta w_1 = \begin{bmatrix}-1\\0.8\end{bmatrix} + \begin{bmatrix}1\\1\end{bmatrix} = \begin{bmatrix}0\\1.8\end{bmatrix}.$$

The new weight vector, $w_1(1)$, is shown in Figure 5.11. As Figure 5.11 illustrates, $w_1(1)$ is closer to the second data record x than $w_1(0)$ and points more to the direction of x since x has $t = 1$ and thus lies on the positive side of the decision boundary.

With the new weights, we present the inputs of the data records to the perceptron again in the second iteration of evaluating and updating the weights if needed. We present the inputs of the first data record:

$$net = w_{1,1}(1)x_1 + w_{1,2}(1)x_2 = 0 \times (-1) + 1.8 \times (-1) = -1.8.$$

Since $net < 0$, we have $o_1 = -1$. Hence, the perceptron with the weight vector (0, 1.8) produces the target output for the inputs of the first data record, $t_1 = -1$. With $(t_1 - o_1) = 0$, there is no need to change the connection weights. Next, we present the inputs of the second data record to the perceptron:

$$net = w_{1,1}(1)x_1 + w_{1,2}(1)x_2 = 0 \times 1 + 1.8 \times 1 = 1.8.$$

Since $net > 0$, we have $o_1 = 1$. Hence, the perceptron with the weight vector (0, 1.8) produces the target output for the inputs of the second data record, $t = 1$. With $(t - o) = 0$, there is no need to change the connection weights. The perceptron with the weight vector (0, 1.8) produces the target outputs for all the data records in the training data set. The learning of the connection weights from the data records in the training data set is finished after one iteration of changing the connection weights with the final weight vector (0, 1.8). The decision boundary is the line, $x_2 = 0$.

The general equations for the learning method of determining connection weights are given as follows:

$$\Delta w_j = \alpha(t_j - o_j)x = \alpha e_j x \qquad (5.13)$$

$$w_j(k+1) = w_j(k) + \Delta w_j \qquad (5.14)$$

or

$$\Delta w_{j,i} = \alpha(t_j - o_j)x_i = \alpha e_j x_i \qquad (5.15)$$

$$w_{j,i}(k+1) = w_{j,i}(k) + \Delta w_{j,i}, \qquad (5.16)$$

Where
  $e_j = t_j - o_j$ represents the output error
  α is the learning rate taking a value usually in the range (0, 1)

In Equation 5.11, α is set to 1/2. Since the bias of processing unit j is the weight of connection from the input $x_0 = 1$ to the processing unit, Equations 5.15 and 5.16 can be extended for changing the bias of processing unit j as follows:

$$\Delta b_j = \alpha(t_j - o_j) \times x_0 = \alpha(t_j - o_j) \times 1 = \alpha e_j \qquad (5.17)$$

$$b_j(k+1) = b_j(k) + \Delta b_j. \qquad (5.18)$$

5.3.5 Limitation of a Perceptron

As described in Sections 5.3.2 and 5.3.3, each processing unit implements a linear decision boundary, that is, a linearly separable function. Even with multiple processing units in one layer, a perceptron is limited to implementing a linearly separable function. For example, the XOR function in Table 5.3 is not a linearly separable function. There is only one output for the XOR function. Using one processing unit to represent the output, we have one decision boundary, which is a straight line representing a linear function. However, there does not exit such a straight line in the input space to separate the two data points with $o = 1$ from the other two data points with $o = -1$. A nonlinear decision boundary such as the one shown in Figure 5.12 is needed to separate the two data points with $o = 1$ from the other two data points with $o = -1$. To use processing units that implement linearly separable functions for constructing an ANN to implement the XOR function, we need two processing units in one layer (the hidden layer) to implement two decision boundaries and one processing unit in another layer (the output layer) to combine the outputs of the two hidden units as shown in Table 5.4 and Figure 5.7. Table 5.5 defines the logical NOT function used in Table 5.4. Hence, we need a two-layer ANN to implement the XOR function, which is a nonlinearly separable function.

The learning method described by Equations 5.13 through 5.18 can be used to learn the connection weights to each output unit using a set of training data because the target value t for each output unit is given in the training data. For each hidden unit, Equations 5.13 through 5.18 are not applicable because we do not know t for the hidden unit. Hence, we encounter a difficulty in learning connection weights and biases from training data for a multilayer ANN. This learning difficulty for multilayer ANNs is overcome by the back-propagation learning method described in the next section.

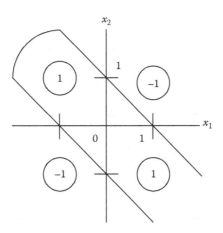

FIGURE 5.12
Four data points of the XOR function.

TABLE 5.4

Function of Each Processing Unit in a Two-Layer ANN to Implement the XOR Function

x_1	x_2	$o_1 = x_1$ OR x_2	$o_2 =$ NOT $(x_1$ OR $x_2)$	$o_3 = o_1$ AND o_2
−1	−1	−1	1	−1
−1	1	1	1	1
1	−1	1	1	1
1	1	1	−1	−1

TABLE 5.5

NOT Function

x	o
−1	1
1	−1

5.4 Back-Propagation Learning Method for a Multilayer Feedforward ANN

The back-propagation learning method for a multilayer ANN (Rumelhart et al., 1986) aims at searching for a set of connection weights (including biases) W that minimizes the output error. The output error for a training data record d is defined as follows:

$$E_d(W) = \frac{1}{2}\sum_j (t_{j,d} - o_{j,d})^2 \qquad (5.19)$$

Where
 $t_{j,d}$ is the target output of output unit j for the training data record d
 $o_{j,d}$ is the actual output produced by output unit j of the ANN with the weights W for the training data record d

The output error for a set of training data records is defined as follows:

$$E(W) = \frac{1}{2}\sum_d \sum_j (t_{j,d} - o_{j,d})^2. \qquad (5.20)$$

Because each $o_{j,d}$ depends on W, E is a function of W. The back-propagation learning method searches in the space of possible weights and evaluates a given set of weights based on their associated E values. The search is called the gradient descent search that changes weights by moving them in the direction of reducing the output error after passing the inputs of the data record d through the ANN with the weights W, as follows:

$$\Delta w_{j,i} = -\alpha \frac{\partial E_d}{\partial w_{j,i}} = -\alpha \frac{\partial E_d}{\partial net_j} \frac{\partial net_j}{\partial w_{j,i}} = \alpha \delta_j \frac{\partial \left(\sum_k w_{j,k} \tilde{o}_k \right)}{\partial w_{j,i}} = \alpha \delta_j \tilde{o}_i \qquad (5.21)$$

where δ_j is defined as

$$\delta_j = -\frac{\partial E_d}{\partial net_j}, \qquad (5.22)$$

Where
  α is the learning rate with a value typically in (0, 1)
  $\tilde{o}_i$ is input i to processing unit j

If unit j directly receives the inputs of the ANN, $\tilde{o}_i$ is x_i; otherwise, $\tilde{o}_i$ is from a unit in the preceding layer feeding its output as an input to unit j. To change a bias for a processing unit, Equation 5.21 is modified by using $\tilde{o}_i = 1$ as follows:

$$\Delta b_j = \alpha \delta_j. \qquad (5.23)$$

If unit j is an output unit,

$$\delta_j = -\frac{\partial E_d}{\partial net_j} = -\frac{\partial E_d}{\partial o_j} \frac{\partial o_j}{\partial net_j} = -\frac{\partial \left(\frac{1}{2} \sum_j (t_{j,d} - o_{j,d})^2 \right)}{\partial o_j} \frac{\partial \left(f_j(net_j) \right)}{\partial net_j}$$

$$= (t_{j,d} - o_{j,d}) f'_j(net_j), \qquad (5.24)$$

where f' denotes the derivative of the function f with regard to net. To obtain a value for the term $f'_j(net_j)$ in Equation 5.24, the transfer function f for unit j must be a semi-linear, nondecreasing, and differentiable function, e.g., linear, sigmoid, and tanh. For the sigmoid transfer function

$$o_j = f_j(net_j) = \frac{1}{1 + e^{-net_j}},$$

we have the following:

$$f'_j(net_j) = \frac{1}{1 + e^{-net_j}} \frac{e^{-net_j}}{1 + e^{-net_j}} = o_j(1 - o_j). \qquad (5.25)$$

If unit j is a hidden unit feeding its output as an input to output units,

$$\delta_j = -\frac{\partial E_d}{\partial net_j} = -\frac{\partial E_d}{\partial o_j}\frac{\partial o_j}{\partial net_j} = -\frac{\partial E_d}{\partial o_j}f'_j(net_j) = -\left(\sum_n \frac{\partial E_d}{\partial net_n}\frac{\partial net_n}{\partial o_j}\right)f'_j(net_j),$$

where net_n is the net sum of output unit n. Using Equation 5.22, we rewrite δ_j as follows:

$$\delta_j = \left(\sum_n \delta_n \frac{\partial net_n}{\partial o_j}\right)f'_j(net_j) = \left(\sum_n \delta_n \frac{\partial\left(\sum_j w_{n,j}o_j\right)}{\partial o_j}\right)f'_j(net_j)$$

$$= \left(\sum_n \delta_n w_{n,j}\right)f'_j(net_j). \tag{5.26}$$

Since we need δ_n in Equation 5.26, which is computed for output unit n, changing the weights of the ANN should start with changing the weights for output units and move on to changing the weights for hidden units in the preceding layer so that δ_n for output unit n can be used in computing δ_j for hidden unit j. In other words, δ_n for output unit n is back-propagated to compute δ_j for hidden unit j, which gives the name of the back-propagation learning.

Changes to weights and biases, as determined by Equations 5.21 and 5.23, are used to update weights and biases of the ANN as follows:

$$w_{j,i}(k+1) = w_{j,i}(k) + \Delta w_{j,i} \tag{5.27}$$

$$b_j(k+1) = b_j(k) + \Delta b_j. \tag{5.28}$$

Example 5.2

Given the ANN for the XOR function and the first data record in Table 5.3 with $x_1 = -1$, $x_2 = -1$, and $t = -1$, use the back-propagation method to update the weights and biases of the ANN. In the ANN, the sigmoid transfer function is used by each of the two hidden units and the linear function is used by the output unit. The ANN starts with the following arbitrarily assigned values of weights and biases in $(-1, 1)$ as shown in Figure 5.13:

$$w_{1,1} = 0.1 \quad w_{2,1} = -0.1 \quad w_{1,2} = 0.2 \quad w_{2,2} = -0.2 \quad b_1 = -0.3$$
$$b_2 = -0.4 \quad w_{3,1} = 0.3 \quad w_{3,2} = 0.4 \quad b_3 = 0.5.$$

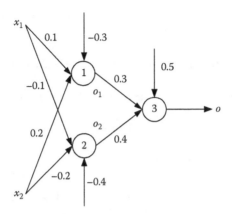

FIGURE 5.13
A set of weights with randomly assigned values in a two-layer feedforward ANN for the XOR function.

Use the learning rate $\alpha = 0.3$.

Passing the inputs of the data record, $x_1 = -1$ and $x_2 = -1$, through the ANN, we obtain the following:

$$o_1 = sig(w_{1,1}x_1 + w_{1,2}x_2 + b_1) = sig(0.1 \times (-1) + 0.2 \times (-1) + (-0.3))$$

$$= sig(-0.6) = \frac{1}{1+e^{-(-0.6)}} = 0.3543$$

$$o_2 = sig(w_{2,1}x_1 + w_{2,2}x_2 + b_2) = sig((-0.1) \times (-1) + (-0.2) \times (-1) + (-0.4))$$

$$= sig(-0.2) = \frac{1}{1+e^{-(-0.2)}} = 0.4502$$

$$o = lin(w_{3,1}o_1 + w_{3,2}o_2 + b_3) = lin(0.3 \times 0.3543 + 0.4 \times 0.4502 + 0.5)$$

$$= lin(0.7864) = 0.7864$$

Since the difference between $o = 0.6871$ and $t = -1$ is large, we need to change the weights and biases of the ANN. Equations 5.21 and 5.23 are used to determine changes to the weights and bias for the output unit as follows:

$$\Delta w_{3,1} = \alpha \delta_3 \bar{o}_1 = 0.3 \times \delta_3 \times o_1 = 0.3 \times \delta_3 \times 0.3543$$

$$\Delta w_{3,2} = \alpha \delta_3 \bar{o}_2 = 0.3 \times \delta_3 \times o_2 = 0.3 \times \delta_3 \times 0.4502$$

$$\Delta b_3 = \alpha \delta_3 = 0.3 \delta_3.$$

Equation 5.24 is used to determine δ_3, and then δ_3 is used to determine $\Delta w_{3,1}$, $\Delta w_{3,2}$, and Δb_3 as follows:

$$\delta_3 = (t-o) f_3'(net_3) = (t_{jd} - o_{jd}) lin'(net_3) = (-1 - 0.7864) \times 1 = -1.7864$$

$$\Delta w_{3,1} = 0.3 \times \delta_3 \times 0.3543 = 0.3 \times (-1.7864) \times 0.3543 = -0.1899$$

$$\Delta w_{3,2} = 0.3 \times \delta_3 \times 0.4502 = 0.3 \times (-1.7864) \times 0.4502 = -0.2413$$

$$\Delta b_3 = 0.3 \delta_3 = 0.3 \times (-1.7864) = -0.5359.$$

Equations 5.21, 5.23, 5.25, and 5.26 are used to determine changes to the weights and bias for each hidden unit as follows:

$$\delta_1 = -\left(\sum_n \delta_n w_{n,1}\right) f_1'(net_1) = -\left(\sum_{n=3}^{n=3} \delta_n w_{n,1}\right) f_1'(net_1)$$

$$= \delta_3 w_{3,1} o_1 (1-o_1) = (-1.7864) \times 0.3 \times 0.3543 \times (1 - 0.3543) = -0.1226$$

$$\delta_2 = -\left(\sum_n \delta_n w_{n,2}\right) f_2'(net_2) = -\left(\sum_{n=3}^{n=3} \delta_n w_{n,2}\right) f_2'(net_2) = \delta_3 w_{3,2} o_2 (1-o_2)$$

$$= (-1.7864) \times 0.4 \times 0.4502 \times (1 - 0.4502) = -0.1769$$

$$\Delta w_{1,1} = \alpha \delta_1 x_1 = 0.3 \times \delta_1 \times x_1 = 0.3 \times (-0.1226) \times (-1) = 0.0368$$

$$\Delta w_{1,2} = \alpha \delta_1 x_2 = 0.3 \times \delta_1 \times x_2 = 0.3 \times (-0.1226) \times (-1) = 0.0368$$

$$\Delta w_{2,1} = \alpha \delta_2 x_1 = 0.3 \times \delta_2 \times x_1 = 0.3 \times (-0.1769) \times (-1) = 0.0531$$

$$\Delta w_{2,2} = \alpha \delta_2 x_2 = 0.3 \times \delta_2 \times x_2 = 0.3 \times (-0.1769) \times (-1) = 0.0531$$

$$\Delta b_1 = \alpha \delta_1 = 0.3 \times (-0.1226) = -0.0368$$

$$\Delta b_2 = \alpha \delta_2 = 0.3 \delta_2 = 0.3 \times (-0.1769) = -0.0531.$$

Using the changes to all the weights and biases of the ANN, Equations 5.27 and 5.28 are used to perform an iteration of updating the weights and biases as follows:

$$w_{1,1}(1) = w_{1,1}(0) + \Delta w_{1,1} = 0.1 + 0.0368 = 0.1368$$

$$w_{1,2}(1) = w_{1,2}(0) + \Delta w_{1,2} = 0.2 + 0.0368 = 0.2368$$

$$w_{2,1}(1) = w_{2,1}(0) + \Delta w_{2,1} = -0.1 + 0.0531 = -0.0469$$

$$w_{2,2}(1) = w_{2,2}(0) + \Delta w_{2,2} = -0.2 + 0.0531 = -0.1469$$

$$w_{3,1}(1) = w_{3,1}(0) + \Delta w_{3,1} = 0.3 - 0.1899 = 0.1101$$

$$w_{3,2}(1) = w_{3,2}(0) + \Delta w_{3,2} = 0.4 - 0.2413 = 0.1587$$

$$b_1(1) = b_1(0) + \Delta b_1 = -0.3 - 0.1226 = -0.4226$$

$$b_2(1) = b_2(0) + \Delta b_2 = -0.4 - 0.0531 = -0.4531$$

$$b_3(1) = b_3(0) + \Delta b_3 = 0.5 - 0.5359 = -0.0359.$$

This new set of weights and biases, $w_{j,i}(1)$ and $b_j(1)$, will be used to pass the inputs of the second data record through the ANN and then update the weights and biases again to obtain $w_{j,i}(2)$ and $b_j(2)$ if necessary. This process repeats again for the third data record, the fourth data record, back to the first data record, and so on, until the measure of the output error E as defined in Equation 5.20 is smaller than a preset threshold, e.g., 0.1.

A measure of the output error, such as E, or the root-mean-squared error over all the training data records can be used to determine when the learning of ANN weights and biases can stop. The number of iterations, e.g., 1000 iterations, is another criterion that can be used to stop the learning.

Updating weights and biases after passing each data record in the training data set is called the incremental learning. In the incremental learning, weights and biases are updated so that they will work better for one data record. Changes based on one data record may go in the different direction where changes made for another data record go, making the learning take a long time to converge to the final set of weights and biases that work for all the data records. The batch learning is to hold the update of weights and biases until all the data records in the training data set are passed through the ANN and their associated changes of weights and biases are computed and averaged. The average of weight and bias changes for all the data records, that is, the overall effect of changes on weights and biases by all the data records, is used to update weights and biases.

The learning rate also affects how well and fast the learning proceeds. As illustrated in Figure 5.14, a small learning rate, e.g., 0.01, produces a small change of weights and biases and thus a small decrease in E, and makes the learning take a long time to reach the global minimum value of E or a local minimum value of E. However, a large learning rate produces a large change of weights and biases, which may cause the search of W for minimizing E not to reach a local or global minimum value of E. Hence, as a tradeoff between

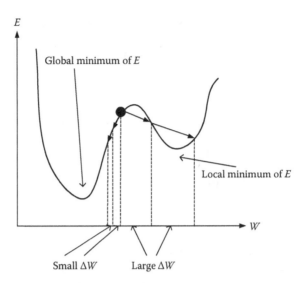

FIGURE 5.14
Effect of the learning rate.

a small learning rate and a large learning rate, a method of adaptive learning rates can be used to start with a large learning rate for speeding up the learning process and then change to a small learning rate for taking small steps to reach a local or global minimum value of E.

Unlike the decision trees in Chapter 4, an ANN does not show an explicit model of the classification and prediction function that the ANN has learned from the training data. The function is implicitly represented through connection weights and biases that cannot be translated into meaningful classification and prediction patterns in the problem domain. Although the knowledge of classification and prediction patterns has been acquired by the ANN, such knowledge is not available in an interpretable form. Hence, ANNs help the task of performing classification and prediction but not the task of discovering knowledge.

5.5 Empirical Selection of an ANN Architecture for a Good Fit to Data

Unlike the regression models in Chapter 2, the learning of a classification and prediction function by a multilayer feedforward ANN does not require defining a specific form of that function, which may be difficult when a data set is large, and we have little prior knowledge about the domain or the data. The complexity of the ANN and the function which the ANN learns and represents depends much on the number of hidden units. The more hidden units the ANN has, the more complex function the ANN can learn and represent. However, if we use a complex ANN to learn a simple function, we may see the function of the ANN over-fit the data as illustrated in Figure 5.15. In Figure 5.15, data points are generated using a linear model:

$$y = x + \varepsilon,$$

where ε denotes a random noise. However, a nonlinear model is fitted to the training data points as illustrated by the filled circles in Figure 5.15, covering every training data point with no difference between the target y value and the predicted y value from the nonlinear model. Although the nonlinear model provides a perfect fit to the training data, the prediction performance of the nonlinear model on new data points in the testing data set as illustrated by the unfilled circles in Figure 5.15 will be poorer than that of the linear model, $y = x$, for the following reasons:

- The nonlinear model captures the random noise ε in the model
- The random noises from new data points behave independently and differently of the random noises from the training data points
- The random noises from the training data points that are captured in the nonlinear model do not match well with the random noises from new data points in the testing data set, causing prediction errors

In general, an over-fitted model does not generalize well to new data points in the testing data set. When we do not have prior knowledge about a given data set (e.g., the form or complexity of the classification and prediction function), we have to empirically try out ANN architectures with varying levels

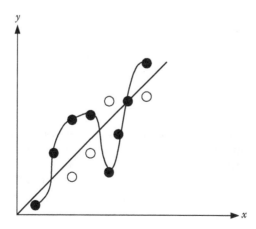

FIGURE 5.15
An illustration of a nonlinear model overfitting to data from a linear model.

of complexity by using different numbers of hidden units. Each ANN architecture is trained to learn weights and biases of connections from a training data set and is tested for the prediction performance on a testing data set. The ANN architecture, which performs well on the testing data, is considered to provide a good fit to the data and is selected.

5.6 Software and Applications

The website http://www.knuggets.com has information about various data mining tools. The following software packages provide software tools for ANNs with back-propagation learning:

- Weka (http://www.cs.waikato.ac.nz/ml/weka/)
- MATLAB® (www.mathworks.com/)

Some applications of ANNs can be found in (Ye et al., 1993; Ye, 1996, 2003, Chapter 3; Ye and Zhao, 1996, 1997).

Exercises

5.1 The training data set for the Boolean function $y = $ NOT x is given next. Use the graphical method to determine the decision boundary, the weight, and the bias of a single-unit perceptron for this Boolean function.

The training data set:

x	y
−1	1
1	−1

5.2 Consider the single-unit perceptron in Exercise 5.1. Assign 0.2 to initial weights and bias and use the learning rate of 0.3. Use the learning method to perform one iteration of the weight and bias update for the two data records of the Boolean function in Exercise 5.1.

5.3 The training data set for a classification function with three attribute variables and one target variable is given below. Use the graphical method to determine the decision boundary, the weight, and the bias of a single-neuron perceptron for this classification function.

The training data set:

x_1	x_2	x_3	y
-1	-1	-1	-1
-1	-1	1	-1
-1	1	-1	-1
-1	1	1	1
1	-1	-1	-1
1	-1	1	1
1	1	-1	1
1	1	1	1

5.4 A single-unit perceptron is used to learn the classification function in Exercise 5.3. Assign 0.4 to the initial weights and 1.5 to the initial bias and use the learning rate of 0.2. Use the learning method to perform one iteration of the weight and bias update for the third and fourth data records of this function.

5.5 Consider a fully connected two-layer feedforward ANN with one input variable, one hidden unit, and two output variables. Assign initial weights and biases of 0.1 and use the learning rate of 0.3. The transfer function is the sigmoid function for each unit. Show the architecture of the ANN and perform one iteration of weight and bias update using the back-propagation learning algorithm and the following training example:

x	y_1	y_2
1	0	1

5.6 The following ANN with the initial weights and biases is used to learn the XOR function given below. The transfer function for units 1 and 4 is the linear function. The transfer function for units 2 and 3 is the sigmoid transfer function. The learning rate is $\alpha = 0.3$. Perform one iteration of the weight and bias update for $w_{1,1}, w_{1,2}, w_{2,1}, w_{3,1}, w_{4,2}, w_{4,3}, b_2$, after feeding $x_1 = 0$ and $x_2 = 1$ to the ANN.

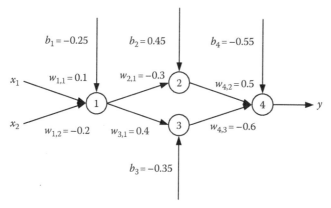

XOR:

x_1	x_2	y
0	0	0
0	1	1
1	0	1
1	1	0

6

Support Vector Machines

A support vector machine (SVM) learns a classification function with two target classes by solving a quadratic programming problem. In this chapter, we briefly review the theoretical foundation of SVM that leads to the formulation of a quadratic programming problem for learning a classifier. We then introduce the SVM formulation for a linear classifier and a linearly separable problem, followed by the SVM formulation for a linear classifier and a non-linearly separable problem and the SVM formulation for a nonlinear classifier and a nonlinearly separable problem based on kernel functions. We also give methods of applying SVM for a classification function with more than two target classes. A list of data mining software packages that support SVM is provided. Some applications of SVM are given with references.

6.1 Theoretical Foundation for Formulating and Solving an Optimization Problem to Learn a Classification Function

Consider a set of n data points, $(x_1, y_1), \ldots, (x_n, y_n)$, and a classification function to fit to data, $y = f_A(x)$, where y takes one of two categorical values $\{-1, 1\}$, x is a p-dimensional vector of variables, and A is a set of parameters in the function f that needs to be learned or determined using the training data. For example, if an artificial neural network (ANN) is used to learn and represent the classification function f, connection weights and biases are the parameters in f. The expected risk of classification using f measures the classification error, and is defined as

$$R(A) = \int |f_A(x) - y| P(x, y) dx dy, \qquad (6.1)$$

where $P(x, y)$ denotes the probability function of x and y. The expected risk of classification depends on A values. A smaller expected risk of classification indicates a better generalization performance of the classification function in that the classification function is capable of classifying more data points correctly. Different sets of A values give different classification functions $f_A(x)$ and thus produce different classification errors and different levels of the expected risk. The empirical risk over a sample of n data points is defined as

$$R_{emp}(A) = \frac{1}{n} \sum_{i=1}^{n} |f_A(x_i) - y_i|. \tag{6.2}$$

Vapnik and Chervonenkis (Vapnik, 1989, 2000) provide the following bound on the expected risk of classification which holds with the probability $1 - \eta$:

$$R(A) \leq R_{emp}(A) + \sqrt{\frac{v\left(\ln\frac{2n}{v} + 1\right) - \ln\frac{\eta}{4}}{n}}, \tag{6.3}$$

where v denotes the VC (Vapnik and Chervonenkis) dimension of f_A and measures the complexity of f_A, which is controlled by the number of parameters A in f for many classification functions. Hence, the expected risk of classification is bound by both the empirical risk of classification and the second term in Equation 6.3 with the second term increasing with the VC-dimension. To minimize the expected risk of classification, we need to minimize both the empirical risk and the VC-dimension of f_A at the same time. This is called the structural risk minimization principle. Minimizing the VC-dimension of f_A, that is, the complexity of f_A, is like looking for a classification function with the minimum description length for a good generalization performance as discussed in Chapter 4. SVM searches for a set of A values that minimize the empirical risk and the VC-dimension at the same time by formulating and solving an optimization problem, specifically, a quadratic programming problem. The following sections provide the SVM formulation of the quadratic programming problem for three types of classification problems: (1) a linear classifier and a linearly separable problem, (2) a linear classifier and a nonlinearly separable problem, and (3) a nonlinear classifier and a nonlinearly separable problem. As discussed in Chapter 5, the logical AND function is a linearly separable classification problem and requires only a linear classifier in Type (1), and the logical XOR function is a nonlinearly separable classification problem and requires a nonlinear classifier in Type (3). Because a linear classifier generally has a lower VC-dimension than a nonlinear classifier, using a linear classifier for a nonlinearly separable problem in Type (2) can sometimes produce a lower bound on the expected risk of classification than using a nonlinear classifier for the nonlinearly separable problem.

6.2 SVM Formulation for a Linear Classifier and a Linearly Separable Problem

Consider the definition of a linear classifier for a perceptron in Chapter 5:

$$f_{w,b}(x) = sign(w'x + b). \tag{6.4}$$

The decision boundary separating two target classes {−1, 1} is

$$w'x + b = 0. \tag{6.5}$$

The linear classifier works in the following way:

$$y = sign(w'x + b) = 1 \quad \text{if } w'x + b > 0 \tag{6.6}$$

$$y = sign(w'x + b) = -1 \quad \text{if } w'x + b \le 0.$$

If we impose a constraint,

$$\|w\| \le M,$$

where M is a constant and $\|w\|$ denotes the norm of the p-dimensional vector w and is defined as

$$\|w\| = \sqrt{w_1^2 + \cdots + w_p^2}.$$

The set of hyperplanes defined by the following:

$$\{f_{w,b} = sign(w'x + b) \mid \|w\| \le M\},$$

has the VC-dimension v that satisfies the bound (Vapnik, 1989, 2000):

$$v \le \min\{M^2, p\} + 1. \tag{6.7}$$

By minimizing $\|w\|$, we will minimize M and thus the VC-dimension v. Hence, to minimize the VC-dimension v as required by the structural risk minimization principle, we want to minimize $\|w\|$, or equivalently:

$$\min \frac{1}{2}\|w\|^2. \tag{6.8}$$

Rescaling w does not change the slope of the hyperplane for the decision boundary. Rescaling b does not change the slope of the decision boundary but moves the hyperplane of the decision boundary in parallel. For example, in the two-dimensional vector space shown in Figure 6.1, the decision boundary is

$$w_1 x_1 + w_2 x_2 + b = 0 \quad \text{or} \quad x_2 = -\frac{w_1}{w_2} x_1 - \frac{b}{w_2}, \tag{6.9}$$

the slope of the line for the decision boundary is $-w_1/w_2$, and the intercept of the line for the decision boundary is $-b/w_2$. Rescaling w to $c_w w$, where c_w is a constant, does not change the slope of the line for the decision boundary as $-c_w w_1/c_w w_2 = -w_1/w_2$. Rescaling b to $c_b b$, where c_b is a constant, does not change the slope of the line for the decision boundary, but changes the intercept of the line to $-c_b b/w_2$ and thus moves the line in parallel.

Figure 6.1 shows examples of data points with the target value of 1 (indicated by small circles) and examples of data points with the target value of –1 (indicated by small squares). Among the data points with the target value of 1, we consider the data point closest to the decision boundary, x_{+1}, as shown by the data point with the solid circle in Figure 6.1. Among the data points with the target value of –1, we consider the data point closest to the decision boundary, x_{-1}, as shown by the data point with the solid square in Figure 6.1. Suppose that for two data points x_{+1} and x_{-1} we have

$$\begin{aligned} w' x_{+1} + b &= c_{+1} \\ w' x_{-1} + b &= c_{-1}. \end{aligned} \tag{6.10}$$

We want to rescale w to $c_w w$ and rescale b to $c_b b$ such that we have

$$\begin{aligned} c_w w' x_{+1} + c_b b &= 1 \\ c_w w' x_{-1} + c_b b &= -1, \end{aligned} \tag{6.11}$$

and still denote the rescaled values by w and b. We have

$$\min\{|w' x_i + b|, \quad i = 1, \dots, n\} = 1,$$

which implies $|w' x + b| = 1$ for the data point in each target class closest to the decision boundary $w'x + b = 0$.

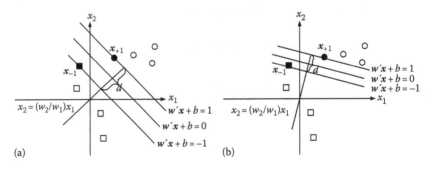

FIGURE 6.1
SVM for a linear classifier and a linearly separable problem. (a) A decision boundary with a large margin. (b) A decision boundary with a small margin.

For example, in the two-dimensional vector space of x, Equations 6.10 and 6.11 become the following:

$$w_1 x_{+1,1} + w_2 x_{+1,2} + b = c_{+1} \tag{6.12}$$

$$w_1 x_{-1,1} + w_2 x_{-1,2} + b = c_{-1} \tag{6.13}$$

$$c_w w_1 x_{+1,1} + c_w w_2 x_{+1,2} + c_b b = 1 \tag{6.14}$$

$$c_w w_1 x_{-1,1} + c_w w_2 x_{-1,2} + c_b b = -1. \tag{6.15}$$

We solve Equations 6.12 through 6.15 to obtain c_w and c_b. We first use Equation 6.14 to obtain

$$c_w = \frac{1 - c_b b}{w_1 x_{+1,1} + w_2 x_{+1,2}}, \tag{6.16}$$

and substitute c_w in Equations 6.16 into 6.15 to obtain

$$\frac{1 - c_b b}{w_1 x_{+1,1} + w_2 x_{+1,2}} (w_1 x_{-1,1} + w_2 x_{-1,2}) + c_b b = -1. \tag{6.17}$$

We then use Equations 6.12 and 6.13 to obtain

$$w_1 x_{+1,1} + w_2 x_{+1,2} = c_{+1} - b \tag{6.18}$$

$$w_1 x_{-1,1} + w_2 x_{-1,2} = c_{-1} - b, \tag{6.19}$$

and substitute Equations 6.18 and 6.19 into Equation 6.17 to obtain

$$\frac{1 - c_b b}{c_{+1} - b}(c_{-1} - b) + c_b b = -1$$

$$\frac{c_{-1} - b}{c_{+1} - b} - \frac{(c_{-1} - b)b}{c_{+1} - b} c_b + b c_b = -1$$

$$c_b = \frac{2b - c_{+1} - c_{-1}}{b^2 + b - c_{-1} b}. \tag{6.20}$$

We finally use Equation 6.14 to compute c_w and substitute Equations 6.18 and 6.20 into the resulting equations to obtain

$$c_w = \frac{1 - c_b b}{w_1 x_{+1,1} + w_2 x_{+1,2}} = \frac{1 - c_b b}{c_{+1} - b} = \frac{1 - (2b - c_{+1} - c_{-1}/b + 1 - c_{-1})}{c_{+1} - b}$$

$$= \frac{1 - b + c_{+1}}{(c_{+1} - b)(b + 1 - c_{-1})}. \tag{6.21}$$

Equations 6.20 and 6.21 show how to rescale w and b in a two-dimensional vector space of x.

Let w and b denote the rescaled values. The hyperplane bisects $w'x + b = 1$ and $w'x + b = -1$ is $w'x + b = 0$, as shown in Figure 6.1. Any data point x with the target class +1 satisfies

$$w'x + b \geq 1$$

since the data point with the target class of +1 closest to $w'x + b = 0$ has $w'x + b = 1$. Any data point x with the target class of −1 satisfies

$$w'x + b \leq -1$$

since the data point with the target class of −1 closest to $w'x + b = 0$ has $w'x + b = -1$. Therefore, the linear classifier can be defined as follows:

$$y = sign(w'x + b) = 1 \quad \text{if } w'x + b \geq 1 \tag{6.22}$$

$$y = sign(w'x + b) = -1 \quad \text{if } w'x + b \leq -1.$$

To minimize the empirical risk R_{emp} or the empirical classification error as required by the structural risk minimization principle defined by Equation 6.3, we require

$$y_i(w'x_i + b) \geq 1, \quad i = 1, \ldots, n. \tag{6.23}$$

If $y_i = 1$, we want $w'x_i + b \geq 1$ so that the linear classifier in Equation 6.22 produces the target class 1. If $y_i = -1$, we want $w'x_i + b \leq -1$ so that the linear classifier in Equation 6.22 produces the target class of −1. Hence, Equation 6.23 specifies the requirement of the correct classification for the sample of data points (x_i, y_i), $i = 1, \ldots, n$.

Therefore, putting Equations 6.8 and 6.23 together allows us to apply the structural risk principle of minimizing both the empirical classification error and the VC-dimension of the classification function. Equations 6.8 and 6.23 are put together by formulating a quadratic programming problem:

$$\min_{w,b} \frac{1}{2}\|w\|^2 \tag{6.24}$$

subject to

$$y_i(w'x_i + b) \geq 1, \quad i = 1, \ldots, n.$$

6.3 Geometric Interpretation of the SVM Formulation for the Linear Classifier

$\|w\|$ in the objective function of the quadratic programming problem in Formulation 6.24 has a geometric interpretation in that $2/\|w\|$ is the distance of the two hyperplanes $w'x + b = 1$ and $w'x + b = -1$. This distance is called the margin of the decision boundary or the margin of the linear classifier, with the $w'x + b = 0$ being the decision boundary. To show this in the two-dimensional vector space of x, let us compute the distance of two parallel lines $w'x + b = 1$ and $w'x + b = -1$ in Figure 6.1. These two parallel lines can be represented as follows:

$$w_1 x_1 + w_2 x_2 + b = 1 \qquad (6.25)$$

$$w_1 x_1 + w_2 x_2 + b = -1. \qquad (6.26)$$

The following line

$$w_2 x_1 - w_1 x_2 = 0 \qquad (6.27)$$

passes through the origin and is perpendicular to the lines defined in Equations 6.25 and 6.26 since the slope of the parallel lines in Equations 6.25 and 6.26 is $-w_1/w_2$ and the slope of the line in Equation 6.27 is $-w_2/w_1$, which is the negative reciprocal to $-w_1/w_2$. By solving Equations 6.25 and 6.27 for x_1 and x_2, we obtain the coordinates of the data point where these two lines are intersected: $\left(\dfrac{1-b}{w_1^2 + w_2^2} w_1, \dfrac{1-b}{w_1^2 + w_2^2} w_2\right)$. By solving Equations 6.26 and 6.27 for x_1 and x_2, we obtain the coordinates of the data point where these two lines are intersected: $\left(\dfrac{-1-b}{w_1^2 + w_2^2} w_1, \dfrac{-1-b}{w_1^2 + w_2^2} w_2\right)$. Then we compute the distance of the two data points, $\left(\dfrac{1-b}{w_1^2 + w_2^2} w_1, \dfrac{1-b}{w_1^2 + w_2^2} w_2\right)$ and $\left(\dfrac{-1-b}{w_1^2 + w_2^2} w_1, \dfrac{-1-b}{w_1^2 + w_2^2} w_2\right)$:

$$\begin{aligned} d &= \sqrt{\left(\dfrac{1-b}{w_1^2 + w_2^2} w_1 - \dfrac{-1-b}{w_1^2 + w_2^2} w_1\right)^2 + \left(\dfrac{1-b}{w_1^2 + w_2^2} w_2 - \dfrac{-1-b}{w_1^2 + w_2^2} w_2\right)^2} \\ &= \dfrac{1}{w_1^2 + w_2^2} \sqrt{2^2 w_1^2 + 2^2 w_2^2} = \dfrac{2}{\sqrt{w_1^2 + w_2^2}} = \dfrac{2}{\|w\|}. \end{aligned} \qquad (6.28)$$

Hence, minimizing $(1/2)\|w\|^2$ in the objective function of the quadratic programming problem in Formulation 6.24 is to maximize the margin of the linear classifier or the generalization performance of the linear classifier. Figure 6.1a and b shows two different linear classifiers with two different decision boundaries that classify the eight data points correctly but have different margins. The linear classifier in Figure 6.1a has a larger margin and is expected to have a better generalization performance than that in Figure 6.1b.

6.4 Solution of the Quadratic Programming Problem for a Linear Classifier

The quadratic programming problem in Formulation 6.24 has a quadratic objective function and a linear constraint with regard to w and b, is a convex optimization problem, and can be solved using the Lagrange multiplier method for the following problem:

$$\min_{w,b} \max_{\alpha \geq 0} L(w,b,\alpha) = \frac{1}{2}\|w\|^2 - \sum_{i=1}^{n} \alpha_i \left[y_i(w'x_i + b) - 1 \right] \quad (6.29)$$

subject to

$$\alpha_i \left[y_i(w'x_i + b) - 1 \right] = 0 \quad i = 1, \ldots, n \quad (6.30)$$

$$\alpha_i \geq 0 \quad i = 1, \ldots, n,$$

where α_i, $i = 1, \ldots, n$ are the non-negative Lagrange multipliers, and the two equations in the constrains are known as the Karush–Kuhn–Tucker condition (Burges, 1998) and are the transformation of the inequality constraint in Equation 6.23. The solution to Formulation 6.29 is at the saddle point of $L(w,b,\alpha)$, where $L(w,b,\alpha)$ is minimized with regard to w and b and maximized with regard to α. Minimizing $(1/2)\|w\|^2$ with regard to w and b covers the objective function in Formulation 6.24. Minimizing $-\sum_{i=1}^{n} \alpha_i \left[y_i(w'x_i + b) - 1 \right]$ is to maximize $\sum_{i=1}^{n} \alpha_i \left[y_i(w'x_i + b) - 1 \right]$ with regard to α and satisfy $y_i(w'x_i + b) \geq 1$—the constraint in Formulation 6.24, since $\alpha_i \geq 0$. At the point where $L(w,b,\alpha)$ is minimized with regard to w and b, we have

$$\frac{\partial L(w,b,\alpha)}{\partial w} = w - \sum_{i=1}^{n} \alpha_i y_i x_i = 0 \quad \text{or} \quad w = \sum_{i=1}^{n} \alpha_i y_i x_i \quad (6.31)$$

$$\frac{\partial L(w,b,\alpha)}{\partial b} = \sum_{i=1}^{n} \alpha_i y_i = 0. \quad (6.32)$$

Note that w is determined by only the training data points (x_i, y_i) for which $\alpha_i > 0$. Those training data vectors with the corresponding $\alpha_i > 0$ are called support vectors. Using the Karush–Kuhn–Tucker condition in Equation 6.30 and any support vector (x_i, y_i) with $\alpha_i > 0$, we have

$$y_i(w'x_i + b) - 1 = 0 \quad (6.33)$$

in order to satisfy Equation 6.32. We also have

$$y_i^2 = 1 \tag{6.34}$$

since y_i takes the value of 1 or −1. We solve Equations 6.33 and 6.34 for b and get

$$b = y_i - w'x_i \tag{6.35}$$

because

$$y_i(w'x_i + b) - 1 = y_i(w'x_i + y_i - w'x_i) - 1 = y_i^2 - 1 = 0.$$

To compute w using Equations 6.31 and 6.32 and compute b using Equation 6.35, we need to know the values of the Lagrange multipliers α. We substitute Equations 6.31 and 6.32 into $L(w,b,\alpha)$ in Formulation 6.29 to obtain $L(\alpha)$

$$\begin{aligned}L(\alpha) &= \frac{1}{2}\sum_{i=1}^{n}\sum_{j=1}^{n}\alpha_i\alpha_j y_i y_j x_i' x_j - \sum_{i=1}^{n}\sum_{j=1}^{n}\alpha_i\alpha_j y_i y_j x_i' x_j - b\sum_{i=1}^{n}\alpha_i y_i + \sum_{i=1}^{n}\alpha_i \\ &= \sum_{i=1}^{n}\alpha_i - \frac{1}{2}\sum_{i=1}^{n}\sum_{j=1}^{n}\alpha_i\alpha_j y_i y_j x_i' x_j.\end{aligned} \tag{6.36}$$

Hence, the dual problem to the quadratic programming problem in Formulation 6.24 is

$$\max_\alpha L(\alpha) = \sum_{i=1}^{n}\alpha_i - \frac{1}{2}\sum_{i=1}^{n}\sum_{j=1}^{n}\alpha_i\alpha_j y_i y_j x_i' x_j \tag{6.37}$$

subject to

$$\sum_{i=1}^{n}\alpha_i y_i = 0$$

$$\alpha_i\left[y_i(w'x_i + b) - 1\right] = 0 \quad \text{or} \quad \sum_{j=1}^{n}\alpha_i\alpha_j y_i y_j x_i' x_j + \alpha_i y_i b - \alpha_i = 0 \quad i = 1,\ldots,n$$

$$\alpha_i \geq 0 \quad i = 1,\ldots,n.$$

In summary, the linear classifier for SVM is solved in the following steps:

1. Solve the optimization problem in Formulation 6.37 to obtain α:

$$\max_\alpha L(\alpha) = \sum_{i=1}^{n} \alpha_i - \frac{1}{2} \sum_{i=1}^{n} \sum_{j=1}^{n} \alpha_i \alpha_j y_i y_j x_i' x_j$$

subject to

$$\sum_{i=1}^{n} \alpha_i y_i = 0$$

$$\sum_{j=1}^{n} \alpha_i \alpha_j y_i y_j x_j' x_i + \alpha_i y_i b - \alpha_i = 0 \quad i = 1, \ldots, n$$

$$\alpha_i \geq 0 \quad i = 1, \ldots, n.$$

2. Use Equation 6.31 to obtain w:

$$w = \sum_{i=1}^{n} \alpha_i y_i x_i.$$

3. Use Equation 6.35 and a support vector (x_i, y_i) to obtain b:

$$b = y_i - w' x_i.$$

and the decision function of the linear classifier is given in Equation 6.22:

$$y = \text{sign}(w'x + b) = 1 \quad \text{if } w'x + b \geq 1$$
$$y = \text{sign}(w'x + b) = -1 \quad \text{if } w'x + b \leq -1,$$

or Equation 6.4:

$$f_{w,b}(x) = \text{sign}(w'x + b) = \text{sign}\left(\sum_{i=1}^{n} \alpha_i y_i x_i' x + b\right).$$

Note that only the support vectors with the corresponding $\alpha_i > 0$ contribute to the computation of w, b and the decision function of the linear classifier.

Example 6.1

Determine the linear classifier of SVM for the AND function in Table 5.1, which is copied here in Table 6.1 with $x = (x_1, x_2)$.

There are four training data points in this problem. We formulate and solve the optimization problem in Formulation 6.24 as follows:

$$\min_{w_1, w_2, b} \frac{1}{2}\left[(w_1)^2 + (w_2)^2\right]$$

subject to

$$w_1 + w_2 - b \geq 1$$
$$w_1 - w_2 - b \geq 1$$
$$-w_1 + w_2 - b \geq 1$$
$$w_1 + w_2 + b \geq 1.$$

Using the optimization toolbox in MATLAB®, we obtain the following optimal solution to the aforementioned optimization problem:

$$w_1 = 1, \quad w_2 = 1, \quad b = -1.$$

That is, we have

$$w = \begin{bmatrix} 1 \\ 1 \end{bmatrix} \quad b = -1.$$

This solution gives the decision function in Equation 6.22 or 6.4 as follows:

$$\begin{cases} y = \text{sign}\left(\begin{bmatrix} 1 & 1 \end{bmatrix} \begin{bmatrix} x_1 \\ x_2 \end{bmatrix} - 1\right) = \text{sign}(x_1 + x_2 - 1) = 1 & \text{if } x_1 + x_2 - 1 \geq 1 \\ y = \text{sign}\left(\begin{bmatrix} 1 & 1 \end{bmatrix} \begin{bmatrix} x_1 \\ x_2 \end{bmatrix} - 1\right) = \text{sign}(x_1 + x_2 - 1) = -1 & \text{if } x_1 + x_2 - 1 \leq -1 \end{cases}$$

or

$$f_{w,b}(x) = \text{sign}(w'x + b) = \text{sign}\left(\begin{bmatrix} 1 & 1 \end{bmatrix} \begin{bmatrix} x_1 \\ x_2 \end{bmatrix} - 1\right) = \text{sign}(x_1 + x_2 - 1).$$

TABLE 6.1

AND Function

Data Point #	Inputs		Output
i	x_1	x_2	y
1	−1	−1	−1
2	−1	1	−1
3	1	−1	−1
4	1	1	1

We can also formulate the optimization problem in Formulation 6.37:

$$\max_\alpha L(\alpha) = \sum_{i=1}^{n}\alpha_i - \frac{1}{2}\sum_{i=1}^{n}\sum_{j=1}^{n}\alpha_i\alpha_j y_i y_j x'_i x_j$$

$$= \alpha_1 + \alpha_2 + \alpha_3 + \alpha_4 - \frac{1}{2}[\alpha_1\alpha_1 y_1 y_1 x'_1 x_1 + \alpha_1\alpha_2 y_1 y_2 x'_1 x_2$$
$$+ \alpha_1\alpha_3 y_1 y_3 x'_1 x_3 + \alpha_1\alpha_4 y_1 y_4 x'_1 x_4 + \alpha_2\alpha_1 y_2 y_1 x'_2 x_1 + \alpha_2\alpha_2 y_2 y_2 x'_2 x_2$$
$$+ \alpha_2\alpha_3 y_2 y_3 x'_2 x_3 + \alpha_2\alpha_4 y_2 y_4 x'_2 x_4 + \alpha_3\alpha_1 y_3 y_1 x'_3 x_1 + \alpha_3\alpha_2 y_3 y_2 x'_3 x_2$$
$$+ \alpha_3\alpha_3 y_3 y_3 x'_3 x_3 + \alpha_3\alpha_4 y_3 y_4 x'_3 x_4 + \alpha_4\alpha_1 y_4 y_1 x'_4 x_1 + \alpha_4\alpha_2 y_4 y_2 x'_4 x_2$$
$$+ \alpha_4\alpha_3 y_4 y_3 x'_4 x_3 + \alpha_4\alpha_4 y_4 y_4 x'_4 x_4]$$

$$= \alpha_1 + \alpha_2 + \alpha_3 + \alpha_4 - \frac{1}{2}\left[\alpha_1\alpha_1(-1)(-1)[-1\ -1]\begin{bmatrix}-1\\-1\end{bmatrix}\right.$$

$$+2\alpha_1\alpha_2(-1)(-1)[-1\ -1]\begin{bmatrix}-1\\1\end{bmatrix} + 2\alpha_1\alpha_3(-1)(-1)[-1\ -1]\begin{bmatrix}1\\-1\end{bmatrix}$$

$$+2\alpha_1\alpha_4(-1)(1)[-1\ -1]\begin{bmatrix}1\\1\end{bmatrix} + \alpha_2\alpha_2(-1)(-1)[-1\ 1]\begin{bmatrix}-1\\1\end{bmatrix}$$

$$+2\alpha_2\alpha_3(-1)(-1)[-1\ 1]\begin{bmatrix}1\\-1\end{bmatrix} + 2\alpha_2\alpha_4(-1)(1)[-1\ 1]\begin{bmatrix}1\\1\end{bmatrix}$$

$$+\alpha_3\alpha_3(-1)(-1)[1\ -1]\begin{bmatrix}1\\-1\end{bmatrix} + 2\alpha_3\alpha_4(-1)(1)[1\ -1]\begin{bmatrix}1\\1\end{bmatrix}$$

$$\left.+\alpha_4\alpha_4(1)(1)[1\ 1]\begin{bmatrix}1\\1\end{bmatrix}\right]$$

$$= \alpha_1 + \alpha_2 + \alpha_3 + \alpha_4 - \frac{1}{2}(2\alpha_1^2 + 2\alpha_2^2 + 2\alpha_3^2 + 2\alpha_4^2 - 4\alpha_1\alpha_4 - 4\alpha_2\alpha_3)$$

$$= -\alpha_1^2 - \alpha_2^2 - \alpha_3^2 - \alpha_4^2 + 2\alpha_1\alpha_4 + 2\alpha_2\alpha_3 + \alpha_1 + \alpha_2 + \alpha_3 + \alpha_4$$

$$= -(\alpha_1 - \alpha_4)^2 - (\alpha_2 - \alpha_3)^2 + \alpha_1 + \alpha_2 + \alpha_3 + \alpha_4$$

subject to

$$\sum_{i=1}^{n}\alpha_i y_i = \alpha_1 y_1 + \alpha_2 y_2 + \alpha_3 y_3 + \alpha_i y_4 = -\alpha_1 - \alpha_2 - \alpha_3 + \alpha_4 = 0$$

$$\left(\sum_{j=1}^{n}\alpha_i\alpha_j y_i y_j x'_j x_i + \alpha_i y_i b - \alpha_i = 0 \quad i = 1, 2, 3, 4 \text{ become:}\right)$$

$$\alpha_1(-1)\left[\alpha_1(-1)[-1\ -1]\begin{bmatrix}-1\\-1\end{bmatrix} + \alpha_2(-1)[-1\ 1]\begin{bmatrix}-1\\-1\end{bmatrix} + \alpha_3(-1)[1\ -1]\begin{bmatrix}-1\\-1\end{bmatrix}\right.$$

$$\left.+\alpha_4(1)[1\ 1]\begin{bmatrix}-1\\-1\end{bmatrix}\right] + \alpha_1(-1)b - \alpha_1 = 0 \quad \text{or} \quad -\alpha_1(-2\alpha_1 - 2\alpha_4) - \alpha_1 b - \alpha_1 = 0$$

$$\alpha_2(-1)\left[\alpha_1(-1)[-1\ \ -1]\begin{bmatrix}-1\\1\end{bmatrix}+\alpha_2(-1)[-1\ \ 1]\begin{bmatrix}-1\\1\end{bmatrix}+\alpha_3(-1)[1\ \ -1]\begin{bmatrix}-1\\1\end{bmatrix}\right.$$

$$\left.\alpha_4(1)[1\ \ 1]\begin{bmatrix}-1\\1\end{bmatrix}\right]+\alpha_2(-1)b-\alpha_2=0\quad\text{or}\quad-\alpha_2(-2\alpha_2+2\alpha_3)-\alpha_2 b-\alpha_2=0$$

$$\alpha_3(-1)\left[\alpha_1(-1)[-1\ \ -1]\begin{bmatrix}1\\-1\end{bmatrix}+\alpha_2(-1)[-1\ \ 1]\begin{bmatrix}1\\-1\end{bmatrix}+\alpha_3(-1)[1\ \ -1]\begin{bmatrix}1\\-1\end{bmatrix}\right.$$

$$\left.+\alpha_4(1)[1\ \ 1]\begin{bmatrix}1\\-1\end{bmatrix}\right]+\alpha_3(-1)b-\alpha_3=0\quad\text{or}\quad-\alpha_3(2\alpha_2-2\alpha_3)-\alpha_3 b-\alpha_3=0$$

$$\alpha_4(1)\left[\alpha_1(-1)[-1\ \ -1]\begin{bmatrix}1\\1\end{bmatrix}+\alpha_2(-1)[-1\ \ 1]\begin{bmatrix}1\\1\end{bmatrix}+\alpha_3(-1)[1\ \ -1]\begin{bmatrix}1\\1\end{bmatrix}\right.$$

$$\left.+\alpha_4(1)[1\ \ 1]\begin{bmatrix}1\\1\end{bmatrix}\right]+\alpha_4(1)b-\alpha_4=0\quad\text{or}\quad\alpha_4(2\alpha_1+2\alpha_4)+\alpha_4 b-\alpha_4=0$$

$$\alpha_i\geq 0,\quad i=1,2,3,4.$$

Using the optimization toolbox in MATLAB to solve the aforementioned optimization problem, we obtain the optimal solution:

$$\alpha_1=0,\quad\alpha_2=0.5,\quad\alpha_3=0.5,\quad\alpha_4=1,\quad b=-1,$$

and the value of the objective function equals to 1.

The values of the Lagrange multipliers indicate that the second, third, and fourth data points in Table 6.1 are the support vectors. We then obtain w using Equation 6.31:

$$w=\sum_{i=1}^{4}\alpha_i y_i x_i$$

$$w_1=\alpha_1 y_1 x_{1,1}+\alpha_2 y_2 x_{2,1}+\alpha_3 y_3 x_{3,1}+\alpha_4 y_4 x_{4,1}$$
$$=(0)(-1)(-1)+(0.5)(-1)(-1)+(0.5)(-1)(1)+(1)(1)(1)=1$$

$$w_2=\alpha_1 y_1 x_{1,2}+\alpha_2 y_2 x_{2,2}+\alpha_3 y_3 x_{3,2}+\alpha_4 y_4 x_{4,2}$$
$$=(0)(-1)(-1)+(0.5)(-1)(1)+(0.5)(-1)(-1)+(1)(1)(1)=1.$$

The optimal solution already includes the value of $b = -1$. We obtain the same value of b using Equation 6.35 and the fourth data point as the support vector:

$$b = y_4 - w'x_4 = 1 - \begin{bmatrix} 1 & 1 \end{bmatrix} \begin{bmatrix} 1 \\ 1 \end{bmatrix} = -1.$$

The optimal solution of the dual problem for SVM gives the same decision function:

$$\begin{cases} y = sign\left(\begin{bmatrix} 1 & 1 \end{bmatrix} \begin{bmatrix} x_1 \\ x_2 \end{bmatrix} - 1\right) = sign(x_1 + x_2 - 1) = 1 & \text{if } x_1 + x_2 - 1 \geq 1 \\ y = sign\left(\begin{bmatrix} 1 & 1 \end{bmatrix} \begin{bmatrix} x_1 \\ x_2 \end{bmatrix} - 1\right) = sign(x_1 + x_2 - 1) = -1 & \text{if } x_1 + x_2 - 1 \leq -1 \end{cases}$$

or

$$f_{w,b}(x) = sign(w'x + b) = sign\left(\begin{bmatrix} 1 & 1 \end{bmatrix} \begin{bmatrix} x_1 \\ x_2 \end{bmatrix} - 1\right) = sign(x_1 + x_2 - 1).$$

Hence, the optimization problem and its dual problem of SVM for this example problem produces the same optimal solution and the decision function. Figure 6.2 illustrates the decision function and the support vectors for this problem. The decision function of SVM is the same as that of ANN for the same problem illustrated in Figure 5.10 in Chapter 5.

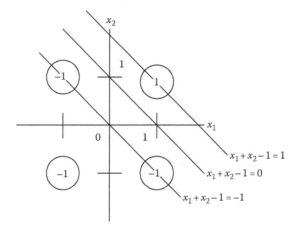

FIGURE 6.2
Decision function and support vectors for the SVM linear classifier in Example 6.1.

Support Vector Machines

Many books and papers in literature introduce SVMs using the dual optimization problem in Formulation 6.37 but without the set of constraints:

$$\sum_{j=1}^{n} \alpha_i \alpha_j y_i y_j x'_j x_i + \alpha_i y_i b - \alpha_i = 0 \quad i = 1, \ldots, n.$$

As seen from Example 6.1, without this set of constraints, the dual problem becomes

$$\max_{\alpha} -(\alpha_1 - \alpha_4)^2 - (\alpha_2 - \alpha_3)^2 + \alpha_1 + \alpha_2 + \alpha_3 + \alpha_4$$

subject to

$$-\alpha_1 - \alpha_2 - \alpha_3 + \alpha_4 = 0$$

$$\alpha_i \geq 0, \quad i = 1, 2, 3, 4.$$

If we let $\alpha_1 = \alpha_4 > 0$ and $\alpha_2 = \alpha_3 = 0$, which satisfy all the constraints, then the objective function becomes $\max \alpha_1 + \alpha_4$, which is unbounded as α_1 and α_4 can keep increasing their value without a bound. Hence, Formulation 6.37 of the dual problem with the full set of constraints should be used.

6.5 SVM Formulation for a Linear Classifier and a Nonlinearly Separable Problem

If a SVM linear classifier is applied to a nonlinearly separable problem (e.g., the logical XOR function described in Chapter 5), it is expected that not every data point in the sample data set can be classified correctly by the SVM linear classifier. The formulation of a SVM for a linear classifier in Formulation 6.24 can be extended to use a soft margin by introducing a set of additional non-negative parameters β_i, $i = 1, \ldots, n$, into the SVM formulation

$$\min_{w,b,\beta} \frac{1}{2} \|w\|^2 + C \left(\sum_{i=1}^{n} \beta_i \right)^k \quad (6.38)$$

subject to

$$y_i (wx_i + b) \geq 1 - \beta_i, \quad i = 1, \ldots, n$$

$$\beta_i \geq 0, \quad i = 1, \ldots, n,$$

where $C > 0$ and $k \geq 1$ are predetermined for giving the penalty of misclassifying the data points. Introducing β_i into the constraint in Formulation 6.38

allows a data point to be misclassified with β_i measuring the level of the misclassification. If a data point is correctly classified, β_i is zero. Minimizing $C\left(\sum_{i=1}^{n}\beta_i\right)^k$ in the objective function is to minimize the misclassification error, while minimizing $(1/2)\|w\|^2$ in the objective function is to minimize the VC-dimension as discussed previously.

Using the Lagrange multiplier method, we transform Formulation 6.38 to

$$\min_{w,b,\beta}\max_{\alpha\geq 0,\gamma\geq 0} L(w,b,\beta,\alpha,\gamma) = \frac{1}{2}\|w\|^2 + C\left(\sum_{i=1}^{n}\beta_i\right)^k$$
$$-\sum_{i=1}^{n}\alpha_i\left[y_i(wx_i+b)-1+\beta_i\right]-\sum_{i=1}^{n}\gamma_i\beta_i, \qquad (6.39)$$

where γ_i, $i = 1, ..., n$, are the non-negative Lagrange multipliers. The solution to Formulation 6.39 is at the saddle point of $L(w,b,\beta,\alpha,\gamma)$, where $L(w,b,\beta,\alpha,\gamma)$ is minimized with regard to w, b, and β and maximized with regard to α and γ. At the point where $L(w,b,\beta,\alpha,\gamma)$ is minimized with regard to w, b, and β, we have

$$\frac{\partial L(w,b,\beta,\alpha,\gamma)}{\partial w} = w - \sum_{i=1}^{n}\alpha_i y_i x_i = 0 \quad \text{or} \quad w = \sum_{i=1}^{n}\alpha_i y_i x_i \qquad (6.40)$$

$$\frac{\partial L(w,b,\beta,\alpha,\gamma)}{\partial b} = \sum_{i=1}^{n}\alpha_i y_i = 0 \qquad (6.41)$$

$$\frac{\partial L(w,b,\beta,\alpha,\gamma)}{\partial \beta} = \begin{cases} pC\left(\sum_{i=1}^{n}\beta_i\right)^{k-1}-\alpha_i-\gamma_i = 0 & i=1,...,n \quad \text{if } k>1 \\ C-\alpha_i-\gamma_i = 0 & i=1,...,n \quad \text{if } k=1 \end{cases}. \qquad (6.42)$$

When $k > 1$, we denote

$$\delta = pC\left(\sum_{i=1}^{n}\beta_i\right)^{k-1} \quad \text{or} \quad \sum_{i=1}^{n}\beta_i = \left(\frac{\delta}{pC}\right)^{1/k-1}. \qquad (6.43)$$

We can rewrite Equation 6.42 to

$$\begin{cases} \delta-\alpha_i-\gamma_i = 0 & \text{or} \quad \gamma_i = \delta-\alpha_i \quad i=1,...,n \quad \text{if } k>1 \\ C-\alpha_i-\gamma_i = 0 & \text{or} \quad \gamma_i = C-\alpha_i \quad i=1,...,n \quad \text{if } k=1 \end{cases}. \qquad (6.44)$$

Support Vector Machines

The Karush–Kuhn–Tucker condition of the optimal solution to Formulation 6.39 gives

$$\alpha_i \left[y_i(wx_i + b) - 1 + \beta_i \right] = 0. \tag{6.45}$$

Using a data point (x_i, y_i) that is correctly classified by the SVM, we have $\beta_i = 0$ and thus the following based on Equation 6.45:

$$b = y_i - w'x_i, \tag{6.46}$$

which is the same as Equation 6.35. Equations 6.40 and 6.46 are used to compute w and b, respectively, if α is known. We use the dual problem of Formulation 6.39 to determine α as follows.

When $k = 1$, substituting w, b, and γ in Equations 6.40, 6.44, and 6.46, respectively, into Formulation 6.39 produces

$$\max_{\alpha \geq 0} L(\alpha) = \frac{1}{2}\|w\|^2 + C\left(\sum_{i=1}^{n}\beta_i\right)^k - \sum_{i=1}^{n}\alpha_i\left[y_i(wx_i + b) - 1 + \beta_i\right] - \sum_{i=1}^{n}\gamma_i\beta_i$$

$$= \frac{1}{2}\sum_{i=1}^{n}\sum_{j=1}^{n}\alpha_i\alpha_j y_i y_j x_i' x_j + C\sum_{i=1}^{n}\beta_i - \sum_{i=1}^{n}\alpha_i\left[y_i\left(\sum_{j=1}^{n}\alpha_j y_j x_j' x_i + b\right) - 1 + \beta_i\right]$$

$$- \sum_{i=1}^{n}(C - \alpha_i)\beta_i = \sum_{i=1}^{n}\alpha_i - \frac{1}{2}\sum_{i=1}^{n}\sum_{j=1}^{n}\alpha_i\alpha_j y_i y_j x_i' x_j \tag{6.47}$$

subject to

$$\sum_{i=1}^{n}\alpha_i y_i = 0$$

$$\alpha_i \leq C \quad i = 1, \ldots, n$$

$$\alpha_i \geq 0 \quad i = 1, \ldots, n.$$

The constraint $\alpha_i \leq C$ comes from Equation 6.44:

$$C - \alpha_i - \gamma_i = 0 \quad \text{or} \quad C - \alpha_i = \gamma_i.$$

Since $\gamma_i \geq 0$, we have $C \geq \alpha_i$.

When $k > 1$, substituting w, b, and γ in Equations 6.40, 6.44, and 6.46, respectively, into Formulation 6.39 produces

$$\max_{\alpha \geq 0, \delta} L(\alpha) = \frac{1}{2}\|w\|^2 + C\left(\sum_{i=1}^{n}\beta_i\right)^k - \sum_{i=1}^{n}\alpha_i\left[y_i(wx_i + b) - 1 + \beta_i\right] - \sum_{i=1}^{n}\gamma_i\beta_i$$

$$= \frac{1}{2}\sum_{i=1}^{n}\sum_{j=1}^{n}\alpha_i\alpha_j y_i y_j x_i'x_j + C\left(\sum_{i=1}^{n}\beta_i\right)^k - \sum_{i=1}^{n}\alpha_i\left[y_i\left(\sum_{j=1}^{n}\alpha_j y_j x_j' x_i + b\right) - 1 + \beta_i\right]$$

$$- \sum_{i=1}^{n}(\delta - \alpha_i)\beta_i = \sum_{i=1}^{n}\alpha_i - \frac{1}{2}\sum_{i=1}^{n}\sum_{j=1}^{n}\alpha_i\alpha_j y_i y_j x_i'x_j - \frac{\delta^{\frac{p}{p-1}}}{(pC)^{\frac{1}{p-1}}}\left(1 - \frac{1}{p}\right) \quad (6.48)$$

subject to

$$\sum_{i=1}^{n}\alpha_i y_i = 0$$

$$\alpha_i \leq \delta \quad i = 1, \ldots, n$$

$$\alpha_i \geq 0 \quad i = 1, \ldots, n.$$

The decision function of the linear classifier is given in Equation 6.22:

$$y = sign(w'x + b) = 1 \quad \text{if } w'x + b \geq 1$$

$$y = sign(w'x + b) = -1 \quad \text{if } w'x + b \leq -1,$$

or Equation 6.4:

$$f_{w,b}(x) = sign(w'x + b) = sign\left(\sum_{i=1}^{n}\alpha_i y_i x_i' x + b\right).$$

Only the support vectors with the corresponding $\alpha_i > 0$ contribute to the computation of w, b, and the decision function of the linear classifier.

6.6 SVM Formulation for a Nonlinear Classifier and a Nonlinearly Separable Problem

The soft margin SVM is extended to a nonlinearly separable problem by transforming the p-dimensional x into a l-dimensional feature space where x can be classified using a linear classifier. The transformation of x is represented as

$$x \to \varphi(x),$$

where

$$\varphi(x) = (h_1\varphi_1(x), \ldots, h_l\varphi_l(x)). \qquad (6.49)$$

The formulation of the soft margin SVM becomes
When $k = 1$,

$$\max_{\alpha \geq 0} L(\alpha) = \sum_{i=1}^{n} \alpha_i - \frac{1}{2} \sum_{i=1}^{n} \sum_{j=1}^{n} \alpha_i \alpha_j y_i y_j \varphi(x_i)' \varphi(x_j) \qquad (6.50)$$

subject to

$$\sum_{i=1}^{n} \alpha_i y_i = 0$$

$$\alpha_i \leq C \quad i = 1, \ldots, n$$

$$\alpha_i \geq 0 \quad i = 1, \ldots, n.$$

When $k > 1$,

$$\max_{\alpha \geq 0, \delta} L(\alpha) = \sum_{i=1}^{n} \alpha_i - \frac{1}{2} \sum_{i=1}^{n} \sum_{j=1}^{n} \alpha_i \alpha_j y_i y_j \varphi(x_i)' \varphi(x_j) - \frac{\delta^{p/p-1}}{(pC)^{1/p-1}} \left(1 - \frac{1}{p}\right) \qquad (6.51)$$

subject to

$$\sum_{i=1}^{n} \alpha_i y_i = 0$$

$$\alpha_i \leq \delta \quad i = 1, \ldots, n$$

$$\alpha_i \geq 0 \quad i = 1, \ldots, n,$$

with the decision function:

$$f_{w,b}(x) = sign\left(\sum_{i=1}^{n} \alpha_i y_i \varphi(x_i)' \varphi(x) + b\right). \qquad (6.52)$$

If we define a kernel function $K(x, y)$ as

$$K(x, y) = \varphi(x)' \varphi(y) = \sum_{i=1}^{l} h_i^2 \varphi_i(x)' \varphi_i(y), \qquad (6.53)$$

the formulation of the soft margin SVM in Equations 6.50 through 6.52 becomes:

When $k = 1$,

$$\max_{\alpha \geq 0} L(\alpha) = \sum_{i=1}^{n} \alpha_i - \frac{1}{2} \sum_{i=1}^{n} \sum_{j=1}^{n} \alpha_i \alpha_j y_i y_j K(x_i, x_j) \qquad (6.54)$$

subject to

$$\sum_{i=1}^{n} \alpha_i y_i = 0$$

$$\alpha_i \leq C \quad i = 1, \ldots, n$$

$$\alpha_i \geq 0 \quad i = 1, \ldots, n.$$

When $k > 1$,

$$\max_{\alpha \geq 0, \delta} L(\alpha) = \sum_{i=1}^{n} \alpha_i - \frac{1}{2} \sum_{i=1}^{n} \sum_{j=1}^{n} \alpha_i \alpha_j y_i y_j K(x_i, x_j) - \frac{\delta^{p/p-1}}{(pC)^{1/p-1}} \left(1 - \frac{1}{p}\right) \qquad (6.55)$$

subject to

$$\sum_{i=1}^{n} \alpha_i y_i = 0$$

$$\alpha_i \leq \delta \quad i = 1, \ldots, n$$

$$\alpha_i \geq 0 \quad i = 1, \ldots, n.$$

with the decision function:

$$f_{w,b}(x) = \text{sign}\left(\sum_{i=1}^{n} \alpha_i y_i K(x_i, x) + b\right). \qquad (6.56)$$

The soft margin SVM in Equations 6.50 through 6.52 requires the transformation $\varphi(x)$ and then solve the SVM in the feature space, while the soft margin SVM in Equations 6.54 through 6.56 uses a kernel function $K(x, y)$ directly.

To work in the feature space using Equations 6.50 through 6.52, some examples of the transformation function for an input vector x in a one-dimensional space are provided next:

$$\varphi(x) = (1, x, \ldots, x^d) \qquad (6.57)$$

$$K(x, y) = \varphi(x)' \varphi(y) = 1 + xy + \cdots + (xy)^d.$$

Support Vector Machines

$$\varphi(x) = \left(\sin x, \frac{1}{\sqrt{2}}\sin(2x), \ldots, \frac{1}{\sqrt{i}}\sin(ix), \ldots\right) \quad (6.58)$$

$$K(x,y) = \varphi(x)'\varphi(y) = \sum_{i=1}^{\infty} \frac{1}{i}\sin(ix)\sin(iy) = \frac{1}{2}\log\left|\frac{\sin(x+y/2)}{\sin(x-y/2)}\right|$$

$$x, y \in [0, \pi].$$

An example of the transformation function for an input vector $x = (x_1, x_2)$ in a two-dimensional space is given next:

$$\varphi(x) = \left(1, \sqrt{2}x_1, \sqrt{2}x_2, x_1^2, x_2^2, \sqrt{2}x_1x_2\right) \quad (6.59)$$

$$K(x,y) = \varphi(x)'\varphi(y) = (1+xy)^2.$$

An example of the transformation function for an input vector $x = (x_1, x_2, x_3)$ in a three-dimensional space is given next:

$$\varphi(x) = \left(1, \sqrt{2}x_1, \sqrt{2}x_2, \sqrt{2}x_3, x_1^2, x_2^2, x_3^2, \sqrt{2}x_1x_2, \sqrt{2}x_1x_3, \sqrt{2}x_2x_3,\right) \quad (6.60)$$

$$K(x,y) = \varphi(x)'\varphi(y) = (1+xy)^2.$$

Principal component analysis described in Chapter 14 can be used to produce the principal components for constructing $\varphi(x)$. However, principal components may not necessarily give appropriate features that lead to a linear classifier in the feature space.

For the transformation functions in Equations 6.57 through 6.60, it is easier to compute the kernel functions directly than starting from computing the transformation functions and working in the feature space since the SVM can be solved using a kernel function directly. Some examples of the kernel functions are provided next:

$$K(x,y) = (1+xy)^d \quad (6.61)$$

$$K(x,y) = e^{-\frac{|x-y|^2}{2\sigma^2}} \quad (6.62)$$

$$K(x,y) = \tanh(\rho xy - \theta). \quad (6.63)$$

The kernel functions in Equations 6.61 through 6.63 produce a polynomial decision function as shown in Figure 6.3, a Gaussian radial basis function as shown in Figure 6.4, and a multi-year perceptron for some values of ρ and θ.

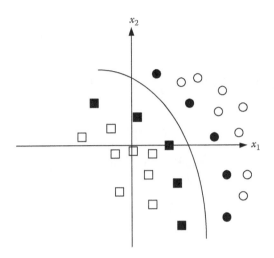

FIGURE 6.3
A polynomial decision function in a two-dimensional space.

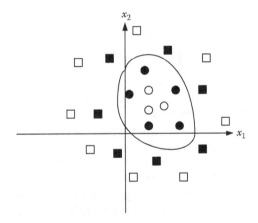

FIGURE 6.4
A Gaussian radial basis function in a two-dimensional space.

The addition and the tensor product of kernel functions are often used to construct more complex kernel functions as follows:

$$K(x,y) = \sum_i K_i(x,y) \tag{6.64}$$

$$K(x,y) = \prod_i K_i(x,y). \tag{6.65}$$

6.7 Methods of Using SVM for Multi-Class Classification Problems

SVM described in the previous sections is for a binary classifier that deals with only two target classes. For a classification problem with more than two target classes, there are several methods that can be used to first build binary classifiers and combine binary classifiers to handle multiple target classes. Suppose that the target classes are $T_1, T_2, ..., T_s$. In the one-versus-one method, a binary classifier is built for every pair of target classes, T_i versus T_j, $i \neq j$. Among the target classes produced by all the binary classifiers for a given input vector, the most dominant target class is taken as the final target class for the input vector. In the one-versus-all method, suppose that a binary classifier is built to distinguish each target class T_i from all the other target classes that are considered together as another target class $NOT\text{-}T_i$. If all the binary classifiers produce the consistent classification results for a given input vector with one binary classifier producing T_i and all the other classifiers producing $NOT\text{-}T_j$, $j \neq i$, the final target class for the input vector is T_i. However, if all the binary classifiers produce inconsistent classification results for a given input vector, it is difficult to determine the final target class for the input vector. For example, there may exist T_i and T_j, $i \neq j$, in the classification results, and it is difficult to determine whether the final target class is T_i or T_j. The error-correcting output coding method generates a unique binary code consisting of binary bits for each target class, builds a binary classifier for each binary bit, and takes the target class with the string of binary bits closest to the resulting string of the binary bits from all the binary classifiers. However, it is not straightforward to generate a unique binary code for each target class so that the resulting set of binary codes for all the target classes leads to the minimum classification error for the training data points.

6.8 Comparison of ANN and SVM

The learning of an ANN, as described in Chapter 5, requires the search for weights and biases of the ANN toward the minimum of the classification error for the training data points, although the search may end up with a local minimum. An SVM is solved to obtain the global optimal solution. However, for a nonlinear classifier and a nonlinearly separable problem, it is often uncertain what kernel function is right to transform the nonlinearly problem into a linearly separable problem since the underlying classification function is unknown. Without an appropriate kernel function, we may end up with using an inappropriate kernel function and thus a solution with the classification error greater than that from a global optimal solution when an appropriate kernel function is used. Hence, using an SVM for a nonlinear classifier and a nonlinearly separable problem involves the search for a good kernel function to classify the training data through trials and errors, just as learning an ANN involves determining an appropriate configuration of the ANN (i.e., the number of hidden units) through trials and errors. Moreover, computing $\sum_{i=1}^{n}\sum_{j=1}^{n}\alpha_i\alpha_j y_i y_j x_i' x_j$ or $\sum_{i=1}^{n}\sum_{j=1}^{n}\alpha_i\alpha_j y_i y_j K(x_i, x_j)$ in the objective function of an SVM for a large set of training data (e.g., one containing 50,000 training data points) requires computing 2.5×10^9 terms and a large memory space, and thus induces a large computational cost. Osuna et al. (1997) apply an SVM to a face detection problem and show that the classification performance of the SVM is close to that of an ANN developed by Sung and Poggio (1998).

6.9 Software and Applications

MATLAB (www.mathworks.com) supports SVM. The optimization toolbox in MATLAB can be used to solve an optimization problem in SVM. Osuna et al. (1997) report an application of SVM to face detection. There are many other SVM applications in literature (www.support-vector-machines.org).

Exercises

6.1 Determine the linear classifier of SVM for the OR function in Table 5.2 using the SVM formulation for a linear classifier in Formulations 6.24 and 6.29.

6.2 Determine the linear classifier of SVM for the NOT function using the SVM formulation for a linear classifier in Formulations 6.24 and 6.29. The training data set for the NOT function, $y = $ NOT x, is given next:

The training data set:

X	Y
−1	1
1	−1

6.3 Determine the linear classifier of SVM for a classification function with the following training data, using the SVM formulation for a linear classifier in Formulations 6.24 and 6.29.

The training data set:

x_1	x_2	x_3	y
−1	−1	−1	0
−1	−1	1	0
−1	1	−1	0
−1	1	1	1
1	−1	−1	0
1	−1	1	1
1	1	−1	1
1	1	1	1

7
k-Nearest Neighbor Classifier and Supervised Clustering

This chapter introduces two classification methods: k-nearest neighbor classifier and supervised clustering, which includes the k-nearest neighbor classifier as a part of the method. Some applications of supervised clustering are given with references.

7.1 k-Nearest Neighbor Classifier

For a data point x_i with p attribute variables:

$$x_i = \begin{bmatrix} x_{i,1} \\ \vdots \\ x_{i,p} \end{bmatrix}$$

and one target variable y whose categorical value needs to be determined, a k-nearest neighbor classifier first locates k data points that are most similar to (i.e., closest to) the data point as the k-nearest neighbors of the data point and then uses the target classes of these k-nearest neighbors to determine the target class of the data point. To determine the k-nearest neighbors of the data point, we need to use a measure of similarity or dissimilarity between data points. Many measures of similarity or dissimilarity exist, including the Euclidean distance, the Minkowski distance, the Hamming distance, Pearson's correlation coefficient, and cosine similarity, which are described in this section.

The Euclidean distance is defined as

$$d(x_i, x_j) = \sqrt{\sum_{l=1}^{p}(x_{i,l} - x_{j,l})^2}, \quad i \neq j. \tag{7.1}$$

The Euclidean distance is a measure of dissimilarity between two data points x_i and x_j. The larger the Euclidean distance is, the more dissimilar the two data points are, and the farther apart the two data points are separated in the p-dimensional data space.

The Minkowski distance is defined as

$$d(x_i, x_j) = \left(\sum_{l=1}^{p} |x_{i,l} - x_{j,l}|^r \right)^{1/r}, \quad i \neq j. \tag{7.2}$$

The Minkowski distance is also a measure of dissimilarity. If we let $r = 2$, the Minkowski distance gives the Euclidean distance. If we let $r = 1$ and each attribute variable takes a binary value, the Minkowski distance gives the Hamming distance which counts the number of bits different between two binary strings.

When the Minkowski distance measure is used, different attribute variables may have different means, variances, and ranges and bring different scales into the distance computation. For example, values of one attribute variable, x_i, may range from 0 to 10, whereas values of another attribute variable, x_j, may range from 0 to 1. Two values of x_i, 1 and 8, produce the absolute difference of 7, whereas two values of x_j, 0.1 and 0.8, produce the absolute difference of 0.7. When both 7 and 0.7 are used in summing up the differences of two data points over all the attribute variables in Equation 7.2, the absolute difference on x_j becomes irrelevant when it is compared with the absolute difference on x_i. Hence, the normalization may be necessary before the Minkowski distance measure is used. Several methods of normalization can be used. One normalization method uses the following formula to normalize a variable x and produce a normalized variable z with the mean of zero and the variance of 1:

$$z = \frac{x - \bar{x}}{s}, \tag{7.3}$$

where $\bar{x}$ and s are the sample average and the sample standard deviation of x. Another normalization method uses the following formula to normalize a variable x and produce a normalized variable z with values in the range of [0, 1]:

$$z = \frac{x_{max} - x}{x_{max} - x_{min}}. \tag{7.4}$$

The normalization is performed by applying the same normalization method to all the attribute variables. The normalized attribute variables are used to compute the Minkowski distance.

The following defines Pearson's correlation coefficient ρ:

$$\rho_{x_i x_j} = \frac{s_{x_i x_j}}{s_{x_i} s_{x_j}}, \tag{7.5}$$

where $s_{x_i x_j}$, s_{x_i}, and s_{x_j} are the estimated covariance of x_i and x_j, the estimated standard deviation of x_i, and the estimated standard deviation of x_j, respectively, and are computed using a sample of n data points as follows:

$$s_{x_i x_j} = \frac{1}{n-1} \sum_{l=1}^{p} (x_{i,l} - \overline{x_i})(x_{j,l} - \overline{x_j}) \tag{7.6}$$

$$s_{x_i} = \sqrt{\frac{1}{n-1} \sum_{l=1}^{p} (x_{i,l} - \overline{x_i})^2} \tag{7.7}$$

$$s_{x_j} = \sqrt{\frac{1}{n-1} \sum_{l=1}^{p} (x_{j,l} - \overline{x_j})^2} \tag{7.8}$$

$$\overline{x_i} = \frac{1}{n} \sum_{l=1}^{p} x_{i,l} \tag{7.9}$$

$$\overline{x_j} = \frac{1}{n} \sum_{l=1}^{p} x_{j,l}. \tag{7.10}$$

Pearson's correlation coefficient falls in the range of [−1, 1] and is a measure of similarity between two data points x_i and x_j. The larger the value of Pearson's correlation coefficient, the more correlated or similar the two data points are. A more detailed description of Pearson's correlation coefficient is given in Chapter 14.

The cosine similarity considers two data points x_i and x_j as two vectors in the p-dimensional space and uses the cosine of the angle θ between the two vectors to measure the similarity of the two data points as follows:

$$\cos(\theta) = \frac{x_i' x_j}{\|x_i\| \|x_j\|}, \tag{7.11}$$

where $\|x_i\|$ and $\|x_j\|$ are the length of the two vectors and are computed as follows:

$$\|x_i\| = \sqrt{x_{i,1}^2 + \cdots + x_{i,p}^2} \tag{7.12}$$

$$\|x_j\| = \sqrt{x_{j,1}^2 + \cdots + x_{j,p}^2}. \tag{7.13}$$

When θ=0°, that is, the two vectors point to the same direction, cos(θ)=1. When θ=180°, that is, the two vectors point to the opposite directions, cos(θ)=−1. When θ = 90° or 270°, that is, the two vectors are orthogonal, cos(θ)=0. Hence, like Pearson's correlation coefficient, the cosine similarity measure gives a value in the range of [−1, 1] and is a measure of similarity between two data points x_i and x_j. The larger the value of the cosine similarity, the more similar the two data points are. A more detailed description of the computation of the angle between two data vectors is given in Chapter 14.

To classify a data point x, the similarity of the data point x to each of n data points in the training data set is computed using a selected measure of similarity or dissimilarity. Among the n data points in the training data set, k data points that are most similar to the data point x are considered as the k-nearest neighbors of x. The dominant target class of the k-nearest neighbors is taken as the target class of x. In other words, the k-nearest neighbors use the majority voting rule to determine the target class of x. For example, suppose that for a data point x to be classified, we have the following:

- k is set to 3
- The target variable takes one of two target classes: A and B
- Two of the 3-nearest neighbors have the target class of A

The 3-nearest neighbor classifier assigns A as the target class of x.

Example 7.1

Use a 3-nearest neighbor classifier and the Euclidean distance measure of dissimilarity to classify whether or not a manufacturing system is faulty using values of the nine quality variables. The training data set in Table 7.1 gives a part of the data set in Table 1.4 and includes nine single-fault cases and the nonfault case in a manufacturing system. For the ith data observation, there are nine attribute variables for the quality of parts, $(x_{i,1}, \ldots, x_{i,9})$, and one target variable y_i for system fault. Table 7.2 gives test cases for some multiple-fault cases.

For the first data point in the testing data set x = (1, 1, 0, 1, 1, 0, 1, 1, 1), the Euclidean distances of this data point to the ten data points in the training data set are 1.73, 2, 2.45, 2.24, 2, 2.65, 2.45, 2.45, 2.45, 2.65, respectively. For example, the Euclidean distance between x and the first data point in the training data set x_1 = (1, 0, 0, 0, 1, 0, 1, 0, 1) is

$$d(x_1, x) = \sqrt{(1-1)^2 + (0-1)^2 + (0-0)^2 + (0-1)^2 + (1-1)^2 + (0-0)^2 + (1-1)^2 + (0-1)^2 + (1-1)^2}$$

$$= \sqrt{3} = 1.73.$$

TABLE 7.1

Training Data Set for System Fault Detection

Instance i (Faulty Machine)	Attribute Variables									Target Variable
	Quality of Parts									
	x_{i1}	x_{i2}	x_{i3}	x_{i4}	x_{i5}	x_{i6}	x_{i7}	x_{i8}	x_{i9}	System Fault y_i
1 (M1)	1	0	0	0	1	0	1	0	1	1
2 (M2)	0	1	0	1	0	0	0	1	0	1
3 (M3)	0	0	1	1	0	1	1	1	0	1
4 (M4)	0	0	0	1	0	0	0	1	0	1
5 (M5)	0	0	0	0	1	0	1	0	1	1
6 (M6)	0	0	0	0	0	1	1	0	0	1
7 (M7)	0	0	0	0	0	0	1	0	0	1
8 (M8)	0	0	0	0	0	0	0	1	0	1
9 (M9)	0	0	0	0	0	0	0	0	1	1
10 (none)	0	0	0	0	0	0	0	0	0	0

TABLE 7.2

Testing Data Set for System Fault Detection and the Classification Results in Examples 7.1 and 7.2

Instance i (Faulty Machine)	Attribute Variables (Quality of Parts)									Target Variable (System Fault y_i)	
	x_{i1}	x_{i2}	x_{i3}	x_{i4}	x_{i5}	x_{i6}	x_{i7}	x_{i8}	x_{i9}	True Value	Classified Value
1 (M1, M2)	1	1	0	1	1	0	1	1	1	1	1
2 (M2, M3)	0	1	1	1	0	1	1	1	0	1	1
3 (M1, M3)	1	0	1	1	1	1	1	1	1	1	1
4 (M1, M4)	1	0	0	1	1	0	1	1	1	1	1
5 (M1, M6)	1	0	0	0	1	1	1	0	1	1	1
6 (M2, M6)	0	1	0	1	0	1	1	1	0	1	1
7 (M2, M5)	0	1	0	1	1	0	1	1	0	1	1
8 (M3, M5)	0	0	1	1	1	1	1	1	1	1	1
9 (M4, M7)	0	0	0	1	0	0	1	1	0	1	1
10 (M5, M8)	0	0	0	0	1	0	1	1	0	1	1
11 (M3, M9)	0	0	1	1	0	1	1	1	1	1	1
12 (M1, M8)	1	0	0	0	1	0	1	1	1	1	1
13 (M1, M2, M3)	1	1	1	1	1	1	1	1	1	1	1
14 (M2, M3, M5)	0	1	1	1	1	1	1	1	1	1	1
15 (M2, M3, M9)	0	1	1	1	0	1	1	1	1	1	1
16 (M1, M6, M8)	1	0	0	0	1	1	1	1	1	1	1

The 3-nearest neighbors of x are x_1, x_2, and x_5 in the training data set which all take the target class of 1 for the system being faulty. Hence, target class of 1 is assigned to the first data point in the testing data set. Since in the training data set there is only one data point with the target class of 0, the 3-nearest neighbors of each data point in the testing data set have at least two data points whose target class is 1, producing the target class of 1 for each data point in the testing data set. If we attempt to classify data point 10 with the true target class of 0 in the training data set, the 3-nearest neighbors of this data point are the data point itself and two other data points with the target class of 1, making the target class of 1 for data point 10 in the training data set, which is different from the true target class of this data point.

However, if we let $k = 1$ for this example, the 1-nearest neighbor classifier assigns the correct target class to each data point in the training data set since each data point in the training data set has itself as its 1-nearest neighbor. The 1-nearest neighbor classifier also assigns the correct target class of 1 to each data point in the testing data set since data point 10 in the training data set is the only data point with the target class of 0 and its attribute variables have the values of zero, making data point 10 not be the 1-nearest neighbor to any data point in the testing data set.

The classification results in Example 7.1 for $k = 3$ in comparison with the classification results for $k = 1$ indicate that the selection of the k value plays an important role in determining the target class of a data point. In Example 7.1, $k = 1$ produces a better classification performance than $k = 3$. In some other examples or applications, if k is too small, e.g., $k = 1$, the 1-nearest neighbor of the data point x may happen to be an outlier or come from noise in the training data set. Letting x take the target class of such a neighbor does not give the outcome that reflects data patterns in the data set. If k is too large, the group of the k-nearest neighbors may include the data points that are located far away from and are not even similar to x. Letting such dissimilar data points vote for the target class of x as its neighbors seems irrational.

The supervised clustering method in the next section extends the k-nearest neighbor classifier by first determining similar data clusters and then using these data clusters to classify a data point. Since data clusters give a more coherent picture of the training data set than individual data points, classifying a data point based on its nearest data clusters and their target classes is expected to give more robust classification performance than a k-nearest neighbor classifier which depends on individual data points.

7.2 Supervised Clustering

The supervised clustering algorithm was developed and applied to cyber attack detection for classifying the observed data of computer and network activities into one of two target classes: attacks and normal use activities

(Li and Ye, 2002, 2005, 2006; Ye, 2008; Ye and Li, 2002). The algorithm can also be applied to other classification problems.

For cyber attack detection, the training data contain large amounts of computer and network data for learning data patterns of attacks and normal use activities. In addition, more training data are added over time to update data patterns of attacks and normal activities. Hence, a scalable, incremental learning algorithm is required so that data patterns of attacks and normal use activities are maintained and updated incrementally with the addition of each new data observation rather than processing all data observations in the training data set in one batch. The supervised clustering algorithm was developed as a scalable, incremental learning algorithm to learn and update data patterns for classification.

During the training, the supervised clustering algorithm takes data points in the training data set one by one to group them into clusters of similar data points based on their attribute values and target values. We start with the first data point in the training data set and let the first cluster to contain this data point and to take the target class of the data point as the target class of the data cluster. Taking the second data point in the training data set, we want to let this data point join the closest cluster that has the same target class as the target class of this data point. In the supervised clustering algorithm, we use the mean vector of all the data points in a data cluster as the centroid of the data cluster that is used to represent the location of the data cluster and compute the distance of a data point to this cluster. The clustering of data points is based on not only values of attribute variables to measure the distance of a data point to a data cluster but also target classes of the data point and the data cluster to make the data point join a data cluster with the same target class. All data points in the same cluster have the same target class, which is also the target class of the cluster. Because the algorithm uses the target class to guide or supervise the clustering of data points, the algorithm is called supervised clustering.

Suppose that the distance of the first data point and the second data point in the training data set is large but the second data point has the same target class as the target class of the first cluster containing the first data point, the second data point still has to join this cluster because this is the only data cluster so far with the same target class. Hence, the clustering results depend on the order in which data points are taken from the training data set, causing the problem called the local bias of the input order. To address this problem, the supervised clustering algorithm sets up an initial data cluster for each target class. For each target class, the centroid of all data points with the target class in the training data set is first computed using the mean vector of the data points. Then an initial cluster for the target class is set up to have the mean vector as the centroid of the cluster and the target class, which is different from any target class of the data points in the training data set. For example, if there are totally two target classes of T_1 and T_2 in the training data, there are two initial clusters. One initial cluster has the mean vector of the data points for T_1 as the centroid. Another initial cluster has the mean vector of the data points for T_2 as the centroid. Both initial clusters are assigned to a target class, e.g., T_3, which is different from T_1 and T_2. Because these initial data clusters do not contain any individual data points, they are called the dummy clusters. All the dummy clusters have the target class that is different from any target class in the training data set. The supervised clustering algorithm requires a data point to form its own cluster if its closest data cluster is a dummy cluster. With the dummy clusters, the first data point from the training data set forms a new cluster since there are only dummy clusters initially and the closest cluster to this data point is a dummy cluster. If the second data point has the same target class of the first data point but is located far away from the first data point, a dummy cluster is more likely the closest cluster to the second data point than the data cluster containing the first data point. This makes the second data point form its own cluster rather than joining the cluster with the first data point, and thus addresses the local bias problem due to the input order of training data points.

During the testing, the supervised clustering algorithm applies a k-nearest neighbor classifier to the data clusters obtained from the training phase by determining the k-nearest cluster neighbors of the data point to be classified and letting these k-nearest data clusters vote for the target class of the data point.

Table 7.3 gives the steps of the supervised clustering algorithm. The following notations are used in the description of the algorithm:

$x_i = (x_{i,1}, \ldots, x_{i,p}, y_i)$: a data point in the training data set with a known value of y_i, for $i = 1, \ldots, n$

$x = (x_1, \ldots, x_p, y)$: a testing data point with the value of y to be determined

T_j: the jth target class, $j = 1, \ldots, s$

C: a data cluster

n_C: the number of data points in the data cluster C

$\overline{x_C}$: the centroid of the data cluster C that is the mean vector of all data points in C

In Step 4 of the training phase, after the data point x_i joins the data cluster C, the centroid of the data cluster C is updated incrementally to produce $\overline{x_C}(t+1)$ (the updated centroid) using x_i, $\overline{x_C}(t)$ (the current cluster centroid), and $n_C(t)$ (the current number of data points in C):

$$\overline{x_C}(t+1) = \begin{bmatrix} \dfrac{n_C(t)\overline{x_{C1}}(t) + x_{i,1}}{n_C(t)+1} \\ \vdots \\ \dfrac{n_C(t)\overline{x_{Cp}}(t) + x_{i,p}}{n_C(t)+1} \end{bmatrix}. \tag{7.14}$$

k-Nearest Neighbor Classifier and Supervised Clustering

TABLE 7.3

Supervised Clustering Algorithm

Step	Description
Training	
1	Set up s dummy clusters for s target classes, respectively, determine the centroid of each dummy cluster by computing the mean vector of all the data points in the training data set with the target class T_j, and assign T_{s+1} as the target class of each dummy cluster where $T_{s+1} \neq T_j, j = 1, \ldots, s$
2	FOR $i = 1$ to n
3	Compute the distance of x_i to each data cluster C including each dummy cluster, $d(x_i, \overline{x_C})$, using a measure of similarity
4	If the nearest cluster to the data point x_i has the same target class as that of the data point, let the data point join this cluster, and update the centroid of this cluster and the number of data points in this cluster
5	If the nearest cluster to the data point x_i has a different target class from that of the data point, form a new cluster containing this data point, use the attribute values of this data point as the centroid of this new cluster, let the number of data points in the cluster be 1, and assign the target class of the data point as the target class of the new cluster
Testing	
1	Compute the distance of the data point x to each data cluster C excluding each dummy cluster, $d(x, \overline{x_C})$
2	Let the k-nearest neighbor clusters of the data point vote for the target class of the data point

During the training, the dummy cluster for a certain target class can be removed if many data clusters have been generated for that target class. Since the centroid of the dummy cluster for a target class is the mean vector of all the training data points with the target class, it is likely that the dummy cluster for the target class is the closest cluster to a data point. Removing the dummy cluster for the target class eliminates this likelihood and stops the creation of a new cluster for the data point because the dummy cluster for the target class is the closest cluster to the data point.

Example 7.2

Use the supervised clustering algorithm with the Euclidean distance measure of dissimilarity and the 1-nearest neighbor classifier to classify whether or not a manufacturing system is faulty using the training data set in Table 7.1 and the testing data set in Table 7.2. Both tables are explained in Example 7.1.

In Step 1 of training, two dummy clusters C_1 and C_2 are set up for two target classes, $y = 1$ and $y = 0$, respectively:

$y_{C_1} = 2$ (indicating that C_1 is a dummy cluster whose target class is different from two target classes in the training and testing data sets)

$y_{C_2} = 2$ (indicating that C_2 is a dummy cluster)

$$\overline{x_{C_1}} = \begin{bmatrix} \dfrac{1+0+0+0+0+0+0+0+0}{9} \\ \dfrac{0+1+0+0+0+0+0+0+0}{9} \\ \dfrac{0+0+1+0+0+0+0+0+0}{9} \\ \dfrac{0+1+1+1+0+0+0+0+0}{9} \\ \dfrac{1+0+0+0+1+0+0+0+0}{9} \\ \dfrac{0+0+1+0+0+1+0+0+0}{9} \\ \dfrac{1+0+1+0+1+1+1+0+0}{9} \\ \dfrac{0+1+1+1+0+0+0+1+0}{9} \\ \dfrac{1+0+0+0+1+0+0+0+1}{9} \end{bmatrix} = \begin{bmatrix} 0.11 \\ 0.11 \\ 0.11 \\ 0.33 \\ 0.22 \\ 0.22 \\ 0.56 \\ 0.44 \\ 0.33 \end{bmatrix}$$

$$\overline{x_{C_2}} = \begin{bmatrix} \dfrac{0}{1} \\ \dfrac{0}{1} \\ \dfrac{0}{1} \\ \dfrac{0}{1} \\ \dfrac{0}{1} \\ \dfrac{0}{1} \\ \dfrac{0}{1} \\ \dfrac{0}{1} \\ \dfrac{0}{1} \end{bmatrix} = \begin{bmatrix} 0 \\ 0 \\ 0 \\ 0 \\ 0 \\ 0 \\ 0 \\ 0 \\ 0 \end{bmatrix}$$

$$n_{C_1} = 9$$

$$n_{C_2} = 1.$$

In Step 2 of training, the first data point x_1 in the training data set is considered:

$$x_1 = \begin{bmatrix} 1 \\ 0 \\ 0 \\ 0 \\ 1 \\ 0 \\ 1 \\ 0 \\ 1 \end{bmatrix} \quad y = 1.$$

In Step 3 of training, the Euclidean distance of x_1 to each of the current clusters C_1 and C_2 is computed:

$$d(x_1, \overline{x_{C_1}}) = \sqrt{\begin{array}{l}(1-0.11)^2 + (0-0.11)^2 + (0-0.11)^2 + (0-0.33)^2 + (1-0.22)^2 \\ + (0-0.22)^2 + (1-0.56)^2 + (0-0.44)^2 + (1-0.33)^2\end{array}} = 1.56$$

$$d(x_1, \overline{x_{C_2}}) = \sqrt{\begin{array}{l}(1-0)^2 + (0-0)^2 + (0-0)^2 + (0-0)^2 + (1-0)^2 \\ + (0-0)^2 + (1-0)^2 + (0-0)^2 + (1-0)^2\end{array}} = 2.$$

Since C_1 is the closest cluster to x_1 and has a different target class from that of x_1, Step 5 of training is executed to form a new data cluster C_3 containing x_1:

$$y_{C_3} = 1$$

$$\overline{x_{C_3}} = \begin{bmatrix} 1 \\ 0 \\ 0 \\ 0 \\ 1 \\ 0 \\ 1 \\ 0 \\ 1 \end{bmatrix}$$

$$n_{C_3} = 1.$$

Going back to Step 2 of training, the second data point x_2 in the training data set is considered:

$$x_2 = \begin{bmatrix} 0 \\ 1 \\ 0 \\ 1 \\ 0 \\ 0 \\ 0 \\ 1 \\ 0 \end{bmatrix} \quad y = 1.$$

In Step 3 of training, the Euclidean distance of x_2 to each of the current clusters C_1, C_2, and C_3 is computed:

$$d(x_2, \overline{x_{C_1}}) = \sqrt{\begin{array}{l}(0-0.11)^2 + (1-0.11)^2 + (0-0.11)^2 + (1-0.33)^2 + (0-0.22)^2 \\ + (0-0.22)^2 + (0-0.56)^2 + (1-0.44)^2 + (0-0.33)^2\end{array}} = 1.44$$

$$d(x_2, \overline{x_{C_2}}) = \sqrt{\begin{array}{l}(0-0)^2 + (1-0)^2 + (0-0)^2 + (1-0)^2 + (0-0)^2 \\ + (0-0)^2 + (0-0)^2 + (1-0)^2 + (0-0)^2\end{array}} = 1.73$$

$$d(x_2, \overline{x_{C_3}}) = \sqrt{\begin{array}{l}(0-1)^2 + (1-0)^2 + (0-0)^2 + (1-0)^2 + (0-1)^2 \\ + (0-0)^2 + (0-1)^2 + (1-0)^2 + (0-1)^2\end{array}} = 2.65.$$

Since C_1 is the closest cluster to x_2 and has a different target class from that of x_2, Step 5 of training is executed to form a new data cluster C_4 containing x_2:

$$y_{C_4} = 1$$

$$\overline{x_{C_4}} = \begin{bmatrix} 0 \\ 1 \\ 0 \\ 1 \\ 0 \\ 0 \\ 0 \\ 1 \\ 0 \end{bmatrix}$$

$n_{C_4} = 1.$

Going back to Step 2 of training, the third data point x_3 in the training data set is considered:

$$x_3 = \begin{bmatrix} 0 \\ 0 \\ 1 \\ 1 \\ 0 \\ 1 \\ 1 \\ 1 \\ 0 \end{bmatrix} \quad y = 1.$$

In Step 3 of training, the Euclidean distance of x_3 to each of the current clusters C_1, C_2, C_3, and C_4 is computed:

$$d(x_3, \overline{x_{C_1}}) = \sqrt{\begin{aligned}&(0-0.11)^2 + (0-0.11)^2 + (1-0.11)^2 + (1-0.33)^2 + (0-0.22)^2 \\ &+ (1-0.22)^2 + (1-0.56)^2 + (1-0.44)^2 + (0-0.33)^2\end{aligned}} = 1.59$$

$$d(x_3, \overline{x_{C_2}}) = \sqrt{\begin{aligned}&(0-0)^2 + (0-0)^2 + (1-0)^2 + (1-0)^2 + (0-0)^2 \\ &+ (1-0)^2 + (1-0)^2 + (1-0)^2 + (0-0)^2\end{aligned}} = 2.24$$

$$d(x_3, \overline{x_{C_3}}) = \sqrt{\begin{aligned}&(0-1)^2 + (0-0)^2 + (1-0)^2 + (1-0)^2 + (0-1)^2 \\ &+ (1-0)^2 + (1-1)^2 + (1-0)^2 + (0-1)^2\end{aligned}} = 2.45$$

$$d(x_3, \overline{x_{C_4}}) = \sqrt{\begin{aligned}&(0-0)^2 + (0-1)^2 + (1-0)^2 + (1-1)^2 + (0-0)^2 \\ &+ (1-0)^2 + (1-0)^2 + (1-1)^2 + (0-0)^2\end{aligned}} = 2.$$

Since C_1 is the closest cluster to x_3 and has a different target class from that of x_3, Step 5 of training is executed to form a new data cluster C_5 containing x_2:

$$y_{C_5} = 1$$

$$\overline{x_{C_5}} = \begin{bmatrix} 0 \\ 0 \\ 1 \\ 1 \\ 0 \\ 1 \\ 1 \\ 1 \\ 0 \end{bmatrix}$$

$$n_{C_5} = 1.$$

Going back to Step 2 of training again, the fourth data point x_4 in the training data set is considered:

$$x_3 = \begin{bmatrix} 0 \\ 0 \\ 0 \\ 1 \\ 0 \\ 0 \\ 0 \\ 1 \\ 0 \end{bmatrix} \quad y = 1.$$

In Step 3 of training, the Euclidean distance of x_4 to each of the current clusters $C_1, C_2, C_3, C_4,$ and C_5 is computed:

$$d(x_4, \overline{x_{C_1}}) = \sqrt{\begin{array}{l}(0-0.11)^2 + (0-0.11)^2 + (0-0.11)^2 + (1-0.33)^2 + (0-0.22)^2 \\ + (0-0.22)^2 + (0-0.56)^2 + (1-0.44)^2 + (0-0.33)^2\end{array}} = 1.14$$

$$d(x_4, \overline{x_{C_2}}) = \sqrt{\begin{array}{l}(0-0)^2 + (0-0)^2 + (0-0)^2 + (1-0)^2 + (0-0)^2 \\ + (0-0)^2 + (0-0)^2 + (1-0)^2 + (0-0)^2\end{array}} = 1.41$$

$$d(x_4, \overline{x_{C_3}}) = \sqrt{\begin{array}{l}(0-1)^2 + (0-0)^2 + (0-0)^2 + (1-0)^2 + (0-1)^2 \\ + (0-0)^2 + (0-1)^2 + (1-0)^2 + (0-1)^2\end{array}} = 2.24$$

$$d(x_4, \overline{x_{C_4}}) = \sqrt{\begin{array}{l}(0-0)^2 + (0-1)^2 + (0-0)^2 + (1-1)^2 + (0-0)^2 \\ + (0-0)^2 + (0-0)^2 + (1-1)^2 + (0-0)^2\end{array}} = 1$$

$$d(x_4, \overline{x_{C_5}}) = \sqrt{\begin{array}{l}(0-0)^2 + (0-0)^2 + (0-1)^2 + (1-1)^2 + (0-0)^2 \\ + (0-1)^2 + (0-1)^2 + (1-1)^2 + (0-0)^2\end{array}} = 1.73.$$

Since C_4 is the closest cluster to x_4 and has the same target class as that of x_4, Step 4 of training is executed to add x_4 into the cluster C_4, which is updated next:

$$y_{C_4} = 1$$

$$\overline{x_{C_4}} = \begin{bmatrix} \frac{0+0}{2} \\ \frac{1+0}{2} \\ \frac{0+0}{2} \\ \frac{1+1}{2} \\ \frac{0+0}{2} \\ \frac{0+0}{2} \\ \frac{0+0}{2} \\ \frac{1+1}{2} \\ \frac{0+0}{2} \end{bmatrix} = \begin{bmatrix} 0 \\ 0.5 \\ 0 \\ 1 \\ 0 \\ 0 \\ 0 \\ 1 \\ 0 \end{bmatrix}$$

$$n_{C_4} = 2.$$

The training continues with the remaining data points $x_5, x_6, x_7, x_8,$ and x_9 and produces the final clusters $C_1, C_2, C_3 = \{x_1, x_5\}, C_4 = \{x_2, x_4\}, C_5 = \{x_3\}, C_6 = \{x_6\}, C_7 = \{x_7\}, C_8 = \{x_8\}, C_9 = \{x_9\},$ and $C_{10} = \{x_{10}\}$:

$$y_{C_1} = 2$$

$$\overline{x_{C_1}} = \begin{bmatrix} 0.11 \\ 0.11 \\ 0.11 \\ 0.33 \\ 0.22 \\ 0.22 \\ 0.56 \\ 0.44 \\ 0.33 \end{bmatrix}$$

$$n_{C_1} = 9$$

$$y_{C_2} = 2$$

$$\overline{x_{C_2}} = \begin{bmatrix} 0 \\ 0 \\ 0 \\ 0 \\ 0 \\ 0 \\ 0 \\ 0 \\ 0 \end{bmatrix}$$

$$n_{C_2} = 1$$

$$y_{C_3} = 1$$

$$\overline{x_{C_3}} = \begin{bmatrix} 1 \\ 0 \\ 0 \\ 0 \\ 1 \\ 0 \\ 1 \\ 0 \\ 1 \end{bmatrix}$$

$$n_{C_3} = 1$$

$$y_{C_4} = 1$$

$$\overline{x_{C_4}} = \begin{bmatrix} 0 \\ 0.5 \\ 0 \\ 1 \\ 0 \\ 0 \\ 0 \\ 1 \\ 0 \end{bmatrix}$$

$$n_{C_4} = 2$$

$$y_{C_5} = 1$$

$$\overline{x_{C_5}} = \begin{bmatrix} 0 \\ 0 \\ 1 \\ 1 \\ 0 \\ 1 \\ 1 \\ 1 \\ 0 \end{bmatrix}$$

$$n_{C_5} = 1$$

$$y_{C_6} = 1$$

$$\overline{x_{C_6}} = \begin{bmatrix} 0 \\ 0 \\ 0 \\ 0 \\ 0 \\ 0 \\ 1 \\ 1 \\ 0 \\ 0 \end{bmatrix}$$

$$n_{C_6} = 1$$

$$y_{C_7} = 1$$

$$\overline{x_{C_7}} = \begin{bmatrix} 0 \\ 0 \\ 0 \\ 0 \\ 0 \\ 0 \\ 1 \\ 0 \\ 0 \end{bmatrix}$$

$$n_{C_7} = 1$$

$$y_{C_8} = 1$$

$$\overline{x_{C_8}} = \begin{bmatrix} 0 \\ 0 \\ 0 \\ 0 \\ 0 \\ 0 \\ 0 \\ 1 \\ 0 \end{bmatrix}$$

$$n_{C_8} = 1$$

$$y_{C_9} = 1$$

$$\overline{x_{C_9}} = \begin{bmatrix} 0 \\ 0 \\ 0 \\ 0 \\ 0 \\ 0 \\ 0 \\ 0 \\ 1 \end{bmatrix}$$

$$n_{C_9} = 1$$

$$y_{C_{10}} = 0$$

$$\overline{x_{C_{10}}} = \begin{bmatrix} 0 \\ 0 \\ 0 \\ 0 \\ 0 \\ 0 \\ 0 \\ 0 \\ 0 \end{bmatrix}$$

$$n_{C_{10}} = 1.$$

In the testing, the first data point in the testing data set,

$$x = \begin{bmatrix} 1 \\ 1 \\ 0 \\ 1 \\ 1 \\ 0 \\ 1 \\ 1 \\ 1 \end{bmatrix},$$

has the Euclidean distances of 1.73, 2.06, 2.45, 2.65, 2.45, 2.45, 2.45, and 2.65 to the nondummy clusters C_3, C_4, C_5, C_6, C_7, C_8, C_9, and C_{10}, respectively.

Hence, the cluster C_3 is the nearest neighbor to x, and the target class of x is assigned to be 1. The closest clusters to the remaining data points 2–16 in the testing data set are C_5, C_3, C_3, C_3, C_5, C_4, C_3/C_5, C_4, $C_3/C_6/C_{10}$, C_5, C_3, C_5, C_5, C_5, and C_3. For data point 8, there is a tie between C_3 and C_5 for the closest cluster. Since both C_3 and C_5 have the target class of 1, the target class of 1 is assigned to data point 8. For data point 10, there also a tie among C_3, C_6, and C_{10} for the closest cluster. Since the majority (two clusters C_3 and C_6) of the three clusters tied have the target class of 1, the target class of 1 is assigned to data point 10. Hence, all the data points in the testing data set are assigned to the target class of 1 and are correctly classified as shown in Table 2.2.

7.3 Software and Applications

A k-nearest neighbor classifier and the supervised clustering algorithm can be easily implemented using computer programs. The application of the supervised clustering algorithm to cyber attack detection is reported in Li and Ye (2002, 2005, 2006), Ye (2008), Ye and Li (2002).

Exercises

7.1 In the space shuttle O-ring data set in Table 1.2, the target variable, the Number of O-rings with Stress, has three values: 0, 1, and 2. Consider these three values as categorical values, Launch-Temperature and Leak-Check Pressure as the attribute variables, instances # 13–23 as the training data, instances # 1–12 as the testing data, and the Euclidean distance as the measure of dissimilarity. Construct a 1-nearest neighbor classifier and a 3-nearest neighbor classifier, and test and compare their classification performance.

7.2 Repeat Exercise 7.1 using the normalized attribute variables from the normalization method in Equation 7.3.

7.3 Repeat Exercise 7.1 using the normalized attribute variables from the normalization method in Equation 7.4.

7.4 Using the same training and testing data sets in Exercise 7.1 and the cosine similarity measure, construct a 1-nearest neighbor classifier and a 3-nearest neighbor classifier, and test and compare their classification performance.

7.5 Using the same training and testing data sets in Exercise 7.1, the supervised clustering algorithm, and the Euclidean distance measure of dissimilarity, construct a 1-nearest neighbor cluster classifier and a 3-nearest neighbor cluster classifier, and test and compare their classification performance.

7.6 Repeat Exercise 7.5 using the normalized attribute variables from the normalization method in Equation 7.3.

7.7 Repeat Exercise 7.5 using the normalized attribute variables from the normalization method in Equation 7.4.

7.8 Using the same training and testing data sets in Exercise 7.1, the supervised clustering algorithm, and the cosine similarity measure, construct a 1-nearest neighbor cluster classifier and a 3-nearest neighbor cluster classifier, and test and compare their classification performance.

Part III

Algorithms for Mining Cluster and Association Patterns

8
Hierarchical Clustering

Hierarchical clustering produces groups of similar data points at different levels of similarity. This chapter introduces a bottom-up procedure of hierarchical clustering, called agglomerative hierarchical clustering. A list of software packages that support hierarchical clustering is provided. Some applications of hierarchical clustering are given with references.

8.1 Procedure of Agglomerative Hierarchical Clustering

Given a number of data records in a data set, the agglomerative hierarchical clustering algorithm produces clusters of similar data records in the following steps:

1. Start with clusters, each of which has one data record.
2. Merge the two closest clusters to form a new cluster that replaces the two original clusters and contains data records from the two original clusters.
3. Repeat Step 2 until there is only one cluster left that contains all the data records.

The next section gives several methods of determining the two closest clusters in Step 2.

8.2 Methods of Determining the Distance between Two Clusters

In order to determine the two closest clusters in Step 2, we need a method to compute the distance between two clusters. There are a number of methods for determining the distance between two clusters. This section describes four methods: average linkage, single linkage, complete linkage, and centroid method.

In the average linkage method, the distance of two clusters (cluster K, C_K, and cluster L, C_L), $D_{K,L}$, is the average of distances between pairs of data records, and each pair has one data record from Cluster K and another data record from Cluster L, as follows:

$$D_{K,L} = \sum_{x_K \in C_K} \sum_{x_L \in C_L} \frac{d(x_K, x_L)}{n_K n_L} \tag{8.1}$$

$$x_K = \begin{bmatrix} x_{K,1} \\ \vdots \\ x_{K,p} \end{bmatrix} \quad x_l = \begin{bmatrix} x_{L,1} \\ \vdots \\ x_{L,p} \end{bmatrix},$$

where
 x_K denotes a data record in C_K
 x_L denotes a data record in C_L
 n_K denotes the number of data records in C_K
 n_L denotes the number of data records in C_L
 $d(x_K, x_L)$ is the distance of two data records that can be computed using the following Euclidean distance:

$$d(x_K, x_L) = \sum_{i=1}^{p} (x_{K,i} - x_{L,i})^2 \tag{8.2}$$

or some other dissimilarity measures of two data points that are described in Chapter 7. As described in Chapter 7, the normalization of the variables, $x_1, \ldots, x_p$, may be necessary before using a measure of dissimilarity or similarity to compute the distance of two data records.

Example 8.1

Compute the distance of the following two clusters using the average linkage method and the squared Euclidean distance of data points:

$$C_K = \{x_1, x_2, x_3\}$$

$$C_L = \{x_4, x_5\}$$

$$x_1 = \begin{bmatrix} 1 \\ 0 \\ 0 \\ 0 \\ 1 \\ 0 \\ 1 \\ 0 \\ 1 \end{bmatrix} \quad x_2 = \begin{bmatrix} 0 \\ 0 \\ 0 \\ 0 \\ 1 \\ 0 \\ 1 \\ 0 \\ 1 \end{bmatrix} \quad x_3 = \begin{bmatrix} 0 \\ 0 \\ 0 \\ 0 \\ 0 \\ 0 \\ 0 \\ 0 \\ 1 \end{bmatrix} \quad x_4 = \begin{bmatrix} 0 \\ 0 \\ 0 \\ 0 \\ 0 \\ 1 \\ 1 \\ 0 \\ 0 \end{bmatrix} \quad x_5 = \begin{bmatrix} 0 \\ 0 \\ 0 \\ 0 \\ 0 \\ 0 \\ 1 \\ 0 \\ 0 \end{bmatrix}.$$

Hierarchical Clustering

There are six pairs of data records between C_K and C_L: (x_1, x_4), (x_1, x_5), (x_2, x_4), (x_2, x_5), (x_3, x_4), (x_3, x_5), and their squared Euclidean distance is computed as

$$d(x_1, x_4) = \sum_{i=1}^{9}(x_{1,i} - x_{4,i})^2$$
$$= (1-0)^2 + (0-0)^2 + (0-0)^2 + (0-0)^2 + (1-0)^2 + (0-1)^2$$
$$+ (1-1)^2 + (0-0)^2 + (1-0)^2 = 4$$

$$d(x_1, x_5) = \sum_{i=1}^{9}(x_{1,i} - x_{4,i})^2$$
$$= (1-0)^2 + (0-0)^2 + (0-0)^2 + (0-0)^2 + (1-0)^2 + (0-0)^2$$
$$+ (1-1)^2 + (0-0)^2 + (1-0)^2 = 3$$

$$d(x_2, x_4) = \sum_{i=1}^{9}(x_{1,i} - x_{4,i})^2$$
$$= (0-0)^2 + (0-0)^2 + (0-0)^2 + (0-0)^2 + (1-0)^2 + (0-1)^2$$
$$+ (1-1)^2 + (0-0)^2 + (1-0)^2 = 3$$

$$d(x_2, x_5) = \sum_{i=1}^{9}(x_{1,i} - x_{4,i})^2$$
$$= (0-0)^2 + (0-0)^2 + (0-0)^2 + (0-0)^2 + (1-0)^2 + (0-0)^2$$
$$+ (1-1)^2 + (0-0)^2 + (1-0)^2 = 2$$

$$d(x_3, x_4) = \sum_{i=1}^{9}(x_{1,i} - x_{4,i})^2$$
$$= (0-0)^2 + (0-0)^2 + (0-0)^2 + (0-0)^2 + (0-0)^2 + (0-1)^2$$
$$+ (0-1)^2 + (0-0)^2 + (1-0)^2 = 3$$

$$d(x_3, x_5) = \sum_{i=1}^{9}(x_{1,i} - x_{4,i})^2$$
$$= (0-0)^2 + (0-0)^2 + (0-0)^2 + (0-0)^2 + (0-0)^2 + (0-0)^2$$
$$+ (0-1)^2 + (0-0)^2 + (1-0)^2 = 2$$

$$D_{K,L} = \sum_{x_K \in C_K} \sum_{x_L \in C_L} \frac{d(x_K, x_L)}{n_K n_L} = \frac{4}{3 \times 2} + \frac{3}{3 \times 2} + \frac{3}{3 \times 2} + \frac{2}{3 \times 2} + \frac{3}{3 \times 2} + \frac{2}{3 \times 2} = 2.8333$$

In the single linkage method, the distance between two clusters is the minimum distance between a data record in one cluster and a data record in the other cluster:

$$D_{K,L} = \min\{d(x_k, x_l), x_k \in C_K, x_l \in C_L\}. \tag{8.3}$$

Using the single linkage method, the distance of clusters C_K and C_L in Example 8.1 is computed as

$$D_{K,L} = \min\{d(x_K, x_L), x_K \in C_K, x_L \in C_L\}$$

$$= \min\{d(x_1, x_4), d(x_1, x_5), d(x_2, x_4), d(x_2, x_5), d(x_3, x_4), d(x_3, x_5)\}$$

$$= \min\{4, 3, 3, 2, 3, 4\} = 2.$$

In the complete linkage method, the distance between two clusters is the maximum distance between a data record in one cluster and a data record in the other cluster:

$$D_{K,L} = \max\{d(x_K, x_L), x_K \in C_K, x_L \in C_L\}. \tag{8.4}$$

Using the complete linkage method, the distance of clusters C_K and C_L in Example 8.1 is computed as

$$D_{K,L} = \max\{d(x_K, x_L), x_K \in C_K, x_L \in C_L\}$$

$$= \max\{d(x_1, x_4), d(x_1, x_5), d(x_2, x_4), d(x_2, x_5), d(x_3, x_4), d(x_3, x_5)\}$$

$$= \max\{4, 3, 3, 2, 3, 4\} = 4.$$

In the centroid method, the distance between two clusters is the distance between the centroids of clusters, and the centroid of a cluster is computed using the mean vector of all data records in the cluster, as follows:

$$D_{K,L} = d(\overline{x_K}, \overline{x_L}) \tag{8.5}$$

$$\overline{x_K} = \begin{bmatrix} \frac{\sum_{k=1}^{n_K} x_{k,1}}{n_K} \\ \vdots \\ \frac{\sum_{k=1}^{n_K} x_{k,p}}{n_K} \end{bmatrix} \quad \overline{x_L} = \begin{bmatrix} \frac{\sum_{l=1}^{n_L} x_{l,1}}{n_L} \\ \vdots \\ \frac{\sum_{l=1}^{n_L} x_{l,p}}{n_L} \end{bmatrix}. \tag{8.6}$$

Hierarchical Clustering

Using the centroid linkage method and the squared Euclidean distance of data points, the distance of clusters C_K and C_L in Example 8.1 is computed as

$$\overline{x_K} = \begin{bmatrix} \dfrac{\sum_{k=1}^{n_K} x_{k,1}}{n_K} \\ \vdots \\ \dfrac{\sum_{k=1}^{n_K} x_{k,p}}{n_K} \end{bmatrix} = \begin{bmatrix} \dfrac{1+0+0}{3} \\ \dfrac{0+0+0}{3} \\ \dfrac{0+0+0}{3} \\ \dfrac{0+0+0}{3} \\ \dfrac{1+1+0}{3} \\ \dfrac{0+0+0}{3} \\ \dfrac{1+1+0}{3} \\ \dfrac{0+0+0}{3} \\ \dfrac{1+1+1}{3} \end{bmatrix} = \begin{bmatrix} \dfrac{1}{3} \\ 0 \\ 0 \\ 0 \\ \dfrac{2}{3} \\ 0 \\ \dfrac{2}{3} \\ 0 \\ 1 \end{bmatrix}$$

$$\overline{x_L} = \begin{bmatrix} \dfrac{\sum_{l=1}^{n_L} x_{l,1}}{n_L} \\ \vdots \\ \dfrac{\sum_{l=1}^{n_L} x_{l,p}}{n_L} \end{bmatrix} = \begin{bmatrix} \dfrac{0+0}{2} \\ \dfrac{0+0}{2} \\ \dfrac{0+0}{2} \\ \dfrac{0+0}{2} \\ \dfrac{0+0}{2} \\ \dfrac{1+0}{2} \\ \dfrac{1+1}{0} \\ \dfrac{0+0}{2} \\ \dfrac{0+0}{2} \end{bmatrix} = \begin{bmatrix} 0 \\ 0 \\ 0 \\ 0 \\ 0 \\ \dfrac{1}{2} \\ 1 \\ 0 \\ 0 \end{bmatrix}$$

$$D_{K,L} = d(\overline{x_K}, \overline{x_L}) = \left(\frac{1}{3}-0\right)^2 + (1-0)^2 + (1-0)^2 + (1-0)^2 + \left(\frac{2}{3}-0\right)^2$$

$$+ \left(0-\frac{1}{2}\right)^2 + \left(\frac{2}{3}-1\right)^2 + (0-0)^2 + (1-0)^2 = 4.9167.$$

Various methods of determining the distance between two clusters have different computational costs and may produce different clustering results. For example, the average linkage method, the single linkage method, and the complete linkage method require the computation of the distance between every pair of data points from two clusters. Although the centroid method does not have such a computation requirement, the centroid method must compute the centroid of every new cluster and the distance of the new cluster with existing clusters. The average linkage method and the centroid method take into account and control the dispersion of data points in each cluster, whereas the single linkage method and the complete linkage method place no constraint on the shape of the cluster.

8.3 Illustration of the Hierarchical Clustering Procedure

The hierarchical clustering procedure is illustrated in Example 8.2.

Example 8.2

Produce a hierarchical clustering of the data for system fault detection in Table 8.1 using the single linkage method.

TABLE 8.1

Data Set for System Fault Detection with Nine Cases of Single-Machine Faults

Instance (Faulty Machine)	Attribute Variables about Quality of Parts								
	x_1	x_2	x_3	x_4	x_5	x_6	x_7	x_8	x_9
1 (M1)	1	0	0	0	1	0	1	0	1
2 (M2)	0	1	0	1	0	0	0	1	0
3 (M3)	0	0	1	1	0	1	1	1	0
4 (M4)	0	0	0	1	0	0	0	1	0
5 (M5)	0	0	0	0	1	0	1	0	1
6 (M6)	0	0	0	0	0	1	1	0	0
7 (M7)	0	0	0	0	0	0	1	0	0
8 (M8)	0	0	0	0	0	0	0	1	0
9 (M9)	0	0	0	0	0	0	0	0	1

Hierarchical Clustering

Table 8.1 contains the data set for system fault detection, including nine instances of single-machine faults. Only the nine attribute variables about the quality of parts are used in the hierarchical clustering. The nine data records in the data set are

$$x_1 = \begin{bmatrix} 1 \\ 0 \\ 0 \\ 0 \\ 1 \\ 0 \\ 1 \\ 0 \\ 1 \end{bmatrix} \quad x_2 = \begin{bmatrix} 0 \\ 1 \\ 0 \\ 1 \\ 0 \\ 0 \\ 0 \\ 1 \\ 0 \end{bmatrix} \quad x_3 = \begin{bmatrix} 0 \\ 0 \\ 1 \\ 1 \\ 0 \\ 1 \\ 1 \\ 1 \\ 0 \end{bmatrix} \quad x_4 = \begin{bmatrix} 0 \\ 0 \\ 0 \\ 1 \\ 0 \\ 0 \\ 1 \\ 1 \\ 0 \end{bmatrix} \quad x_5 = \begin{bmatrix} 0 \\ 0 \\ 0 \\ 0 \\ 1 \\ 0 \\ 1 \\ 0 \\ 1 \end{bmatrix} \quad x_6 = \begin{bmatrix} 0 \\ 0 \\ 0 \\ 0 \\ 0 \\ 1 \\ 1 \\ 0 \\ 0 \end{bmatrix}$$

$$x_7 = \begin{bmatrix} 0 \\ 0 \\ 0 \\ 0 \\ 0 \\ 0 \\ 1 \\ 0 \\ 0 \end{bmatrix} \quad x_8 = \begin{bmatrix} 0 \\ 0 \\ 0 \\ 0 \\ 0 \\ 0 \\ 0 \\ 1 \\ 0 \end{bmatrix} \quad x_9 = \begin{bmatrix} 0 \\ 0 \\ 0 \\ 0 \\ 0 \\ 0 \\ 0 \\ 0 \\ 1 \end{bmatrix}.$$

The clustering results will show which single-machine faults have similar symptoms of the part quality problem.

Figure 8.1 shows the hierarchical clustering procedure that starts with the following nine clusters with one data record in each cluster:

$$C_1 = \{x_1\} \quad C_2 = \{x_2\} \quad C_3 = \{x_3\} \quad C_4 = \{x_4\} \quad C_5 = \{x_5\}$$
$$C_6 = \{x_6\} \quad C_7 = \{x_7\} \quad C_8 = \{x_8\} \quad C_9 = \{x_9\}.$$

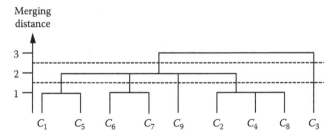

FIGURE 8.1
Result of hierarchical clustering for the data set of system fault detection.

TABLE 8.2

The Distance for Each Pair of Clusters: C_1, C_2, C_3, C_4, C_5, C_6, C_7, C_8, and C_9

	$C_1 =$ $\{x_1\}$	$C_2 =$ $\{x_2\}$	$C_3 =$ $\{x_3\}$	$C_4 =$ $\{x_4\}$	$C_5 =$ $\{x_5\}$	$C_6 =$ $\{x_6\}$	$C_7 =$ $\{x_7\}$	$C_8 =$ $\{x_8\}$	$C_9 =$ $\{x_9\}$
$C_1 = \{x_1\}$		7	7	6	1	4	3	5	3
$C_2 = \{x_2\}$			4	1	6	5	4	2	4
$C_3 = \{x_3\}$				3	6	3	4	4	6
$C_4 = \{x_4\}$					5	4	4	1	3
$C_5 = \{x_5\}$						3	2	4	2
$C_6 = \{x_6\}$							1	3	3
$C_7 = \{x_7\}$								2	2
$C_8 = \{x_8\}$									2
$C_9 = \{x_9\}$									

Since each cluster has only one data record, the distance between two clusters is the distance between two data records in two clusters, respectively. Table 8.2 gives the distance for each pair of data records, which also gives the distance for each pair of clusters.

There are four pairs of clusters that produce the smallest distance of 1: (C_1, C_5), (C_2, C_4), (C_4, C_8), and (C_6, C_7). We merge (C_1, C_5) to form a new cluster $C_{1,5}$, and merge (C_6, C_7) to form a new cluster $C_{6,7}$. Since the cluster C_4 is involved in two pairs of clusters (C_2, C_4) and (C_4, C_8), we can merge only one pair of clusters. We arbitrarily choose to merge (C_2, C_4) to form a new cluster $C_{2,4}$. Figure 8.1 shows these new clusters, in a new set of clusters, $C_{1,5}$, $C_{2,4}$, C_3, $C_{6,7}$, C_8, and C_9.

Table 8.3 gives the distance for each pair of the clusters, $C_{1,5}$, $C_{2,4}$, C_3, $C_{6,7}$, C_8, and C_9, using the single linkage method. For example, there are four pairs of data records between $C_{1,5}$ and $C_{2,4}$: (x_1, x_2), (x_1, x_4), (x_5, x_2), and (x_5, x_4), with their distance being 7, 6, 6, and 5, respectively, from Table 8.2. Hence, the minimum distance is 5, which is taken as the distance of

TABLE 8.3

Distance for Each Pair of Clusters: $C_{1,5}$, $C_{2,4}$, C_3, $C_{6,7}$, C_8, and C_9

	$C_{1,5} =$ $\{x_1, x_5\}$	$C_{2,4} =$ $\{x_2, x_4\}$	$C_3 = \{x_3\}$	$C_{6,7} =$ $\{x_6, x_7\}$	$C_8 = \{x_8\}$	$C_9 = \{x_9\}$
$C_{1,5} = \{x_1, x_5\}$		5 = min $\{7, 6, 6, 5\}$	6 = min $\{7, 6\}$	2 = min $\{4, 3, 3, 2\}$	4 = min $\{5, 4\}$	2 = min $\{3, 2\}$
$C_{2,4} = \{x_2, x_4\}$			3 = min $\{4, 3\}$	4 = min $\{5, 4, 4, 4\}$	1 = min $\{2, 1\}$	3 = min $\{4, 3\}$
$C_3 = \{x_3\}$				3 = min $\{3, 4\}$	4 = min $\{4\}$	6 = min $\{6\}$
$C_{6,7} = \{x_6, x_7\}$					2 = min $\{3, 2\}$	2 = min $\{3, 2\}$
$C_8 = \{x_8\}$						2 = min $\{2\}$
$C_9 = \{x_9\}$						

Hierarchical Clustering

TABLE 8.4

Distance for Each Pair of Clusters: $C_{1,5}$, $C_{2,4,8}$, C_3, $C_{6,7}$, and C_9

	$C_{1,5} = \{x_1, x_5\}$	$C_{2,4,8} = \{x_2, x_4, x_8\}$	$C_3 = \{x_3\}$	$C_{6,7} = \{x_6, x_7\}$	$C_9 = \{x_9\}$
$C_{1,5} = \{x_1, x_5\}$		4 = min {7, 6, 5, 6, 5, 4}	6 = min {7, 6}	2 = min {4, 3, 3, 2}	2 = min {3, 2}
$C_{2,4,8} = \{x_2, x_4, x_8\}$			3 = min {4, 3, 4}	2 = min {5, 4, 4, 4, 3, 2}	3 = min {4, 3, 2}
$C_3 = \{x_3\}$				3 = min {3, 4}	6 = min {6}
$C_{6,7} = \{x_6, x_7\}$					2 = min {3, 2}
$C_9 = \{x_9\}$					

$C_{1,5}$ and $C_{2,4}$. The closest pair of clusters is ($C_{2,4}$, C_8) with the distance of 1. Merging clusters $C_{2,4}$ and C_8 produces a new cluster $C_{2,4,8}$. We have a new set of clusters, $C_{1,5}$, $C_{2,4,8}$, C_3, $C_{6,7}$, and C_9.

Table 8.4 gives the distance for each pair of the clusters, $C_{1,5}$, $C_{2,4,8}$, C_3, $C_{6,7}$, and C_9, using the single linkage method. Four pairs of clusters, ($C_{1,5}$, $C_{6,7}$), ($C_{1,5}$, C_9), ($C_{2,4,8}$, $C_{6,7}$), and ($C_{6,7}$, C_9), produce the smallest distance of 2. Since three clusters, $C_{1,5}$, $C_{6,7}$, and C_9, have the same distance from one another, we merge the three clusters together to form a new cluster, $C_{1,5,6,7,9}$. $C_{6,7}$ is not merged with $C_{2,4,8}$ since $C_{6,7}$ is merged with $C_{1,5}$ and C_9. We have a new set of clusters, $C_{1,5,6,7,9}$, $C_{2,4,8}$, and C_3.

Table 8.5 gives the distance for each pair of the clusters, $C_{1,5,6,7,9}$, $C_{2,4,8}$, and C_3, using the single linkage method. The pair of clusters, ($C_{1,5,6,7,9}$, $C_{2,4,8}$), produces the smallest distance of 2. Merging the clusters, $C_{1,5,6,7,9}$ and $C_{2,4,8}$, forms a new cluster, $C_{1,2,4,5,6,7,8,9}$. We have a new set of clusters, $C_{1,2,4,5,6,7,8,9}$ and C_3, which have the distance of 3 and are merged into one cluster, $C_{1,2,3,4,5,6,7,8,9}$.

Figure 8.1 also shows the merging distance, which is the distance of two clusters when they are merged together. The hierarchical clustering tree shown in Figure 8.1 is called the dendrogram.

Hierarchical clustering allows us to obtain different sets of clusters by setting different thresholds of the merging distance threshold for different levels of data similarity. For example, if we set the threshold of the merging distance to 1.5 as shown by the dash line in Figure 8.1, we obtain the clusters, $C_{1,5}$, $C_{6,7}$, C_9, $C_{2,4,8}$, and C_3, which are considered as the clusters of similar data because each cluster's merging distance

TABLE 8.5

Distance for Each Pair of Clusters: $C_{1,5,6,7,9}$, $C_{2,4,8}$, and C_3

	$C_{1,5,6,7,9} = \{x_1, x_5, x_6, x_7, x_9\}$	$C_{2,4,8} = \{x_2, x_4, x_8\}$	$C_3 = \{x_3\}$
$C_{1,5,6,7,9} = \{x_1, x_5, x_6, x_7, x_9\}$		2 = min{7, 6, 5, 6, 5, 4, 5, 4, 3, 4, 4, 2, 4, 3, 2}	3 = min{7, 6, 3, 4, 6}
$C_{2,4,8} = \{x_2, x_4, x_8\}$			3 = min{4, 3, 4}
$C_3 = \{x_3\}$			

is smaller than or equal to the threshold of 1.5. This set of clusters indicates which machine faults produce similar symptoms of the part quality problem. For instance, the cluster $C_{1,5}$ indicates that the M1 fault and the M5 fault produce similar symptoms of the part quality problem. The production flow of parts in Figure 1.1 shows that parts pass through M1 and M5 consecutively and thus explains why the M1 fault and M5 fault produce similar symptoms of the part quality problem. Hence, the clusters obtained by setting the threshold of the merging distance to 1.5 give a meaningful clustering result that reveals the inherent structure of the system. If we set the threshold of the merging distance to 2.5 as shown by another dash line in Figure 8.1, we obtain the set of clusters, $C_{1,2,4,5,6,7,8,9}$ and C_3, which is not as useful as the set of clusters, $C_{1,5}$, $C_{6,7}$, C_9, $C_{2,4,8}$, and C_3, for revealing the system structure.

This example shows that obtaining a data mining result is not the end of data mining. It is crucial that we can explain the data mining result in a meaningful way in the problem context to make the data mining result useful in the problem domain. Many real-world data sets do not come with prior knowledge of a system generating such data sets. Therefore, after obtaining the hierarchical clustering result, it is important to examine different sets of clusters at different levels of data similarity and determine which set of clusters can be interpreted in a meaningful manner to help reveal the system and generate useful knowledge about the system.

8.4 Nonmonotonic Tree of Hierarchical Clustering

In Figure 8.1, the merging distance of a new cluster is not smaller than the merging distance of any cluster that was formed before the new cluster. Such a hierarchical clustering tree is monotonic. For example, in Figure 8.1, the merging distance of the cluster $C_{2,4}$ is 1, which is equal to the merging distance of $C_{2,4,8}$, and the merging distance of $C_{1,2,4,5,6,7,8,9}$ is 2, which is smaller than the merging distance of $C_{2,4,8}$.

The centroid linkage method can produce a nonmonotonic tree in which the merging distance for a new cluster can be smaller than the merging distance for a cluster that is formed before the new cluster. Figure 8.2 shows three data points, x_1, x_2, and x_3, for which the centroid method produces a nonmonotonic tree of hierarchical clustering. The distance between each pair of the three data points is 2. We start with three initial clusters, C_1, C_2, and C_3, containing the three data points, x_1, x_2, and x_3, respectively. Because the three clusters have the same distance between each other, we arbitrarily choose to merge C_1 and C_2 into a new cluster $C_{1,2}$. As shown in Figure 8.2, the distance between the centroid of $C_{1,2}$ and x_3 is $\sqrt{2^2 - 1^2} = 1.73$, which is smaller than the merging distance of 2 for $C_{1,2}$. Hence, when $C_{1,2}$ is merged

Hierarchical Clustering

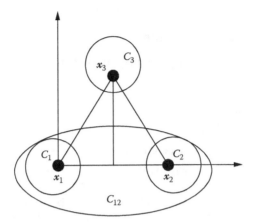

FIGURE 8.2
An example of three data points for which the centroid linkage method produces a nonmonotonic tree of hierarchical clustering.

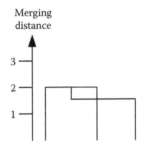

FIGURE 8.3
Nonmonotonic tree of hierarchical clustering for the data points in Figure 8.2.

with C_3 next to produce a new cluster $C_{1,2,3}$, the merging distance of 1.73 for $C_{1,2,3}$ is smaller than the merging distance of 2 for $C_{1,2}$. Figure 8.3 shows the non-monotonic tree of hierarchical clustering for these three data points using the centroid method.

The single linkage method, which is used in Example 8.2, computes the distance between two clusters using the smallest distance between two data points, one data point in one cluster, and another data point in the other cluster. The smallest distance between two data points is used to form a new cluster. The distance used to form a cluster earlier cannot be used again to form a new cluster later, because the distance is already inside a cluster and a distance with a data point outside a cluster is needed to form a new cluster later. Hence, the distance to form a new cluster later must come from a distance not used before, which must be greater than or equal to a distance selected and used earlier. Hence, the hierarchical clustering tree from the single linkage method is always monotonic.

8.5 Software and Applications

Hierarchical clustering is supported by many statistical software packages, including:

- SAS (www.sas.com)
- SPSS (www.spss.com)
- Statistica (www.statistica.com)
- MATLAB® (www.matworks.com)

Some applications of hierarchical clustering can be found in (Ye, 1997, 2003, Chapter 10; Ye and Salvendy, 1991, 1994; Ye and Zhao, 1996). In the work by Ye and Salvendy (1994), the hierarchical clustering is used to reveal the knowledge structure of C programming from expert programmers and novice programmers.

Exercises

8.1 Produce a hierarchical clustering of 23 data points in the space shuttle O-ring data set in Table 1.2. Use Launch-Temperature and Leak-Check Pressure as the attribute variables, the normalization method in Equation 7.4 to obtain the normalized Launch-Temperature and Leak-Check Pressure, the Euclidean distance of data points, and the single linkage method.

8.2 Repeat Exercise 8.1 using the complete linkage method.

8.3 Repeat Exercise 8.1 using the cosine similarity measure to compute the distance of data points.

8.4 Repeat Exercise 8.3 using the complete linkage method.

8.5 Discuss whether or not it is possible for the complete linkage method to produce a nonmonotonic tree of hierarchical clustering.

8.6 Discuss whether or not it is possible for the average linkage method to produce a nonmonotonic tree of hierarchical clustering.

9

K-Means Clustering and Density-Based Clustering

This chapter introduces K-means and density-based clustering algorithms that produce nonhierarchical groups of similar data points based on the centroid and density of a cluster, respectively. A list of software packages that support these clustering algorithms is provided. Some applications of these clustering algorithms are given with references.

9.1 K-Means Clustering

Table 9.1 lists the steps of the K-means clustering algorithm. The K-means clustering algorithm starts with a given K value and the initially assigned centroids of the K clusters. The algorithm proceeds by having each of n data points in the data set join its closest cluster and updating the centroids of the clusters until the centroids of the clusters do not change any more and consequently each data point does not move from its current cluster to another cluster. In Step 7 of the algorithm, if there is any change of cluster centroids in Steps 3–6, we have to check if the change of cluster centroids causes the further movement of any data point by going back to Step 2.

To determine the closest cluster to a data point, the distance of a data point to a data cluster needs to be computed. The mean vector of data points in a cluster is often used as the centroid of the cluster. Using a measure of similarity or dissimilarity, we compute the distance of a data point to the centroid of the cluster as the distance of a data point to the cluster. Measures of similarity or dissimilarity are described in Chapter 7.

One method of assigning the initial centroids of the K clusters is to randomly select K data points from the data set and use these data points to set up the centroids of the K clusters. Although this method uses specific data points to set up the initial centroids of the K clusters, the K clusters have no data point in each of them initially. There are also other methods of setting up the initial centroids of the K clusters, such as using the result of a hierarchical clustering to obtain the K clusters and using the centroids of these clusters as the initial centroids of the K clusters for the K-means clustering algorithm.

TABLE 9.1

K-Means Clustering Algorithm

Step	Description
1	Set up the initial centroids of the K clusters
2	REPEAT
3	FOR $i = 1$ to n
4	Compute the distance of the data point x_i to each of the K clusters using a measure of similarity or dissimilarity
5	IF x_i is not in any cluster or its closest cluster is not its current cluster
6	Move x_i to its closest cluster and update the centroid of the cluster
7	UNTIL no change of cluster centroids occurs in Steps 3–6

For a large data set, the stopping criterion for the REPEAT-UNTIL loop in Step 7 of the algorithm can be relaxed so that the REPEAT-UNTIL loop stops when the amount of changes to the cluster centroids is less than a threshold, e.g., less than 5% of the data points changing their clusters.

The K-means clustering algorithm minimizes the following sum of squared errors (SSE) or distances between data points and their cluster centroids (Ye, 2003, Chapter 10):

$$\text{SSE} = \sum_{i=1}^{K} \sum_{x \in C_i} d(x, \overline{x_{C_i}})^2. \tag{9.1}$$

In Equation 9.1, the mean vector of data points in the cluster C_i is used as the cluster centroid to compute the distance between a data point in the cluster C_i and the centroid of the cluster C_i.

Since K-means clustering depends on the parameter K, knowledge in the application domain may help the selection of an appropriate K value to produce a K-means clustering result that is meaningful in the application domain. Different K-means clustering results using different K values may be obtained so that different results can be compared and evaluated.

Example 9.1

Produce the 5-means clusters for the data set of system fault detection in Table 9.2 using the Euclidean distance as the measure of dissimilarity. This is the same data set for Example 8.1. The data set includes nine instances of single-machine faults, and the data point for each instance has the nine attribute variables about the quality of parts.

In Step 1 of the K-means clustering algorithm, we arbitrarily select data points 1, 3, 5, 7, and 9 to set up the initial centroids of the five clusters, C_1, C_2, C_3, C_4, and C_5, respectively:

K-Means Clustering and Density-Based Clustering

TABLE 9.2

Data Set for System Fault Detection with Nine Cases of Single-Machine Faults

Instance (Faulty Machine)	Attribute Variables about Quality of Parts								
	x_1	x_2	x_3	x_4	x_5	x_6	x_7	x_8	x_9
1 (M1)	1	0	0	0	1	0	1	0	1
2 (M2)	0	1	0	1	0	0	0	1	0
3 (M3)	0	0	1	1	0	1	1	1	0
4 (M4)	0	0	0	1	0	0	0	1	0
5 (M5)	0	0	0	0	1	0	1	0	1
6 (M6)	0	0	0	0	0	1	1	0	0
7 (M7)	0	0	0	0	0	0	1	0	0
8 (M8)	0	0	0	0	0	0	0	1	0
9 (M9)	0	0	0	0	0	0	0	0	1

$$\overline{x_{C_1}} = x_1 = \begin{bmatrix}1\\0\\0\\0\\1\\0\\1\\0\\1\end{bmatrix} \quad \overline{x_{C_2}} = x_3 = \begin{bmatrix}0\\0\\1\\1\\0\\1\\1\\1\\0\end{bmatrix} \quad \overline{x_{C_3}} = x_5 = \begin{bmatrix}0\\0\\0\\0\\1\\0\\1\\0\\1\end{bmatrix} \quad \overline{x_{C_4}} = x_7 = \begin{bmatrix}0\\0\\0\\0\\0\\0\\1\\0\\0\end{bmatrix} \quad \overline{x_{C_5}} = x_9 = \begin{bmatrix}0\\0\\0\\0\\0\\0\\0\\0\\1\end{bmatrix}.$$

The five clusters have no data point in each of them initially. Hence, we have $C_1 = \{\}$, $C_2 = \{\}$, $C_3 = \{\}$, $C_4 = \{\}$, and $C_5 = \{\}$.

In Steps 2 and 3 of the algorithm, we take the first data point x_1 in the data set. In Step 4 of the algorithm, we compute the Euclidean distance of the data point x_1 to each of the five clusters:

$d(x_1, \overline{x_{C_1}})$
$= \sqrt{(1-1)^2 + (0-0)^2 + (0-0)^2 + (0-0)^2 + (1-1)^2 + (0-0)^2 + (1-1)^2 + (0-0)^2 + (1-1)^2}$
$= 0$

$d(x_1, \overline{x_{C_2}})$
$= \sqrt{(1-0)^2 + (0-0)^2 + (0-1)^2 + (0-1)^2 + (1-0)^2 + (0-1)^2 + (1-1)^2 + (0-1)^2 + (1-0)^2}$
$= 2.65$

$d(x_1, \overline{x_{C_3}})$

$= \sqrt{(1-0)^2 + (0-0)^2 + (0-0)^2 + (0-0)^2 + (1-1)^2 + (0-0)^2 + (1-1)^2 + (0-0)^2 + (1-1)^2}$

$= 1$

$d(x_1, \overline{x_{C_4}})$

$= \sqrt{(1-0)^2 + (0-0)^2 + (0-0)^2 + (0-0)^2 + (1-0)^2 + (0-0)^2 + (1-1)^2 + (0-0)^2 + (1-0)^2}$

$= 1.73$

$d(x_1, \overline{x_{C_5}})$

$= \sqrt{(1-0)^2 + (0-0)^2 + (0-0)^2 + (0-0)^2 + (1-0)^2 + (0-0)^2 + (1-0)^2 + (0-0)^2 + (1-1)^2}$

$= 1.73$

In Step 5 of the algorithm, x_1 is not in any cluster. Step 6 of the algorithm is executed to move x_1 to its closest cluster C_1 whose centroid remains the same since its centroid is set up using x_1. We have $C_1 = \{x_1\}$, $C_2 = \{\}$, $C_3 = \{\}$, $C_4 = \{\}$, and $C_5 = \{\}$.

Going back to Step 3, we take the second data point x_2 in the data set. In Step 4, we compute the Euclidean distance of the data point x_2 to each of the five clusters:

$d(x_2, \overline{x_{C_1}})$

$= \sqrt{(0-1)^2 + (1-0)^2 + (0-0)^2 + (1-0)^2 + (0-1)^2 + (0-0)^2 + (0-1)^2 + (1-0)^2 + (0-1)^2}$

$= 2.65$

$d(x_2, \overline{x_{C_2}})$

$= \sqrt{(0-0)^2 + (1-0)^2 + (0-1)^2 + (1-1)^2 + (0-0)^2 + (0-1)^2 + (0-1)^2 + (1-1)^2 + (0-0)^2}$

$= 2$

$d(x_2, \overline{x_{C_3}})$

$= \sqrt{(0-0)^2 + (1-0)^2 + (0-0)^2 + (1-0)^2 + (0-1)^2 + (0-0)^2 + (0-1)^2 + (1-0)^2 + (0-1)^2}$

$= 2.45$

$$d(x_2, \overline{x_{C_4}})$$
$$= \sqrt{(0-0)^2 + (1-0)^2 + (0-0)^2 + (1-0)^2 + (0-0)^2 + (0-0)^2 + (0-1)^2 + (1-0)^2 + (0-0)^2}$$
$$= 2$$

$$d(x_2, \overline{x_{C_5}})$$
$$= \sqrt{(0-0)^2 + (1-0)^2 + (0-0)^2 + (1-0)^2 + (0-0)^2 + (0-0)^2 + (0-0)^2 + (1-0)^2 + (0-1)^2}$$
$$= 2$$

In Step 5, x_2 is not in any cluster. Step 6 of the algorithm is executed. Among the three clusters, C_2, C_4, and C_5, which produce the smallest distance to x_2, we arbitrarily select C_2 and move x_2 to C_2. C_2 has only one data point x_2, and the centroid of C_2 is updated by taking x_2 as its centroid:

$$\overline{x_{C_2}} = \begin{bmatrix} 0 \\ 1 \\ 0 \\ 1 \\ 0 \\ 0 \\ 0 \\ 1 \\ 0 \end{bmatrix}.$$

We have $C_1 = \{x_1\}$, $C_2 = \{x_2\}$, $C_3 = \{\}$, $C_4 = \{\}$, and $C_5 = \{\}$.

Going back to Step 3, we take the third data point x_3 in the data set. In Step 4, we compute the Euclidean distance of the data point x_3 to each of the five clusters:

$$d(x_3, \overline{x_{C_1}})$$
$$= \sqrt{(0-1)^2 + (0-0)^2 + (1-0)^2 + (1-0)^2 + (0-1)^2 + (1-0)^2 + (1-1)^2 + (1-0)^2 + (0-1)^2}$$
$$= 2.65$$

$$d(x_3, \overline{x_{C_2}})$$
$$= \sqrt{(0-0)^2 + (0-1)^2 + (1-0)^2 + (1-1)^2 + (0-0)^2 + (1-0)^2 + (1-0)^2 + (1-1)^2 + (0-0)^2}$$
$$= 2$$

$d(x_3, \overline{x_{C_3}})$

$= \sqrt{(0-0)^2 + (0-0)^2 + (1-0)^2 + (1-0)^2 + (0-1)^2 + (1-0)^2 + (1-1)^2 + (1-0)^2 + (0-1)^2}$

$= 2.45$

$d(x_3, \overline{x_{C_4}})$

$= \sqrt{(0-0)^2 + (0-0)^2 + (1-0)^2 + (1-0)^2 + (0-0)^2 + (1-0)^2 + (1-1)^2 + (1-0)^2 + (0-0)^2}$

$= 2$

$d(x_3, \overline{x_{C_5}})$

$= \sqrt{(0-0)^2 + (0-0)^2 + (1-0)^2 + (1-0)^2 + (0-0)^2 + (1-0)^2 + (1-0)^2 + (1-0)^2 + (0-1)^2}$

$= 2.45$

In Step 5, x_3 is not in any cluster. Step 6 of the algorithm is executed. Between the two clusters, C_2 and C_4, which produce the smallest distance to x_3, we arbitrarily select C_2 and move x_3 to C_2. C_2 has two data points, x_2 and x_3, and the centroid of C_2 is updated:

$$\overline{x_{C_2}} = \begin{bmatrix} \frac{0+0}{2} \\ \frac{1+0}{2} \\ \frac{0+1}{2} \\ \frac{1+1}{2} \\ \frac{0+0}{2} \\ \frac{0+1}{2} \\ \frac{0+1}{2} \\ \frac{1+1}{2} \\ \frac{0+0}{2} \end{bmatrix} = \begin{bmatrix} 0 \\ 0.5 \\ 0.5 \\ 1 \\ 0 \\ 0.5 \\ 0.5 \\ 1 \\ 0 \end{bmatrix}.$$

We have $C_1 = \{x_1\}$, $C_2 = \{x_2, x_3\}$, $C_3 = \{\}$, $C_4 = \{\}$, and $C_5 = \{\}$.

Going back to Step 3, we take the fourth data point x_4 in the data set. In Step 4, we compute the Euclidean distance of the data point x_4 to each of the five clusters:

$$d\left(x_4, \overline{x_{C_1}}\right)$$
$$= \sqrt{(0-1)^2 + (0-0)^2 + (0-0)^2 + (1-0)^2 + (0-1)^2 + (0-0)^2 + (0-1)^2 + (1-0)^2 + (0-1)^2}$$
$$= 2.45$$

$$d\left(x_4, \overline{x_{C_2}}\right)$$
$$= \sqrt{(0-0)^2 + (0-0.5)^2 + (0-0.5)^2 + (1-1)^2 + (0-0)^2 + (0-0.5)^2 + (0-0.5)^2 + (1-1)^2 + (0-0)^2}$$
$$= 1$$

$$d\left(x_4, \overline{x_{C_3}}\right)$$
$$= \sqrt{(0-0)^2 + (0-0)^2 + (0-0)^2 + (1-0)^2 + (0-1)^2 + (0-0)^2 + (0-1)^2 + (1-0)^2 + (0-1)^2}$$
$$= 2.24$$

$$d\left(x_4, \overline{x_{C_4}}\right)$$
$$= \sqrt{(0-0)^2 + (0-0)^2 + (0-0)^2 + (1-0)^2 + (0-0)^2 + (0-0)^2 + (0-1)^2 + (1-0)^2 + (0-0)^2}$$
$$= 1.73$$

$$d\left(x_4, \overline{x_{C_5}}\right)$$
$$= \sqrt{(0-0)^2 + (0-0)^2 + (0-0)^2 + (1-0)^2 + (0-0)^2 + (0-0)^2 + (0-0)^2 + (1-0)^2 + (0-1)^2}$$
$$= 1.73$$

In Step 5, x_4 is not in any cluster. Step 6 of the algorithm is executed to move x_4 to its closest cluster C_2, and the centroid of C_2 is updated:

$$\overline{x_{C_2}} = \begin{bmatrix} \frac{0+0+0}{3} \\ \frac{1+0+0}{3} \\ \frac{0+1+0}{3} \\ \frac{1+1+1}{3} \\ \frac{0+0+0}{3} \\ \frac{0+1+0}{3} \\ \frac{0+1+0}{3} \\ \frac{1+1+1}{3} \\ \frac{0+0+0}{3} \end{bmatrix} = \begin{bmatrix} 0 \\ 0.33 \\ 0.33 \\ 1 \\ 0 \\ 0.33 \\ 0.33 \\ 1 \\ 0 \end{bmatrix}.$$

We have $C_1 = \{x_1\}$, $C_2 = \{x_2, x_3, x_4\}$, $C_3 = \{\}$, $C_4 = \{\}$, and $C_5 = \{\}$.

Going back to Step 3, we take the fifth data point x_5 in the data set. In Step 4, we know that x_5 is closest to C_3 since C_3 is initially set up using x_5 and is not updated since then. In Step 5, x_5 is not in any cluster. Step 6 of the algorithm is executed to move x_5 to its closest cluster C_3 whose centroid remains the same. We have $C_1 = \{x_1\}$, $C_2 = \{x_2, x_3, x_4\}$, $C_3 = \{x_5\}$, $C_4 = \{\}$, and $C_5 = \{\}$.

Going back to Step 3, we take the sixth data point x_6 in the data set. In Step 4, we compute the Euclidean distance of the data point x_6 to each of the five clusters:

$d(x_6, \overline{x_{C_1}})$

$= \sqrt{(0-1)^2 + (0-0)^2 + (0-0)^2 + (0-0)^2 + (0-1)^2 + (1-0)^2 + (1-1)^2 + (0-0)^2 + (0-1)^2}$

$= 2$

$d(x_6, \overline{x_{C_2}})$

$= \sqrt{(0-0)^2 + (0-0.33)^2 + (0-0.33)^2 + (0-1)^2 + (0-0)^2 + (1-0.33)^2 + (1-0.33)^2 + (0-1)^2 + (0-0)^2}$

$= 1.77$

$d(x_6, \overline{x_{C_3}})$

$= \sqrt{(0-0)^2 + (0-0)^2 + (0-0)^2 + (0-0)^2 + (0-1)^2 + (1-0)^2 + (1-1)^2 + (0-0)^2 + (0-1)^2}$

$= 1.73$

$$d(x_6, \overline{x_{C_4}})$$
$$= \sqrt{(0-0)^2 + (0-0)^2 + (0-0)^2 + (0-0)^2 + (0-0)^2 + (1-0)^2 + (1-1)^2 + (0-0)^2 + (0-0)^2}$$
$$= 1$$

$$d(x_6, \overline{x_{C_5}})$$
$$= \sqrt{(0-0)^2 + (0-0)^2 + (0-0)^2 + (0-0)^2 + (0-0)^2 + (1-0)^2 + (1-0)^2 + (0-0)^2 + (0-1)^2}$$
$$= 1.73$$

In Step 5, x_6 is not in any cluster. Step 6 of the algorithm is executed to move x_6 to its closest cluster C_4, and the centroid of C_4 is updated:

$$\overline{x_{C_4}} = \begin{bmatrix} 0 \\ 0 \\ 0 \\ 0 \\ 0 \\ 1 \\ 1 \\ 0 \\ 0 \end{bmatrix}.$$

We have $C_1 = \{x_1\}$, $C_2 = \{x_2, x_3, x_4\}$, $C_3 = \{x_5\}$, $C_4 = \{x_6\}$, and $C_5 = \{\}$.

Going back to Step 3, we take the sixth data point x_7 in the data set. In Step 4, we compute the Euclidean distance of the data point x_7 to each of the five clusters:

$$d(x_7, \overline{x_{C_1}})$$
$$= \sqrt{(0-1)^2 + (0-0)^2 + (0-0)^2 + (0-0)^2 + (0-1)^2 + (0-0)^2 + (1-1)^2 + (0-0)^2 + (0-1)^2}$$
$$= 1.73$$

$$d(x_7, \overline{x_{C_2}})$$
$$= \sqrt{(0-0)^2 + (0-0.33)^2 + (0-0.33)^2 + (0-1)^2 + (0-0)^2 + (0-0.33)^2 + (1-0.33)^2 + (0-1)^2 + (0-0)^2}$$
$$= 1.67$$

$$d(x_7, \overline{x_{C_3}})$$

$$= \sqrt{(0-0)^2 + (0-0)^2 + (0-0)^2 + (0-0)^2 + (0-1)^2 + (0-0)^2 + (1-1)^2 + (0-0)^2 + (0-1)^2}$$

$$= 1.41$$

$$d(x_7, \overline{x_{C_4}})$$

$$= \sqrt{(0-0)^2 + (0-0)^2 + (0-0)^2 + (0-0)^2 + (0-0)^2 + (0-1)^2 + (1-1)^2 + (0-0)^2 + (0-0)^2}$$

$$= 1$$

$$d(x_7, \overline{x_{C_5}})$$

$$= \sqrt{(0-0)^2 + (0-0)^2 + (0-0)^2 + (0-0)^2 + (0-0)^2 + (0-0)^2 + (1-0)^2 + (0-0)^2 + (0-1)^2}$$

$$= 1.41$$

In Step 5, x_7 is not in any cluster. Step 6 of the algorithm is executed to move x_7 to its closest cluster C_4, and the centroid of C_4 is updated:

$$\overline{x_{C_4}} = \begin{bmatrix} \frac{0+0}{2} \\ \frac{0+0}{2} \\ \frac{0+0}{2} \\ \frac{0+0}{2} \\ \frac{0+0}{2} \\ \frac{1+0}{2} \\ \frac{1+1}{2} \\ \frac{0+0}{2} \\ \frac{0+0}{2} \end{bmatrix} = \begin{bmatrix} 0 \\ 0 \\ 0 \\ 0 \\ 0 \\ 0.5 \\ 1 \\ 0 \\ 0 \end{bmatrix}$$

We have $C_1 = \{x_1\}$, $C_2 = \{x_2, x_3, x_4\}$, $C_3 = \{x_5\}$, $C_4 = \{x_6, x_7\}$, and $C_5 = \{\}$.

K-Means Clustering and Density-Based Clustering

Going back to Step 3, we take the eighth data point x_8 in the data set. In Step 4, we compute the Euclidean distance of the data point x_8 to each of the five clusters:

$$d\left(x_8, \overline{x_{C_1}}\right)$$
$$= \sqrt{(0-1)^2 + (0-0)^2 + (0-0)^2 + (0-0)^2 + (0-1)^2 + (0-0)^2 + (0-1)^2 + (1-0)^2 + (0-1)^2}$$
$$= 2.27$$

$$d\left(x_8, \overline{x_{C_2}}\right)$$
$$= \sqrt{(0-0)^2 + (0-0.33)^2 + (0-0.33)^2 + (0-1)^2 + (0-0)^2 + (0-0.33)^2 + (0-0.33)^2 + (1-1)^2 + (0-0)^2}$$
$$= 1.20$$

$$d\left(x_8, \overline{x_{C_3}}\right)$$
$$= \sqrt{(0-0)^2 + (0-0)^2 + (0-0)^2 + (0-0)^2 + (0-1)^2 + (0-0)^2 + (0-1)^2 + (1-0)^2 + (0-1)^2}$$
$$= 2$$

$$d\left(x_8, \overline{x_{C_4}}\right)$$
$$= \sqrt{(0-0)^2 + (0-0)^2 + (0-0)^2 + (0-0)^2 + (0-0)^2 + (0-0.5)^2 + (0-1)^2 + (1-0)^2 + (0-0)^2}$$
$$= 1.5$$

$$d\left(x_8, \overline{x_{C_5}}\right)$$
$$= \sqrt{(0-0)^2 + (0-0)^2 + (0-0)^2 + (0-0)^2 + (0-0)^2 + (0-0)^2 + (0-0)^2 + (1-0)^2 + (0-1)^2}$$
$$= 1.41$$

In Step 5, x_8 is not in any cluster. Step 6 of the algorithm is executed to move x_8 to its closest cluster C_2, and the centroid of C_2 is updated:

$$\overline{x_{C_2}} = \begin{bmatrix} \frac{0+0+0+0}{4} \\ \frac{1+0+0+0}{4} \\ \frac{0+1+0+0}{4} \\ \frac{1+1+1+0}{4} \\ \frac{0+0+0+0}{4} \\ \frac{0+1+0+0}{4} \\ \frac{0+1+0+0}{3} \\ \frac{1+1+1+1}{4} \\ \frac{0+0+0+0}{4} \end{bmatrix} = \begin{bmatrix} 0 \\ 0.25 \\ 0.25 \\ 0.75 \\ 0 \\ 0.25 \\ 0.25 \\ 1 \\ 0 \end{bmatrix}.$$

We have $C_1 = \{x_1\}$, $C_2 = \{x_2, x_3, x_4, x_8\}$, $C_3 = \{x_5\}$, $C_4 = \{x_6, x_7\}$, and $C_5 = \{\}$.

Going back to Step 3, we take the ninth data point x_9 in the data set. In Step 4, we know that x_9 is closest to C_5 since C_5 is initially set up using x_9 and is not updated since then. In Step 5, x_9 is not in any cluster. Step 6 of the algorithm is executed to move x_9 to its closest cluster C_5 whose centroid remains the same. We have $C_1 = \{x_1\}$, $C_2 = \{x_2, x_3, x_4, x_8\}$, $C_3 = \{x_5\}$, $C_4 = \{x_6, x_7\}$, and $C_5 = \{x_9\}$.

After finishing the FOR loop in Steps 3–6, we go down to Step 7. Since there are changes of cluster centroids in Steps 3–6, we go back to Step 2 and then Step 3 to start another FOR loop. In this FOR loop, the current cluster of each data point is the closest cluster of the data point. Hence, none of the nine data points move from its current cluster to another cluster, and no change of the cluster centroids occurs in this FOR loop. The 5-means clustering for this example produces five clusters, $C_1 = \{x_1\}$, $C_2 = \{x_2, x_3, x_4, x_8\}$, $C_3 = \{x_5\}$, $C_4 = \{x_6, x_7\}$, and $C_5 = \{x_9\}$. The hierarchical clustering for the same data set shown in Figure 8.1 produces five clusters, $\{x_1, x_5\}$, $\{x_2, x_4, x_8\}$, $\{x_3\}$, $\{x_6, x_7\}$, and $\{x_9\}$, when we set the threshold of the merging distance to 1.5. Hence, the 5-means clustering results are similar but not exactly the same as the hierarchical clustering result.

9.2 Density-Based Clustering

Density-based clustering considers data clusters as regions of data points with high density, which is measured using the number of data points within a given radius (Li and Ye, 2002). Clusters are separated by regions of data points with low density. DBSCAN (Ester et al., 1996) is a density-based clustering algorithm that starts with a set of data points and two parameters: the radius and the minimum number of data points required to form a cluster. The density of a data point x is computed by counting the number of data points within the radius of the data point x. The region of x is the area within the radius of x, which has a dense region if the number of data points in the region of x is greater than or equal to the minimum number of data points. At first, all the data points in the data set are considered unmarked. DBSCAN arbitrarily selects an unmarked data point x from the data set. If the region of the data point x is not dense, x is marked as a noise point. If the region of x is dense, a new cluster is formed containing x, and x is marked as a member of this new cluster. Moreover, each of the data points in the region of x joins the cluster and is marked as a member of this cluster if the data point has not yet joined a cluster. This new cluster is further expanded to include all the data points that have not yet joined a cluster and are in the region of any data point z already in the cluster if the region of z is dense. The expansion of the cluster continues until all the data points connected through the dense regions of data points join the cluster if they have not yet joined a cluster. Note that a noise point may later be found in the dense region of a data point in another cluster and thus may be converted as a member of that cluster. After completing a cluster, DBSCAN selects another unmarked data point and evaluates if the data point is a noise point or a data point to start a new cluster. This process continues until all the data points in the data set are marked as either a noise point or a member of a cluster.

Since density-based clustering depends on two parameters of the radius and the minimum number of data points, knowledge in the application domain may help the selection of appropriate parameter values to produce a clustering result that is meaningful in the application domain. Different clustering results using different parameter values may be obtained so that different results can be compared and evaluated.

9.3 Software and Applications

K-means clustering is supported in:

- Weka (http://www.cs.waikato.ac.nz/ml/weka/)
- MATLAB® (www.matworks.com)
- SAS (www.sas.com)

The application of DBSCAN to spatial data can be found in Ester et al. (1996).

Exercises

9.1 Produce the 2-means clustering of the data points in Table 9.2 using the Euclidean distance as the measure of dissimilarity and using the first and third data points to set up the initial centroids of the two clusters.

9.2 Produce the density-based clustering of the data points in Table 9.2 using the Euclidean distance as the measure of dissimilarity, 1.5 as the radius and 2 as the minimum number of data points required to form a cluster.

9.3 Produce the density-based clustering of the data points in Table 9.2 using the Euclidean distance as the measure of dissimilarity, 2 as the radius and 2 as the minimum number of data points required to form a cluster.

9.4 Produce the 3-means clustering of 23 data points in the space shuttle O-ring data set in Table 1.2. Use Launch-Temperature and Leak-Check Pressure as the attribute variables and the normalization method in Equation 7.4 to obtain the normalized Launch-Temperature and Leak-Check Pressure, the Euclidean distance as the measure of dissimilarity.

9.5 Repeat Exercise 9.4 using the cosine similarity measure.

10
Self-Organizing Map

This chapter describes the self-organizing map (SOM), which is based on the architecture of artificial neural networks and is used for data clustering and visualization. A list of software packages for SOM is provided along with references for applications.

10.1 Algorithm of Self-Organizing Map

SOM was developed by Kohonen (1982). SOM is an artificial neural network with output nodes arranged in a q-dimensional space, called the output map or graph. The one-, two-, or three-dimensional space or arrangement of output nodes, as shown in Figure 10.1, is usually used so that clusters of data points can be visualized, because similar data points are represented by nodes that are close to each other in the output map.

In an SOM, each input variable x_i, $i = 1, \ldots, p$, is connected to each SOM node j, $j = 1, \ldots, k$, with the connection weight w_{ji}. The output vector o of the SOM for a given input vector x is computed as follows:

$$o = \begin{bmatrix} o_1 \\ \vdots \\ o_j \\ \vdots \\ o_k \end{bmatrix} = \begin{bmatrix} w_1' x \\ \vdots \\ w_j' x \\ \vdots \\ w_k' x \end{bmatrix}, \qquad (10.1)$$

where

$$x = \begin{bmatrix} x_1 \\ \vdots \\ x_i \\ \vdots \\ x_p \end{bmatrix}$$

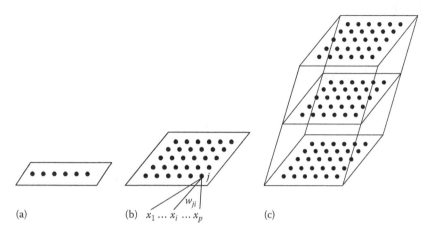

FIGURE 10.1
Architectures of SOM with a (a) one-, (b) two-, and (c) three-dimensional output map.

$$w_j = \begin{bmatrix} w_{j1} \\ \vdots \\ w_{ji} \\ \vdots \\ w_{jp} \end{bmatrix}.$$

Among all the output nodes, the output node producing the largest value for a given input vector x is called the winner node. The winner node of the input vector has the weight vector that is most similar to the input vector. The learning algorithm of SOM determines the connection weights so that the winner nodes of similar input vectors are close together. Table 10.1 lists the steps of the SOM learning algorithm, given a training data set with n data points, x_i, $i = 1, \ldots, n$.

In Step 5 of the algorithm, the connection weights of the winner node for the input vector x_i and the nearby nodes of the winner node are updated to make the weights of the winner node and its nearby nodes more similar to the input vector and thus make these nodes produce larger outputs for the input vector. The neighborhood function $f(j, c)$, which determines the closeness of node j to the winner node c and thus eligibility of node j for the weight change, can be defined in many ways. One example of $f(j, c)$ is

$$f(j, c) = \begin{cases} 1 & \text{if } \|r_j - r_c\| \le B_c(t) \\ 0 & \text{otherwise} \end{cases}, \tag{10.2}$$

where r_j and r_c are the coordinates of node j and the winner node c in the output map, and $B_c(t)$ gives the threshold value that bounds the neighborhood

Self-Organizing Map

TABLE 10.1
Learning Algorithm of SOM

Step	Description
1	Initialize the connection weights of nodes with random positive or negative values, $w'_j(t) = [w_{j1}(t) \quad \cdots \quad w_{jp}(t)], t = 0, j = 1, \ldots, k$
2	REPEAT
3	FOR $i = 1$ to n
4	Determine the winner node c for x_i: $c = \text{argmax}_j w'_j(t) x_i$
5	Update the connection weights of the winner node and its nearby nodes: $w_j(t+1) = w_j(t) + \alpha f(j,c)[x_i - w_j(t)]$, where α is the learning rate and $f(j,c)$ defines whether or not node j is close enough to c to be considered for the weight update
6	$w_j(t+1) = w_j(t)$ for other nodes without the weight update
7	$t = t + 1$
8	UNTIL the sum of weight changes for all the nodes, $E(t)$, is not greater than a threshold ε

of the winner c. $B_c(t)$ is defined as a function of t to have an adaptive learning process that uses a large threshold value at the beginning of the learning process and then decreases the threshold values over iterations. Another example of $f(j, c)$ is

$$f(j,c) = \frac{1}{e^{\frac{\|r_j - r_c\|^2}{2B_c^2(t)}}}. \tag{10.3}$$

In Step 8 of the algorithm, the sum of weight changes for all the nodes is computed:

$$E(t) = \sum_j \|w_j(t+1) - w_j(t)\|. \tag{10.4}$$

After the SOM is learned, clusters of data points are identified by marking each node with the data point(s) that makes the node the winner node. A cluster of data points are located and identified in a close neighborhood in the output map.

Example 10.1

Use the SOM with nine nodes in a one-dimensional chain and their coordinates of 1, 2, 3, 4, 5, 6, 7, 8, and 9 in Figure 10.2 to cluster the nine data points in the data set for system fault detection in Table 10.2, which is the same data set in Tables 8.1 and 9.2. The data set includes nine instances of single-machine faults, and the data point for each

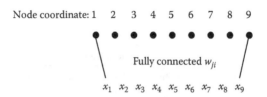

FIGURE 10.2
Architecture of SOM for Example 10.1.

TABLE 10.2

Data Set for System Fault Detection with Nine Cases of Single-Machine Faults

Instance (Faulty Machine)	Attribute Variables about Quality of Parts								
	x_1	x_2	x_3	x_4	x_5	x_6	x_7	x_8	x_9
1 (M1)	1	0	0	0	1	0	1	0	1
2 (M2)	0	1	0	1	0	0	0	1	0
3 (M3)	0	0	1	1	0	1	1	1	0
4 (M4)	0	0	0	1	0	0	0	1	0
5 (M5)	0	0	0	0	1	0	1	0	1
6 (M6)	0	0	0	0	0	1	1	0	0
7 (M7)	0	0	0	0	0	0	1	0	0
8 (M8)	0	0	0	0	0	0	0	1	0
9 (M9)	0	0	0	0	0	0	0	0	1

instance has the nine attribute variables about the quality of parts. The learning rate α is 0.3. The neighborhood function $f(j, c)$ is

$$f(j,c) = \begin{cases} 1 & \text{for } j = c-1, c, c+1 \\ 0 & \text{otherwise} \end{cases}.$$

In Step 1 of the learning process, we initialize the connection weights to the following:

$$w_1(0) = \begin{bmatrix} -0.24 \\ -0.41 \\ 0.46 \\ 0.27 \\ 0.88 \\ -0.09 \\ 0.78 \\ -0.39 \\ 0.91 \end{bmatrix} \quad w_2(0) = \begin{bmatrix} 0.44 \\ 0.44 \\ 0.93 \\ -0.15 \\ 0.84 \\ -0.36 \\ -0.16 \\ 0.55 \\ 0.93 \end{bmatrix} \quad w_3(0) = \begin{bmatrix} 0.96 \\ -0.45 \\ -0.75 \\ 0.35 \\ 0.05 \\ 0.86 \\ 0.12 \\ -0.49 \\ 0.98 \end{bmatrix} \quad w_4(0) = \begin{bmatrix} 0.82 \\ -0.22 \\ 0.60 \\ -0.56 \\ 0.91 \\ -0.80 \\ 0.33 \\ -0.54 \\ 0.47 \end{bmatrix}$$

Self-Organizing Map

$$w_5(0) = \begin{bmatrix} 0.62 \\ 0.44 \\ 0.33 \\ 0.46 \\ -0.25 \\ -0.26 \\ -0.71 \\ -0.61 \\ 0.38 \end{bmatrix} \quad w_6(0) = \begin{bmatrix} -0.47 \\ -0.62 \\ -0.96 \\ -0.43 \\ 0.32 \\ 0.96 \\ 0.70 \\ -0.04 \\ -0.84 \end{bmatrix} \quad w_7(0) = \begin{bmatrix} -0.87 \\ 0.23 \\ 0.37 \\ 0.49 \\ 0.04 \\ 0.33 \\ -0.10 \\ 0.45 \\ -0.96 \end{bmatrix}$$

$$w_8(0) = \begin{bmatrix} -0.95 \\ -0.21 \\ -0.48 \\ 0.05 \\ -0.54 \\ 0.23 \\ -0.37 \\ 0.61 \\ -0.76 \end{bmatrix} \quad w_9(0) = \begin{bmatrix} 0.69 \\ 0.23 \\ -0.69 \\ 0.86 \\ 0.22 \\ -0.91 \\ 0.82 \\ 0.31 \\ 0.31 \end{bmatrix}.$$

Using these initial weights to compute the SOM outputs for the nine data points makes nodes 4, 9, 7, 9, 1, 6, 9, 8, and 3 the winner nodes for $x_1, x_2, x_3, x_4, x_5, x_6, x_7, x_8,$ and x_9, respectively. For example, the output of each node for x_1 is computed to determine the winner node:

$$o = \begin{bmatrix} o_1 \\ o_2 \\ o_3 \\ o_4 \\ o_5 \\ o_6 \\ o_7 \\ o_8 \\ o_9 \end{bmatrix} = \begin{bmatrix} w'_1(0)x_1 \\ w'_2(0)x_1 \\ w'_3(0)x_1 \\ w'_4(0)x_1 \\ w'_5(0)x_1 \\ w'_6(0)x_1 \\ w'_7(0)x_1 \\ w'_8(0)x_1 \\ w'_9(0)x_1 \end{bmatrix}$$

$$
\begin{aligned}
&= \begin{bmatrix}
(-0.24)(1)+(-0.41)(0)+(0.46)(0)+(0.27)(0)+(0.88)(1)+(-0.09)(0) \\
\quad +(0.78)(1)+(-0.39)(0)+(0.91)(1) \\
(0.44)(1)+(0.44)(0)+(0.93)(0)+(-0.15)(0)+(0.84)(1)+(-0.36)(0) \\
\quad +(-0.16)(1)+(0.55)(0)+(0.93)(1) \\
(0.96)(1)+(-0.45)(0)+(-0.75)(0)+(0.75)(0)+(0.05)(1)+(0.86)(0) \\
\quad +(0.12)(1)+(-0.49)(0)+(0.98)(1) \\
(0.82)(1)+(-0.22)(0)+(0.60)(0)+(-0.56)(0)+(0.91)(1)+(-0.89)(0) \\
\quad +(0.33)(1)+(-0.54)(0)+(0.47)(1) \\
(0.62)(1)+(0.44)(0)+(0.33)(0)+(0.46)(0)+(-0.25)(1)+(-0.26)(0) \\
\quad +(-0.71)(1)+(-0.61)(0)+(0.38)(1) \\
(-0.47)(1)+(-0.62)(0)+(-0.96)(0)+(-0.43)(0)+(0.32)(1)+(0.96)(0) \\
\quad +(0.70)(1)+(-0.04)(0)+(-0.84)(1) \\
(-0.87)(1)+(0.23)(0)+(0.37)(0)+(0.49)(0)+(0.04)(1)+(0.33)(0) \\
\quad +(-0.10)(1)+(0.45)(0)+(-0.96)(1) \\
(-0.95)(1)+(-0.21)(0)+(-0.48)(0)+(0.05)(0)+(-0.54)(1)+(0.23)(0) \\
\quad +(-0.37)(1)+(0.61)(0)+(-0.76)(1) \\
(0.69)(1)+(0.23)(0)+(-0.69)(0)+(0.86)(0)+(0.22)(1)+(-0.91)(0) \\
\quad +(0.82)(1)+(0.31)(0)+(0.31)(1)
\end{bmatrix} \\[1em]
&= \begin{bmatrix} 2.33 \\ 2.04 \\ 2.11 \\ 2.53 \\ 0.04 \\ -0.29 \\ -1.9 \\ -2.62 \\ 2.04 \end{bmatrix}.
\end{aligned}
$$

Self-Organizing Map

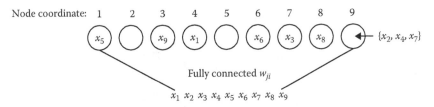

FIGURE 10.3
The winner nodes for the nine data points in Example 10.1 using initial weight values.

Since node 4 has the largest output value $o_4 = 2.53$, node 4 is the winner node for x_1. Figure 10.3 illustrates the output map to indicate the winner nodes for the nine data points and thus initial clusters of the data points based on the initial weights.

In Steps 2 and 3, x_1 is considered. In Step 4, the output of each node for x_1 is computed to determine the winner node. As described earlier, node 4 is the winner node for x_1, and thus $c = 4$. In Step 5, the connection weights to the winner node $c = 4$ and its neighbors $c - 1 = 3$ and $c + 1 = 5$ are updated:

$$w_4(1) = w_4(0) + (0.3)[x_1 - w_4(0)] = (0.7)w_4(0) + (0.3)x_1$$

$$= (0.7)\begin{bmatrix} 0.82 \\ -0.22 \\ 0.60 \\ -0.56 \\ 0.91 \\ -0.80 \\ 0.33 \\ -0.54 \\ 0.47 \end{bmatrix} + (0.3)\begin{bmatrix} 1 \\ 0 \\ 0 \\ 0 \\ 1 \\ 0 \\ 1 \\ 0 \\ 1 \end{bmatrix} = \begin{bmatrix} 0.87 \\ -0.15 \\ 0.42 \\ -0.39 \\ 0.94 \\ -0.56 \\ 0.53 \\ -0.38 \\ 0.63 \end{bmatrix}.$$

$$w_3(1) = w_3(0) + (0.3)[x_1 - w_3(0)] = (0.7)w_3(0) + (0.3)x_1$$

$$= (0.7)\begin{bmatrix} 0.96 \\ -0.45 \\ -0.75 \\ 0.35 \\ 0.05 \\ 0.86 \\ 0.12 \\ -0.49 \\ 0.98 \end{bmatrix} + (0.3)\begin{bmatrix} 1 \\ 0 \\ 0 \\ 0 \\ 1 \\ 0 \\ 1 \\ 0 \\ 1 \end{bmatrix} = \begin{bmatrix} 1.96 \\ -0.32 \\ 0.53 \\ 0.25 \\ 0.34 \\ 0.60 \\ 0.38 \\ -0.34 \\ 0.99 \end{bmatrix}.$$

$$w_5(1) = w_5(0) + (0.3)[x_1 - w_5(0)] = (0.7)w_5(0) + (0.3)x_1$$

$$= (0.7)\begin{bmatrix} 0.62 \\ 0.44 \\ 0.33 \\ 0.46 \\ -0.25 \\ -0.26 \\ -0.71 \\ -0.61 \\ 0.38 \end{bmatrix} + (0.3)\begin{bmatrix} 1 \\ 0 \\ 0 \\ 0 \\ 1 \\ 0 \\ 1 \\ 0 \\ 1 \end{bmatrix} = \begin{bmatrix} 0.73 \\ 0.31 \\ 0.23 \\ 0.32 \\ 0.13 \\ -0.18 \\ 0.80 \\ -0.43 \\ 0.57 \end{bmatrix}.$$

In Step 6, the weights for the other nodes remain the same. In Step 7, t is increased to 1, and the weights of the nine nodes are

$$w_1(1) = \begin{bmatrix} -0.24 \\ -0.41 \\ 0.46 \\ 0.27 \\ 0.88 \\ -0.09 \\ 0.78 \\ -0.39 \\ 0.91 \end{bmatrix} \quad w_2(1) = \begin{bmatrix} 0.44 \\ 0.44 \\ 0.93 \\ -0.15 \\ 0.84 \\ -0.36 \\ -0.16 \\ 0.55 \\ 0.93 \end{bmatrix} \quad w_3(1) = \begin{bmatrix} 1.96 \\ -0.32 \\ 0.53 \\ 0.25 \\ 0.34 \\ 0.60 \\ 0.38 \\ -0.34 \\ 0.99 \end{bmatrix} \quad w_4(1) = \begin{bmatrix} 0.87 \\ -0.15 \\ 0.42 \\ -0.39 \\ 0.94 \\ -0.56 \\ 0.53 \\ -0.38 \\ 0.63 \end{bmatrix}$$

$$w_5(1) = \begin{bmatrix} 0.73 \\ 0.31 \\ 0.23 \\ 0.32 \\ 0.13 \\ -0.18 \\ 0.80 \\ -0.43 \\ 0.57 \end{bmatrix} \quad w_6(1) = \begin{bmatrix} -0.47 \\ -0.62 \\ -0.96 \\ -0.43 \\ 0.32 \\ 0.96 \\ 0.70 \\ -0.04 \\ -0.84 \end{bmatrix} \quad w_7(1) = \begin{bmatrix} -0.87 \\ 0.23 \\ 0.37 \\ 0.49 \\ 0.04 \\ 0.33 \\ -0.10 \\ 0.45 \\ -0.96 \end{bmatrix}$$

$$w_8(1) = \begin{bmatrix} -0.95 \\ -0.21 \\ -0.48 \\ 0.05 \\ -0.54 \\ 0.23 \\ -0.37 \\ 0.61 \\ -0.76 \end{bmatrix} \quad w_9(1) = \begin{bmatrix} 0.69 \\ 0.23 \\ -0.69 \\ 0.86 \\ 0.22 \\ -0.91 \\ 0.82 \\ 0.31 \\ 0.31 \end{bmatrix}.$$

Next, we go back to Steps 2 and 3, and x_2 is considered. The learning process continues until the sum of consecutive weight changes initiated by all the nine data points is small enough.

10.2 Software and Applications

SOM is supported by:

- Weka (http://www.cs.waikato.ac.nz/ml/weka/)
- MATLAB® (www.matworks.com)

Liu and Weisberg (2005) describe the application of SOM to analyze ocean current variability. The application of SOM to brain activity data of monkey in relation to movement directions is reported in Ye (2003, Chapter 3).

Exercises

10.1 Continue the learning process in Example 10.1 to perform the weight updates when x_2 is presented to the SOM.

10.2 Use the software Weka to produce the SOM for Example 10.1.

10.3 Define a two-dimensional SOM and the neighborhood function in Equation 10.2 for Example 10.1 and perform one iteration of the weight update when x_1 is presented to the SOM.

10.4 Use the software Weka to produce a two-dimensional SOM for Example 10.1.

10.5 Produce a one-dimensional SOM with the same neighborhood function in Example 10.1 for the space shuttle O-ring data set in Table 1.2. Use Launch-Temperature and Leak-Check Pressure as the attribute variables and the normalization method in Equation 7.4 to obtain the normalized Launch-Temperature and Leak-Check Pressure.

11
Probability Distributions of Univariate Data

The clustering algorithms in Chapters 8 through 10 can be applied to data with one or more attribute variables. If there is only one attribute variable, we have univariate data. For univariate data, the probability distribution of data points captures not only clusters of data points but also many other characteristics concerning the distribution of data points. Many specific data patterns of univariate data can be identified through their corresponding types of probability distribution. This chapter introduces the concept and characteristics of the probability distribution and the use of the probability distribution characteristics to identify certain univariate data patterns. A list of software packages for identifying the probability distribution characteristics of univariate data is provided along with references for applications.

11.1 Probability Distribution of Univariate Data and Probability Distribution Characteristics of Various Data Patterns

Given an attribute variable x and its data observations, $x_1, \ldots, x_n$, the frequency histogram of data observations is often used to show the frequencies of all the x values. Table 11.1 gives the values of launch temperature in the space shuttle O-ring data set, which is taken from Table 1.2. Figure 11.1 gives a histogram of the launch temperature values in Table 11.1 using an interval width of 5 units. Changing the interval width changes the frequency of data observations in each interval and thus the histogram.

In the histogram in Figure 11.1, the frequency of data observations for each interval can be replaced with the probability density, which can be estimated using the ratio of that frequency to the total number of data observations. Fitting a curve to the histogram of the probability density, we obtain a fitted curve for the probability density function $f(x)$ that gives the probability density for any value of x. A common type of the probability distribution is a normal distribution with the following probability density function:

$$f(x) = \frac{1}{\sqrt{2\pi}\sigma} e^{-\frac{1}{2}\left(\frac{x-\mu}{\sigma}\right)^2}, \qquad (11.1)$$

TABLE 11.1

Values of Launch Temperature in the Space Shuttle O-Ring Data Set

Instance	Launch Temperature
1	66
2	70
3	69
4	68
5	67
6	72
7	73
8	70
9	57
10	63
11	70
12	78
13	67
14	53
15	67
16	75
17	70
18	81
19	76
20	79
21	75
22	76
23	58

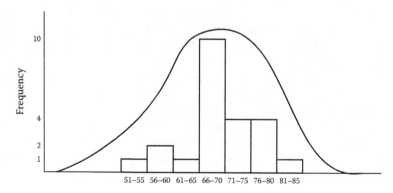

FIGURE 11.1
Frequency histogram of the Launch Temperature data.

Where
  μ is the mean
  σ is the standard deviation

A normal distribution is symmetric with the highest probability density at the mean $x = \mu$ and the same probability density at $x = \mu + a$ and $x = \mu - a$.

Many data patterns manifest special characteristics of their probability distributions. For example, we study time series data of computer and network activities (Ye, 2008, Chapter 9). Time series data consist of data observations over time. From computer and network data, we observe the following data patterns that are illustrated in Figure 11.2:

- Spike
- Random fluctuation
- Step change
- Steady change

The probability distributions of time series data with the spike, random fluctuation, step change, and steady change patterns have special characteristics. Time series data with a spike pattern as shown in Figure 11.2a have the majority of data points with similar values and few data points with higher values producing upward spikes or with lower values producing downward spikes. The high frequency of data points with similar values determines where the mean with a high probability density is located, and few data points with lower (higher) values than the mean for downward (upward) spikes produce a long tail on the left (right) side of the mean and thus a left (right) skewed distribution. Hence, time series data with spikes produce a skewed probability distribution that is asymmetric with most data points having values near the mean and few data points having values spreading over one side of the mean and creating a long tail, as shown in Figure 11.2a. Time series data with a random fluctuation pattern produce a normal distribution that is symmetric, as shown in Figure 11.2b. Time series data with one step change, as shown in Figure 11.2c, produce two clusters of data points with two different centroids and thus a bimodal distribution. Time series data with multiple step changes create multiple clusters of data points with their different centroids and thus a multimodal distribution. Time series data with the steady change (i.e., a steady increase of values or a steady decrease of values) have values evenly distributed and thus produce a uniform distribution, as shown in Figure 11.2d. Therefore, the four patterns of time series data produce four different types of probability distribution:

- Left or right skewed distribution
- Normal distribution
- Multimodal distribution
- Uniform distribution

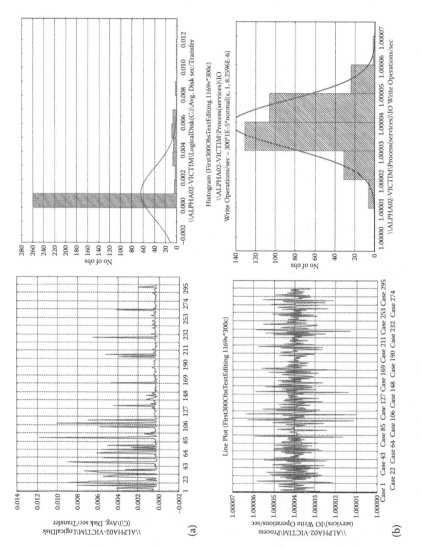

FIGURE 11.2
Time series data patterns and their probability distributions. (a) The data plot and histogram of spike pattern, (b) the data plot and histogram of random fluctuation pattern.

Probability Distributions of Univariate Data

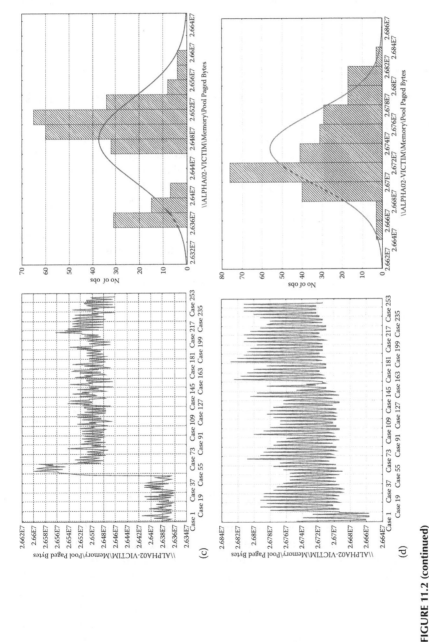

FIGURE 11.2 (continued)
Time series data patterns and their probability distributions. (c) the data plot and histogram of a step change pattern, and (d) the data plot and histogram of a steady change pattern.

As described in Ye (2008, Chapter 9), the four data patterns and their corresponding probability distributions can be used to identify whether or not there are attack activities underway in computer and network systems since computer and network data under attack or normal use conditions may demonstrate different data patterns. Cyber attack detection is an important part of protecting computer and network systems from cyber attacks.

11.2 Method of Distinguishing Four Probability Distributions

We may distinguish these four data patterns by identifying which of four different types of probability distribution data have. Although there are normality tests to determine whether or not data have a normal distribution (Bryc, 1995), statistical tests for identifying one of the other probability distributions are not common. Although the histogram can be plotted to let us first visualize and then determine the probability distribution, we need a test that can be programmed and run on computer without the manual visual inspection, especially when the data set is large and the real-time monitoring of data is required as for the application to cyber attack detection. A method of distinguishing the four probability distributions using a combination of skewness and mode tests is developed in Ye (2008, Chapter 9) and is described in the next section.

The method of distinguishing four probability distributions is based on skewness and mode tests. Skewness is defined as

$$\text{skewness} = E\left(\frac{(x-\mu)^3}{\sigma^3}\right), \tag{11.2}$$

where μ and σ are the mean and standard deviation of data population for the variable x. Given a sample of n data points, $x_1, \ldots, x_n$, the sample skewness is computed:

$$\text{skewness} = \frac{n \sum_{i=1}^{n} (x_i - \bar{x})^3}{(n-1)(n-2)s^3}, \tag{11.3}$$

where $\bar{x}$ and s are the average and standard deviation of the data sample. Unlike the variance that squares both positive and negative deviations from the mean to make both positive and negative deviations from the mean contribute to the variance in the same way, the skewness measures how much data deviations from the mean are symmetric on both sides of the mean. A left-skewed distribution with a long tail on the left side of the mean has a negative value of the skewness. A right-skewed distribution with a long tail on the right side of the mean has a positive value of the skewness.

TABLE 11.2

Combinations of Skewness and Mode Test Results for Distinguishing Four Probability Distributions

Probability Distribution	Dip Test	Mode Test	Skewness Test
Multimodal distribution	Unimodality is rejected	Number of significant modes ≥ 2	Any result
Uniform distribution	Unimodality is not rejected	Number of significant modes > 2	Symmetric
Normal distribution	Unimodality is not rejected	Number of significant modes < 2	Symmetric
Skewed distribution	Unimodality is not rejected	Number of significant modes < 2	Skewed

The mode of a probability distribution for a variable x is located at the value of x that has the maximum probability density. When a probability density function has multiple local maxima, the probability distribution has multiple modes. A large probability density indicates a cluster of similar data points. Hence, the mode is related to the clustering of data points. A normal distribution and a skewed distribution are examples of unimodal distributions with only one mode, in contrast to multimodal distributions with multiple modes. A uniform distribution has no significant mode since data are evenly distributed and are not formed into clusters. The dip test (Hartigan and Hartigan, 1985) determines whether or not a probability distribution is unimodal. The mode test in the R statistical software (www.r-project.org) determines the significance of each potential mode in a probability distribution and gives the number of significant modes.

Table 11.2 describes the special combinations of the skewness and mode test results that are used to distinguish four probability distributions: a multimodal distribution including a bimodal distribution, a uniform distribution, a normal distribution, and a skewed distribution. Therefore, if we know that the data have one of these four probability distributions, we can check the combination of results from the dip test, the mode test, and the skewness test and identify which probability distribution the data have.

11.3 Software and Applications

Statistica (www.statsoft.com) supports the skewness test. R statistical software (www.r-project.org, www.cran.r-project.org/doc/packages/diptest.pdf) supports the dip test and the mode test. In Ye (2008, Chapter 9), computer and network data under attack and normal use conditions can be characterized by different probability distributions of the data under different conditions.

Cyber attack detection is performed by monitoring the observed computer and network data and determining whether or not the change of probability distribution from the normal use condition to an attack condition occurs.

Exercises

11.1 Select and use the software to perform the skewness test, the mode test, and the dip test on the Launch Temperature Data in Table 11.1 and use the test results to determine whether or not the probability distribution of the Launch Temperature data falls into one of the four probability distributions in Table 11.2.

11.2 Select a numeric variable in the data set you obtain in Problem 1.2 and select an interval width to plot a histogram of the data for the variable. Select and use the software to perform the skewness test, the mode test, and the dip test on the data of the variable, and use the test results to determine whether or not the probability distribution of the Launch Temperature data falls into one of the four probability distributions in Table 11.2.

11.3 Select a numeric variable in the data set you obtain in Problem 1.3 and select an interval width to plot a histogram of the data for the variable. Select and use the software to perform the skewness test, the mode test, and the dip test on the data of the variable, and use the test results to determine whether or not the probability distribution of the Launch Temperature data falls into one of the four probability distributions in Table 11.2.

12
Association Rules

Association rules uncover items that are frequently associated together. The algorithm of association rules was initially developed in the context of market basket analysis for studying customer purchasing behaviors that can be used for marketing. Association rules uncover what items customers often purchase together. Items that are frequently purchased together can be placed together in stores or can be associated together at e-commerce websites for promoting the sale of the items or for other marketing purposes. There are many other applications of association rules, for example, text analysis for document classification and retrieval. This chapter introduces the algorithm of mining association rules. A list of software packages that support association rules is provided. Some applications of association rules are given with references.

12.1 Definition of Association Rules and Measures of Association

An item set contains a set of items. For example, a customer's purchase transaction at a grocery store is an item set or a set of grocery items such as eggs, tomatoes, and apples. The data set for system fault detection with nine cases of single-machine faults in Table 8.1 contains nine data records, which can be considered as nine sets of items by taking $x_1, x_2, x_3, x_4, x_5, x_6, x_7, x_8, x_9$ as nine different quality problems with the value of 1 indicating the presence of the given quality problem. Table 12.1 shows the nine item sets obtained from the data set for system fault detection. A frequent association of items in Table 12.1 reveals which quality problems often occur together.

An association rule takes the form of

$$A \to C,$$

Where
  A is an item set called the antecedent
  C is an item set called the consequent

A and C have no common items, that is, $A \cap C = \emptyset$ (an empty set). The relationship of A and C in the association rule means that the presence of the item set

TABLE 12.1

Data Set for System Fault Detection with Nine Cases of Single-Machine Faults and Item Sets Obtained from This Data Set

Instance (Faulty Machine)	Attribute Variables about Quality of Parts									Items in Each Data Record
	x_1	x_2	x_3	x_4	x_5	x_6	x_7	x_8	x_9	
1 (M1)	1	0	0	0	1	0	1	0	1	$\{x_1, x_5, x_7, x_9\}$
2 (M2)	0	1	0	1	0	0	0	1	0	$\{x_2, x_4, x_8\}$
3 (M3)	0	0	1	1	0	1	1	1	0	$\{x_3, x_4, x_6, x_7, x_8\}$
4 (M4)	0	0	0	1	0	0	0	1	0	$\{x_4, x_8\}$
5 (M5)	0	0	0	0	1	0	1	0	1	$\{x_5, x_7, x_9\}$
6 (M6)	0	0	0	0	0	1	1	0	0	$\{x_6, x_7\}$
7 (M7)	0	0	0	0	0	0	1	0	0	$\{x_7\}$
8 (M8)	0	0	0	0	0	0	0	1	0	$\{x_8\}$
9 (M9)	0	0	0	0	0	0	0	0	1	$\{x_9\}$

A in a data record implies the presence of the item set C in the data record, that is, the item set C is associated with the item set A.

The measures of *support*, *confidence*, and *lift* are defined and used to discover item sets A and C that are frequently associated together. $Support(X)$ measures the proportion of data records that contain the item set X, and is defined as

$$support(X) = \frac{|\{S | S \in D \text{ and } S \supseteq X\}|}{N}, \qquad (12.1)$$

Where
  D denotes the data set containing data records
  S is a data record in the data set D (indicated by $S \in D$) and contains the items in X (indicated by $S \supseteq X$)
  $|\ |$ denotes the number of such data records S
  N is the number of the data records in D

Based on the definition, we have

$$support(\emptyset) = \frac{|\{S | S \in D \text{ and } S \supseteq \emptyset\}|}{N} = \frac{N}{N} = 1.$$

For example, for the data set with the nine data records in Table 12.1,

$$support(\{x_5\}) = \frac{2}{9} = 0.22$$

$$support(\{x_7\}) = \frac{5}{9} = 0.56$$

Association Rules

$$support(\{x_9\}) = \frac{3}{9} = 0.33$$

$$support(\{x_5, x_7\}) = \frac{2}{9} = 0.22$$

$$support(\{x_5, x_9\}) = \frac{2}{9} = 0.22.$$

Support($A \rightarrow C$) measures the proportion of data records that contain both the antecedent A and the consequent C in the association rule $A \rightarrow C$, and is defined as

$$support(A \rightarrow C) = support(A \cup C), \quad (12.2)$$

where $A \cup C$ is the union of the item set A and the item set C and contains items from both A and C. Based on the definition, we have

$$support(\emptyset \rightarrow C) = support(C)$$

$$support(A \rightarrow \emptyset) = support(A).$$

For example,

$$support(\{x_5\} \rightarrow \{x_7\}) = support(\{x_5\} \cup \{x_7\}) = support(\{x_5, x_7\}) = 0.22$$

$$support(\{x_5\} \rightarrow \{x_9\}) = support(\{x_5\} \cup \{x_9\}) = support(\{x_5, x_9\}) = 0.22.$$

Confidence($A \rightarrow C$) measures the proportion of data records containing the antecedent A that also contain the consequent C, and is defined as

$$confidence(A \rightarrow C) = \frac{support(A \cup C)}{support(A)}. \quad (12.3)$$

Based on the definition, we have

$$confidence(\emptyset \rightarrow C) = \frac{support(C)}{support(\emptyset)} = \frac{support(C)}{1} = support(C)$$

$$confidence(A \rightarrow \emptyset) = \frac{support(A)}{support(A)} = 1.$$

For example,

$$\text{confidence}(\{x_5\} \rightarrow \{x_7\}) = \frac{\text{support}(\{x_5\} \cup \{x_7\})}{\text{support}(\{x_5\})} = \frac{0.22}{0.22} = 1$$

$$\text{confidence}(\{x_5\} \rightarrow \{x_9\}) = \frac{\text{support}(\{x_5\} \cup \{x_9\})}{\text{support}(\{x_9\})} = \frac{0.22}{0.22} = 1.$$

If the antecedent A and the consequent C are independent and $\text{support}(C)$ is high (the consequent C is contained in many data records in the data set), $\text{support}(A \cup C)$ has a high value because C is contained in many data records that also contain A. As a result, we get a high value of $\text{support}(A \rightarrow C)$ and $\text{confidence}(A \rightarrow C)$ even though A and C are independent and the association of $A \rightarrow C$ is of little interest. For example, if the item set C is contained in every data record in the data, we have

$$\text{support}(A \rightarrow C) = \text{support}(A \cup C) = \text{support}(A)$$

$$\text{confidence}(A \rightarrow C) = \frac{\text{support}(A \cup C)}{\text{support}(A)} = \frac{\text{support}(A)}{\text{support}(A)} = 1.$$

However, the association rule of $A \rightarrow C$ is of little interest to us, because the item set C is in every data record and thus any item set including A is associated with C. To address this issue, $\text{lift}(A \rightarrow C)$ is defined:

$$\text{lift}(A \rightarrow C) = \frac{\text{confidence}(A \rightarrow C)}{\text{support}(C)} = \frac{\text{support}(A \cup C)}{\text{support}(A) \times \text{support}(C)}. \quad (12.4)$$

If the antecedent A and the consequent C are independent but $\text{support}(C)$ is high, the high value of $\text{support}(C)$ produces a low value of $\text{lift}(A \rightarrow C)$. For example,

$$\text{lift}(\{x_5\} \rightarrow \{x_7\}) = \frac{\text{confidence}(\{x_5\} \rightarrow \{x_7\})}{\text{support}(\{x_7\})} = \frac{1}{0.56} = 1.79$$

$$\text{lift}(\{x_5\} \rightarrow \{x_9\}) = \frac{\text{confidence}(\{x_5\} \rightarrow \{x_9\})}{\text{support}(\{x_9\})} = \frac{1}{0.33} = 3.03.$$

Association Rules

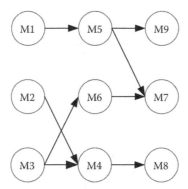

FIGURE 12.1
A manufacturing system with nine machines and production flows of parts.

The association rules, $\{x_5\} \to \{x_7\}$ and $\{x_5\} \to \{x_9\}$, have the same values of *support* and *confidence* but different values of *lift*. Hence, x_5 appears to have a greater impact on the frequency of x_9 than the frequency of x_7. Figure 1.1, which is copied in Figure 12.1, gives the production flows of parts for the data set in Table 12.1. As shown in Figure 12.1, parts flowing through M5 go to M7 and M9. Hence, x_5 should have the same impact on x_7 and x_9. However, parts flowing through M6 also go to M7, x_7 is more frequent than x_9 in the data set, producing a lower lift value for $\{x_5\} \to \{x_7\}$ than that of $\{x_5\} \to \{x_9\}$. In other words, x_7 is impacted by not only x_5 but also x_6 and x_3 as shown in Figure 12.1, which makes x_7 appear less dependent on x_5 since *lift* addresses the independence issue of the antecedent and the consequent by a low value of *lift*.

12.2 Association Rule Discovery

Association rule discovery is used to find all association rules that exceed the minimum thresholds on certain measures of association, typically *support* and *confidence*. Association rules are constructed using frequent item sets that satisfy the minimum support. Given a data set of data records that are made of p items at maximum, an item set can be represented as $(x_1, ..., x_p)$, $x_i = 0$ or 1, $i = 1, ..., p$, with $x_i = 1$ indicating the presence of the ith item in the item set. Since there are 2^p possible combinations of different values for $(x_1, ..., x_p)$, there are $(2^p - 1)$ possible different item sets with 1 to p items, excluding the empty set represented by $(0, ..., 0)$. It is impractical to exhaustively examine the *support* value of every one of $(2^p - 1)$ possible different item sets.

TABLE 12.2

Apriori Algorithm

Step	Description of the Step
1	$F_1 = \{$frequent one-item sets$\}$
2	$i = 1$
3	while $F_i \neq \emptyset$
4	$\quad i = i + 1$
5	$\quad C_i = \{\{x_1, ..., x_{i-2}, x_{i-1}, x_i\} \mid \{x_1, ..., x_{i-2}, x_{i-1}\} \in F_{i-1}$ and $\{x_1, ..., x_{i-2}, x_i\} \in F_{i-1}\}$
6	$\quad$ for all data records $S \in D$
7	$\quad\quad$ for all candidate sets $C \in C_i$
8	$\quad\quad\quad$ if $S \supseteq C$
9	$\quad\quad\quad\quad$ C.count = C.count + 1
10	$\quad F_i = \{C \mid C \in C_i$ and C.count $\geq$ minimum support$\}$
11	return all $F_j, j = 1, ..., i-1$

The Apriori algorithm (Agrawal and Srikant, 1994) provides an efficient procedure of generating frequent item sets by considering that an item set can be a frequent item set only if all of its subsets are frequent item sets. Table 12.2 gives the steps of the Apriori algorithm for a given data set D.

In Step 5 of the Apriori algorithm, the two item sets from F_{i-1} have the same items of $x_1, ..., x_{i-2}$, and the two item sets differ only in only one item with x_{i-1} in one item set and x_i in another item set. A candidate item set for F_i is constructed by including $x_1, ..., x_{i-2}$ (the common items of the two item sets from F_{i-1}), x_{i-1} and x_i. For example, if $\{x_1, x_2, x_3\}$ is a frequent three-item set, any combination of two items from this frequent three-item set, $\{x_1, x_2\}$, $\{x_2, x_3\}$ or $\{x_1, x_3\}$, must be frequent two-item sets. That is, if support($\{x_1, x_2, x_3\}$) is greater than or equal to the minimum support, support($\{x_1, x_2\}$), support($\{x_2, x_3\}$), and support $\{x_1, x_3\}$ must be greater than or equal to the minimum support. Hence, the frequent three-item set, $\{x_1, x_2, x_3\}$, can be constructed using two of its two-item subsets that differ in only one item, $\{x_1, x_2\}$ and $\{x_1, x_3\}$, $\{x_1, x_2\}$ and $\{x_2, x_3\}$, or $\{x_1, x_3\}$ and $\{x_2, x_3\}$. Similarly, a frequent i-item set must come from frequent $(i-1)$-item sets that differ in only one item. This method of constructing a candidate item set for F_i significantly reduces the number of candidate item sets for F_i to be evaluated in Step 7 of the algorithm.

Example 12.1 illustrates the Apriori algorithm. When the data are sparse with each item being relatively infrequent in the data set, the Apriori algorithm is efficient in that it produces a small number of frequent item sets, few of which contain large numbers of items. When the data are dense, the Apriori algorithm is less efficient and produces a large number of frequent item sets.

Association Rules

Example 12.1

From the data set in Table 12.1, find all frequent item sets with *min-support* (minimum support) = 0.2.

Examining the *support* of each one-item set, we obtain

$$F_1 = \left\{ \{x_4\}, \; support = \frac{3}{9} = 0.33, \right.$$

$$\{x_5\}, \; support = \frac{2}{9} = 0.22,$$

$$\{x_6\}, \; support = \frac{2}{9} = 0.22,$$

$$\{x_7\}, \; support = \frac{5}{9} = 0.56,$$

$$\{x_8\}, \; support = \frac{4}{9} = 0.44,$$

$$\left. \{x_9\}, \; support = \frac{3}{9} = 0.33 \right\}.$$

Using the frequent one-item sets to put together the candidate two-item sets and examine their *support*, we obtain

$$F_2 = \left\{ \{x_4, x_8\}, \; support = \frac{3}{9} = 0.33, \right.$$

$$\{x_5, x_7\}, \; support = \frac{2}{9} = 0.22,$$

$$\{x_5, x_9\}, \; support = \frac{2}{9} = 0.22,$$

$$\{x_6, x_7\}, \; support = \frac{2}{9} = 0.22,$$

$$\left. \{x_7, x_9\}, \; support = \frac{2}{9} = 0.22 \right\}.$$

Since $\{x_5, x_7\}$, $\{x_5, x_9\}$, and $\{x_7, x_9\}$ differ from each other in only one item, they are used to construct the three-item set $\{x_5, x_7, x_9\}$—the only three-item set that can be constructed:

$$F_3 = \left\{ \{x_5, x_7, x_9\}, \; support = \frac{2}{9} = 0.22 \right\}.$$

Note that constructing a three-item set from two-item sets that differ in more than one item does not produce a frequent three-item set.

For example, $\{x_4, x_8\}$ and $\{x_5, x_7\}$ are frequent two-item sets that differ in two items. $\{x_4, x_5\}$, $\{x_4, x_7\}$, $\{x_8, x_5\}$, and $\{x_8, x_7\}$ are not frequent two-item sets. A three-item set constructed using $\{x_4, x_8\}$ and $\{x_5, x_7\}$, e.g., $\{x_4, x_5, x_8\}$, is not a frequent three-item set because not every pair of two items from $\{x_4, x_5, x_8\}$ is a frequent two-item set. Specifically, $\{x_4, x_5\}$ and $\{x_8, x_5\}$ are not frequent two-item sets.

Since there is only one frequent three-item set, we cannot generate a candidate four-item set in Step 5 of the Apriori algorithm. That is, $C_4 = \emptyset$. As a result, $F_4 = \emptyset$ in Step 3 of the Apriori algorithm, and we exit the WHILE loop. In Step 11 of the algorithm, we collect all the frequent item sets that satisfy *min-support* = 0.2:

$\{x_4\}, \{x_5\}, \{x_6\}, \{x_7\}, \{x_8\}, \{x_9\}, \{x_4, x_8\}, \{x_5, x_7\}, \{x_5, x_9\}, \{x_6, x_7\}, \{x_7, x_9\}, \{x_5, x_7, x_9\}.$

Example 12.2

Use the frequent item sets from Example 12.1 to generate all the association rules that satisfy *min-support* = 0.2 and *min-confidence* (minimum confidence) = 0.5.

Using each frequent item set F obtained from Example 12.1, we generate each of the following association rules, $A \rightarrow C$, which satisfies

$$A \cup C = F,$$
$$A \cap C = \emptyset,$$

the criteria of the *min-support* and the *min-confidence*:

- $\emptyset \rightarrow \{x_4\}$, support = 0.33, confidence = 0.33
- $\emptyset \rightarrow \{x_5\}$, support = 0.22, confidence = 0.22
- $\emptyset \rightarrow \{x_6\}$, support = 0.22, confidence = 0.22
- $\emptyset \rightarrow \{x_7\}$, support = 0.56, confidence = 0.56
- $\emptyset \rightarrow \{x_8\}$, support = 0.44, confidence = 0.44
- $\emptyset \rightarrow \{x_9\}$, support = 0.33, confidence = 0.33
- $\emptyset \rightarrow \{x_4, x_8\}$, support = 0.33, confidence = 0.33
- $\emptyset \rightarrow \{x_5, x_7\}$, support = 0.22, confidence = 0.22
- $\emptyset \rightarrow \{x_5, x_9\}$, support = 0.22, confidence = 0.22
- $\emptyset \rightarrow \{x_6, x_7\}$, support = 0.22, confidence = 0.22
- $\emptyset \rightarrow \{x_7, x_9\}$, support = 0.22, confidence = 0.22
- $\emptyset \rightarrow \{x_5, x_7, x_9\}$, support = 0.22, confidence = 0.22
- $\{x_4\} \rightarrow \emptyset$, support = 0.33, confidence = 1
- $\{x_5\} \rightarrow \emptyset$, support = 0.22, confidence = 1
- $\{x_6\} \rightarrow \emptyset$, support = 0.22, confidence = 1
- $\{x_7\} \rightarrow \emptyset$, support = 0.56, confidence = 1
- $\{x_8\} \rightarrow \emptyset$, support = 0.44, confidence = 1
- $\{x_9\} \rightarrow \emptyset$, support = 0.33, confidence = 1
- $\{x_4, x_8\} \rightarrow \emptyset$, support = 0.33, confidence = 1
- $\{x_5, x_7\} \rightarrow \emptyset$, support = 0.22, confidence = 1
- $\{x_5, x_9\} \rightarrow \emptyset$, support = 0.22, confidence = 1
- $\{x_6, x_7\} \rightarrow \emptyset$, support = 0.22, confidence = 1

$\{x_7, x_9\} \to \emptyset$, support = 0.22, confidence = 1
$\{x_5, x_7, x_9\} \to \emptyset$, support = 0.22, confidence = 1
$\{x_4\} \to \{x_8\}$, support = 0.33, confidence = 1
$\{x_5\} \to \{x_7\}$, support = 0.22, confidence = 1
$\{x_5\} \to \{x_9\}$, support = 0.22, confidence = 1
$\{x_6\} \to \{x_7\}$, support = 0.22, confidence = 1
$\{x_7\} \to \{x_9\}$, support = 0.22, confidence = 0.39
$\{x_8\} \to \{x_4\}$, support = 0.33, confidence = 0.75
$\{x_7\} \to \{x_5\}$, support = 0.22, confidence = 0.39
$\{x_9\} \to \{x_5\}$, support = 0.22, confidence = 0.67
$\{x_7\} \to \{x_6\}$, support = 0.22, confidence = 0.39
$\{x_9\} \to \{x_7\}$, support = 0.22, confidence = 0.67
$\{x_5\} \to \{x_7, x_9\}$, support = 0.22, confidence = 1
$\{x_7\} \to \{x_5, x_9\}$, support = 0.22, confidence = 0.39
$\{x_9\} \to \{x_5, x_7\}$, support = 0.22, confidence = 0.67
$\{x_7, x_9\} \to \{x_5\}$, support = 0.22, confidence = 1
$\{x_5, x_9\} \to \{x_7\}$, support = 0.22, confidence = 1
$\{x_5, x_7\} \to \{x_9\}$, support = 0.22, confidence = 1.

Removing each association rule in the form of $F \to \emptyset$, we obtain the final set of association rules:

$\{x_4\} \to \emptyset$, support = 0.33, confidence = 1
$\{x_5\} \to \emptyset$, support = 0.22, confidence = 1
$\{x_6\} \to \emptyset$, support = 0.22, confidence = 1
$\{x_7\} \to \emptyset$, support = 0.56, confidence = 1
$\{x_8\} \to \emptyset$, support = 0.44, confidence = 1
$\{x_9\} \to \emptyset$, support = 0.33, confidence = 1
$\{x_4, x_8\} \to \emptyset$, support = 0.33, confidence = 1
$\{x_5, x_7\} \to \emptyset$, support = 0.22, confidence = 1
$\{x_5, x_9\} \to \emptyset$, support = 0.22, confidence = 1
$\{x_6, x_7\} \to \emptyset$, support = 0.22, confidence = 1
$\{x_7, x_9\} \to \emptyset$, support = 0.22, confidence = 1
$\{x_5, x_7, x_9\} \to \emptyset$, support = 0.22, confidence = 1
$\{x_4\} \to \{x_8\}$, support = 0.33, confidence = 1
$\{x_8\} \to \{x_4\}$, support = 0.33, confidence = 0.75
$\{x_5\} \to \{x_7\}$, support = 0.22, confidence = 1
$\{x_5\} \to \{x_9\}$, support = 0.22, confidence = 1
$\{x_5\} \to \{x_7, x_9\}$, support = 0.22, confidence = 1
$\{x_5, x_9\} \to \{x_7\}$, support = 0.22, confidence = 1
$\{x_5, x_7\} \to \{x_9\}$, support = 0.22, confidence = 1
$\{x_9\} \to \{x_5\}$, support = 0.22, confidence = 0.67
$\{x_9\} \to \{x_7\}$, support = 0.22, confidence = 0.67
$\{x_9\} \to \{x_5, x_7\}$, support = 0.22, confidence = 0.67
$\{x_7, x_9\} \to \{x_5\}$, support = 0.22, confidence = 1
$\{x_6\} \to \{x_7\}$, support = 0.22, confidence = 1.

In this final set of association rules, each association rule in the form of $F \to \emptyset$ does not tell the association of two item sets but the presence of the item set F in the data set and can thus be ignored. The remaining association rules reveal the close association of x_4 with x_8, x_5 with x_7, and x_9, and x_6 with x_7, which are consistent with the production flows in Figure 12.1. However, the production flows from M1, M2, and M3 are not captured in the frequent item sets and in the final set of association rules because of the way in which the data set is sampled by considering all the single-machine faults. Since M1, M2, and M3 are at the beginning of the production flows and affected by themselves only, x_1, x_2, and x_3 appear less frequently in the data set than x_4 to x_9. For the same reason, the *confidence* value of the association rule $\{x_4\} \rightarrow \{x_8\}$ is higher than that of the association rule $\{x_8\} \rightarrow \{x_4\}$.

Association rule discovery is not applicable to numeric data. To apply association rule discovery, numeric data need to be converted into categorical data by defining ranges of data values as discussed in Section 4.3 of Chapter 4 and treating values in the same range as the same item.

12.3 Software and Applications

Association rule discovery is supported by Weka (http://www.cs.waikato.ac.nz/ml/weka/) and Statistica (www.statistica.com). Some applications of association rules can be found in Ye (2003, Chapter 2).

Exercises

12.1 Consider 16 data records in the testing data set of system fault detection in Table 3.2 as 16 sets of items by taking x_1, x_2, x_3, x_4, x_5, x_6, x_7, x_8, x_9 as nine different quality problems with the value of 1 indicating the presence of the given quality problem. Find all frequent item sets with *min-support* = 0.5.

12.2 Use the frequent item sets from Exercise 12.1 to generate all the association rules that satisfy *min-support* = 0.5 and *min-confidence* = 0.5.

12.3 Repeat Exercise 12.1 for all 25 data records from Table 12.1 and Table 3.2 as the data set.

12.4 Repeat Exercise 12.2 for all 25 data records from Table 12.1 and Table 3.2 as the data set.

12.5 To illustrate the Apriori algorithm is efficient for a sparse data set, find or create a sparse data set with each item being relatively infrequent in the data set, and apply the Apriori algorithm to the data set to produce frequent item sets with an appropriate value of *min-support*.

12.6 To illustrate the Apriori algorithm is less efficient for a dense data set, find or create a dense data set with each item being relatively frequent in the data records of the data set, and apply the Apriori algorithm to the data set to produce frequent item sets with an appropriate value of *min-support*.

13
Bayesian Network

Bayes classifier in Chapter 3 requires all the attribute variables are independent of each other. Bayesian network in this chapter allows associations among the attribute variables themselves and associations between attribute variables and target variables. Bayesian network uses associations of variables to infer information about any variable in Bayesian network. In this chapter, we first introduce the structure of a Bayesian network and the probability information of variables in a Bayesian network. Then we describe the probabilistic inference that is conducted within a Bayesian network. Finally, we introduce methods of learning the structure and probability information of a Bayesian network. A list of software packages that support Bayesian network is provided. Some applications of Bayesian network are given with references.

13.1 Structure of a Bayesian Network and Probability Distributions of Variables

In Chapter 3, a naive Bayes classifier uses Equation 3.5 (shown next) to classify the value of the target variable y based on the assumption that the attribute variables, $x_1, \ldots, x_p$, are independent of each other:

$$y_{MAP} \approx \arg\max_{y \in Y} p(y) \prod_{i=1}^{p} P(x_i|y).$$

However, in many applications, some attribute variables are associated in a certain way. For example, in the data set for a system fault detection shown in Table 3.1 and copied here in Table 13.1, x_1 is associated with x_5, x_7, and x_9. As shown in Figure 1.1, which is copied here as Figure 13.1, M5, M7, and M9 are on the production path of parts that are processed at M1. The faulty M1 causes the failed part quality after M1 for $x_1 = 1$, which in turn cause $x_5 = 1$, then $x_7 = 1$, and finally $x_9 = 1$. Although x_1 affects x_5, x_7, and x_9, we do not have x_5, x_7, and x_9 affecting x_1. Hence, the cause–effect association of x_1 with x_5, x_7, and x_9 goes in one direction only. Moreover, x_1 is not associated with other variables, x_2, x_3, x_4, x_6, and x_8.

197

TABLE 13.1
Training Data Set for System Fault Detection

Instance (Faulty Machine)	Attribute Variables									Target Variable
	Quality of Parts									
	x_1	x_2	x_3	x_4	x_5	x_6	x_7	x_8	x_9	System Fault y
1 (M1)	1	0	0	0	1	0	1	0	1	1
2 (M2)	0	1	0	1	0	0	0	1	0	1
3 (M3)	0	0	1	1	0	1	1	1	0	1
4 (M4)	0	0	0	1	0	0	0	1	0	1
5 (M5)	0	0	0	0	1	0	1	0	1	1
6 (M6)	0	0	0	0	0	1	1	0	0	1
7 (M7)	0	0	0	0	0	0	1	0	0	1
8 (M8)	0	0	0	0	0	0	0	1	0	1
9 (M9)	0	0	0	0	0	0	0	0	1	1
10 (none)	0	0	0	0	0	0	0	0	0	0

A Bayesian network contains nodes to represent variables (including both attribute variables and target variables) and directed links between nodes to represent directed associations between variables. Each variable is assumed to have a finite set of states or values. There is a directed link from a node representing the variable x_i to a node representing a variable x_j if x_i has a direct impact on x_j, i.e., x_i causes x_j, or x_i influences x_j in some way. In a directed link from x_i to x_j, x_i is a parent of x_j, and x_j is a child of x_i. No directed cycles, e.g., $x_1 \rightarrow x_2 \rightarrow x_3 \rightarrow x_1$, are allowed in a Bayesian network. Hence, the structure of a Bayesian network is a directed, acyclic graph.

Domain knowledge is usually used to determine how variables are linked. For example, the production flow of parts in Figure 13.1 can be used to determine the structure of a Bayesian network shown in Figure 13.2 that includes

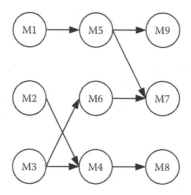

FIGURE 13.1
Manufacturing system with nine machines and production flows of parts.

Bayesian Network

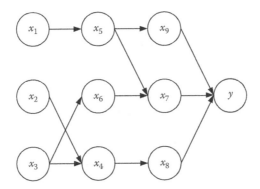

FIGURE 13.2
Structure of a Bayesian network for the data set of system fault detection.

nine attribute variables for the quality of parts at various stages of production, $x_1, x_2, x_3, x_4, x_5, x_6, x_7, x_8$, and x_9, and the target variable for the presence of a system fault, y. In Figure 13.2, x_5 has one parent x_1, x_6 has one parent x_3, x_4 has two parents x_2 and x_3, x_9 has one parent x_5, x_7 has two parents x_5 and x_6, x_8 has one parent x_4, and y has three parents x_7, x_8, and x_9. Instead of drawing a directed link from each of the nine quality variables, $x_1, x_2, x_3, x_4, x_5, x_6, x_7, x_8$, and x_9, to the system fault variable y, we have a directed link from each of three quality variables, x_7, x_8, and x_9, to the system fault variable y, because x_7, x_8, and x_9 are at the last stage of the production flow and capture the effects of x_1, x_2, x_3, x_4, x_5, and x_6 on y.

Given that the variable x has parents $z_1, ..., z_k$, a Bayesian network uses a conditional probability distribution for $P(x|z_1, ..., z_k)$ to quantify the effects of parents $z_1, ..., z_k$ on the child x. For example, we suppose that the device for inspecting the quality of parts in the data set of system fault detection is not 100% reliable, producing data uncertainties and conditional probability distributions in Tables 13.2 through 13.10 for the nodes with

TABLE 13.2

$P(x_5|x_1)$

	$x_1 = 0$	$x_1 = 1$
$x_5 = 0$	$P(x_5 = 0 \mid x_1 = 0) = 0.7$	$P(x_5 = 0 \mid x_1 = 1) = 0.1$
$x_5 = 1$	$P(x_5 = 1 \mid x_1 = 0) = 0.3$	$P(x_5 = 1 \mid x_1 = 1) = 0.9$

TABLE 13.3

$P(x_6|x_3)$

	$x_3 = 0$	$x_3 = 1$
$x_6 = 0$	$P(x_6 = 0 \mid x_3 = 0) = 0.7$	$P(x_6 = 0 \mid x_3 = 1) = 0.1$
$x_6 = 1$	$P(x_6 = 1 \mid x_3 = 0) = 0.3$	$P(x_6 = 1 \mid x_3 = 1) = 0.9$

TABLE 13.4

$P(x_4 | x_3, x_2)$

	$x_2 = 0$	
	$x_3 = 0$	$x_3 = 1$
$x_4 = 0$	$P(x_4 = 0 | x_2 = 0, x_3 = 0) = 0.7$	$P(x_4 = 0 | x_2 = 0, x_3 = 1) = 0.1$
$x_4 = 1$	$P(x_4 = 1 | x_2 = 0, x_3 = 0) = 0.3$	$P(x_4 = 1 | x_2 = 0, x_3 = 1) = 0.9$
	$x_2 = 1$	
	$x_3 = 0$	$x_3 = 1$
$x_4 = 0$	$P(x_4 = 0 | x_2 = 1, x_3 = 0) = 0.1$	$P(x_4 = 0 | x_2 = 1, x_3 = 1) = 0.1$
$x_4 = 1$	$P(x_4 = 1 | x_2 = 1, x_3 = 0) = 0.9$	$P(x_4 = 1 | x_2 = 1, x_3 = 1) = 0.9$

TABLE 13.5

$P(x_9 | x_5)$

	$x_5 = 0$	$x_5 = 1$
$x_9 = 0$	$P(x_9 = 0 | x_5 = 0) = 0.7$	$P(x_9 = 0 | x_5 = 1) = 0.1$
$x_9 = 1$	$P(x_9 = 1 | x_5 = 0) = 0.3$	$P(x_9 = 1 | x_5 = 1) = 0.9$

TABLE 13.6

$P(x_7 | x_5, x_6)$

	$x_5 = 0$	
	$x_6 = 0$	$x_6 = 1$
$x_7 = 0$	$P(x_7 = 0 | x_5 = 0, x_6 = 0) = 0.7$	$P(x_7 = 0 | x_5 = 0, x_6 = 1) = 0.1$
$x_7 = 1$	$P(x_7 = 1 | x_5 = 0, x_6 = 0) = 0.3$	$P(x_7 = 1 | x_5 = 0, x_6 = 1) = 0.9$
	$x_5 = 1$	
	$x_6 = 0$	$x_6 = 1$
$x_7 = 0$	$P(x_7 = 0 | x_5 = 1, x_6 = 0) = 0.1$	$P(x_7 = 0 | x_5 = 1, x_6 = 1) = 0.1$
$x_7 = 1$	$P(x_7 = 1 | x_5 = 1, x_6 = 0) = 0.9$	$P(x_7 = 1 | x_5 = 1, x_6 = 1) = 0.9$

TABLE 13.7

$P(x_8 | x_4)$

	$x_4 = 0$	$x_4 = 1$
$x_8 = 0$	$P(x_8 = 0 | x_4 = 0) = 0.7$	$P(x_8 = 0 | x_4 = 1) = 0.1$
$x_8 = 1$	$P(x_8 = 1 | x_4 = 0) = 0.3$	$P(x_8 = 1 | x_4 = 1) = 0.9$

TABLE 13.8
$P(y|x_9)$

	$x_9 = 0$	$x_9 = 1$
$y = 0$	$P(y=0\|x_9=0) = 0.9$	$P(y=0\|x_9=1) = 0.1$
$y = 1$	$P(y=1\|x_9=0) = 0.1$	$P(y=1\|x_9=1) = 0.9$

TABLE 13.9
$P(y|x_7)$

	$x_7 = 0$	$x_7 = 1$
$y = 0$	$P(y=0\|x_7=0) = 0.9$	$P(y=0\|x_7=1) = 0.1$
$y = 1$	$P(y=1\|x_7=0) = 0.1$	$P(y=1\|x_7=1) = 0.9$

TABLE 13.10
$P(y|x_8)$

	$x_8 = 0$	$x_8 = 1$
$y = 0$	$P(y=0\|x_8=0) = 0.9$	$P(y=0\|x_8=1) = 0.1$
$y = 1$	$P(y=1\|x_8=0) = 0.1$	$P(y=1\|x_{98}=1) = 0.9$

parent(s) in Figure 13.2. For example, in Table 13.2, $P(x_5 = 0|x_1 = 1) = 0.1$ and $P(x_5 = 1|x_1 = 1) = 0.9$ mean that if $x_1 = 1$ then the probability of $x_5 = 0$ is 0.1, the probability of $x_5 = 1$ is 0.9, and the probability of having either value (0 or 1) of x_5 is 0.1 + 0.9 = 1. The reason for not having the probability of 1 for $x_5 = 1$ if $x_1 = 1$ is that the inspection device for x_1 has a small probability of failure. Although the inspection devices tell $x_1 = 1$, there is a small probability that x_1 should be 0. In addition, the inspection device for x_5 also has a small probability of failure, meaning that the inspection device may tell $x_5 = 0$ although x_5 should be 1. The probabilities of failure in the inspection devices produce data uncertainties and thus the conditional probabilities in Tables 13.2 through 13.10.

For the node of a variable x in a Bayesian network that has no parents, the prior probability distribution of x is needed. For example, in the Bayesian network in Figure 13.2, x_1, x_2, and x_3 have no parents, and their prior probability distributions are given in Tables 13.11 through 13.13, respectively.

The prior probability distributions of nodes without parent(s) and the conditional probability distributions of nodes with parent(s) allow computing the joint probability distribution of all the variables in a Bayesian network.

TABLE 13.11
$P(x_1)$

$x_1 = 0$	$x_1 = 1$
$P(x_1 = 0) = 0.8$	$P(x_1 = 1) = 0.2$

TABLE 13.12
$P(x_2)$

$x_2 = 0$	$x_2 = 1$
$P(x_2 = 0) = 0.8$	$P(x_2 = 1) = 0.2$

TABLE 13.13
$P(x_3)$

$x_3 = 0$	$x_3 = 1$
$P(x_3 = 0) = 0.8$	$P(x_3 = 1) = 0.2$

For example, the joint probability distribution of the 10 variables in the Bayesian network in Figure 13.2 is computed next:

$P(x_1, x_2, x_3, x_4, x_5, x_6, x_7, x_8, x_9, y)$

$= P(y|x_1, x_2, x_3, x_4, x_5, x_6, x_7, x_8, x_9)P(x_1, x_2, x_3, x_4, x_5, x_6, x_7, x_8, x_9)$

$= P(y|x_7, x_8, x_9)P(x_1, x_2, x_3, x_4, x_5, x_6, x_7, x_8, x_9)$

$= P(y|x_7, x_8, x_9)P(x_9|x_1, x_2, x_3, x_4, x_5, x_6, x_7, x_8)P(x_1, x_2, x_3, x_4, x_5, x_6, x_7, x_8)$

$= P(y|x_7, x_8, x_9)P(x_9|x_5)P(x_1, x_2, x_3, x_4, x_5, x_6, x_7, x_8)$

$= P(y|x_7, x_8, x_9)P(x_9|x_5)P(x_7|x_1, x_2, x_3, x_4, x_5, x_6, x_8)P(x_1, x_2, x_3, x_4, x_5, x_6, x_8)$

$= P(y|x_7, x_8, x_9)P(x_9|x_5)P(x_7|x_5, x_6)P(x_1, x_2, x_3, x_4, x_5, x_6, x_8) = \cdots$

$= P(y|x_7, x_8, x_9)P(x_9|x_5)P(x_7|x_5, x_6)P(x_8|x_4)P(x_5|x_1)P(x_6|x_3)P(x_4|x_2, x_3)P(x_1, x_2, x_3)$

$= P(y|x_7, x_8, x_9)P(x_9|x_5)P(x_7|x_5, x_6)P(x_8|x_4)P(x_5|x_1)P(x_6|x_3)P(x_4|x_2, x_3)P(x_1)P(x_2)P(x_3).$

In the aforementioned computation, we use the following equations:

$$P(x_1, \ldots, x_i | z_1, \ldots, z_k, v_1, \ldots, v_j) = P(x_1, \ldots, x_i | z_1, \ldots, z_k) \qquad (13.1)$$

$$P(x_1,\ldots,x_i) = \prod_{j=1}^{i} P(x_i), \qquad (13.2)$$

where in Equation 13.1 we have $x_1, \ldots, x_i$ conditionally independent of $v_1, \ldots, v_j$ given $z_1, \ldots, z_k$, and in Equation 13.2 we have $x_1, \ldots, x_i$ independent of each other.

Therefore, the conditional independences and independences among certain variables allow us to express the joint probability distribution of all the variables using the conditional probability distributions of nodes with parent(s) and the prior probability distributions of nodes without parent(s). In other words, a Bayesian network gives a decomposed, simplified representation of the joint probability distribution.

The joint probability distribution of all the variables gives the complete description of all the variables and allows us to answer any questions about all the variables. For example, given the joint probability distribution of two variables x and z, $P(x, z)$, and x takes one of values $a_1, \ldots, a_i$, and z takes one of values $b_1, \ldots, b_j$, we can compute the probabilities for any questions about these two variables:

$$P(x) = \sum_{k=1}^{j} P(x, z = b_k) \qquad (13.3)$$

$$P(z) = \sum_{k=1}^{i} P(x = a_k, z) \qquad (13.4)$$

$$P(x|z) = \frac{P(x,z)}{P(z)} \qquad (13.5)$$

$$P(z|x) = \frac{P(x,z)}{P(x)}. \qquad (13.6)$$

In Equation 13.3, we marginalize z out of $P(x, z)$ to obtain $P(x)$. In Equation 13.4, we marginalize x out of $P(x, z)$ to obtain $P(z)$.

Example 13.1

Given the following joint probability distribution $P(x, z)$:

$$P(x=0, z=0) = 0.2$$

$$P(x=0, z=1) = 0.4$$

$$P(x=1, z=0) = 0.3$$

$$P(x=1, z=1) = 0.1,$$

which sum up to 1, compute $P(x)$, $P(z)$, $P(x|z)$, and $P(x|z)$:

$$P(x=0) = P(x=0, z=0) + P(x=0, z=1) = 0.2 + 0.4 = 0.6$$

$$P(x=1) = P(x=1, z=0) + P(x=1, z=1) = 0.3 + 0.1 = 0.4$$

$$P(z=0) = P(x=0, z=0) + P(x=1, z=0) = 0.2 + 0.3 = 0.5$$

$$P(z=1) = P(x=0, z=1) + P(x=1, z=1) = 0.4 + 0.1 = 0.5$$

$$P(x=0|z=0) = \frac{P(x=0, z=0)}{P(z=0)} = \frac{0.2}{0.5} = 0.4$$

$$P(x=1|z=0) = \frac{P(x=1, z=0)}{P(z=0)} = \frac{0.3}{0.5} = 0.6$$

$$P(x=0|z=1) = \frac{P(x=0, z=1)}{P(z=1)} = \frac{0.4}{0.5} = 0.8$$

$$P(x=1|z=1) = \frac{P(x=1, z=1)}{P(z=1)} = \frac{0.1}{0.5} = 0.2$$

$$P(z=0|x=0) = \frac{P(x=0, z=0)}{P(x=0)} = \frac{0.2}{0.6} = 0.33$$

$$P(z=1|x=0) = \frac{P(x=0, z=1)}{P(x=0)} = \frac{0.4}{0.6} = 0.67$$

$$P(z=0|x=1) = \frac{P(x=1, z=0)}{P(x=1)} = \frac{0.3}{0.4} = 0.75$$

$$P(z=1|x=1) = \frac{P(x=1, z=1)}{P(x=1)} = \frac{0.1}{0.4} = 0.25.$$

13.2 Probabilistic Inference

The probability distributions captured in a Bayesian network represent our prior knowledge about the domain of all the variables. After obtaining evidences for specific values of some variables (evidence variables), we want to use the probabilistic inference for determining posterior probability distributions of certain variables of interest (query variables). That is, we want to see how probabilities of values for the query variables change after knowing specific values of evidence variables. For example, in the Bayesian network in Figure 13.2, we want to know what the probability of $y = 1$ is and what the probability of x_7 is if we have the evidence to confirm $x_9 = 1$. In some applications, evidence variables are variables that can be easily observed, and query variables are variables that are not observable. We give some examples of probability inference next.

Example 13.2

Consider the Bayesian network in Figure 13.2 and the probability distributions in Tables 13.2 through 13.13. Given $x_6 = 1$, what are the probabilities of $x_4 = 1$, $x_3 = 1$, and $x_2 = 1$? In other words, what are $P(x_4 = 1 | x_6 = 1)$, $P(x_3 = 1 | x_6 = 1)$, and $P(x_2 = 1 | x_6 = 1)$? Note that the given condition $x_6 = 1$ does not imply $P(x_6 = 1) = 1$.

To get $P(x_3 = 1 | x_6 = 1)$, we need to obtain $P(x_3, x_6)$.

$$P(x_6, x_3) = P(x_6 | x_3) P(x_3)$$

	$x_3 = 0$	$x_3 = 1$
$x_6 = 0$	$P(x_6 = 0, x_3 = 0) = P(x_6 = 0 | x_3 = 0)$ $P(x_3 = 0) = (0.7)(0.8) = 0.56$	$P(x_6 = 0, x_3 = 1) = P(x_6 = 0 | x_3 = 1)$ $P(x_3 = 1) = (0.1)(0.2) = 0.02$
$x_6 = 1$	$P(x_6 = 1, x_3 = 0) = P(x_6 = 1 | x_3 = 0)$ $P(x_3 = 0) = (0.3)(0.8) = 0.24$	$P(x_6 = 1, x_3 = 1) = P(x_6 = 1 | x_3 = 1)$ $P(x_3 = 1) = (0.9)(0.2) = 0.18$

By marginalizing x_3 out of $P(x_6, x_3)$, we obtain $P(x_6)$:

$$P(x_6 = 0) = P(x_6 = 0, x_3 = 0) + P(x_6 = 0, x_3 = 1) = 0.56 + 0.02 = 0.58$$

$$P(x_6 = 1) = P(x_6 = 1, x_3 = 0) + P(x_6 = 1, x_3 = 1) = 0.24 + 0.18 = 0.42.$$

$$P(x_3 = 1 | x_6 = 1) = \frac{P(x_6 = 1 | x_3 = 1) P(x_3 = 1)}{P(x_6 = 1)} = \frac{(0.9)(0.2)}{0.42} = 0.429$$

Hence, the evidence $x_6 = 1$ changes the probability of $x_3 = 1$ from 0.2 to 0.429.

To obtain $P(x_4=1|x_6=1)$, we need to get $P(x_4, x_6)$. x_4 and x_6 are associated through x_3. Moreover, the association of x_4 and x_3 involves x_2. Hence, we want to marginalize x_3 and x_2 out of $P(x_4, x_3, x_2|x_6 = 1)$ where

$$P(x_4, x_3, x_2|x_6 = 1) = P(x_4|x_3, x_2)P(x_3|x_6 = 1)P(x_2)$$

$$= P(x_4|x_3, x_2)\frac{P(x_6 = 1|x_3)P(x_3)}{P(x_6 = 1)}P(x_2).$$

Although $P(x_4|x_3, x_2)$, $P(x_6|x_3)$, $P(x_3)$, and $P(x_2)$ are given in Tables 13.3, 13.4, 13.12, and 13.13, respectively, $P(x_6)$ needs to be computed. In addition to computing $P(x_6)$, we also compute $P(x_4)$ so we can compare $P(x_4 = 1|x_6 = 1)$ with $P(x_4)$.

To obtain $P(x_4)$ and $P(x_6)$, we first compute the joint probabilities $P(x_4, x_3, x_2)$ and $P(x_6, x_3)$ and then marginalize x_3 and x_2 out of $P(x_4, x_3, x_2)$ and x_3 out of $P(x_6, x_3)$ as follows:

$$P(x_4, x_3, x_2) = P(x_4|x_3, x_2)P(x_3)P(x_2)$$

		$x_2 = 0$	
		$x_3 = 0$	$x_3 = 1$
$x_4 = 0$		$P(x_4 = 0, x_3 = 0, x_2 = 0) = P(x_4 = 0|x_3 = 0, x_2 = 0)$ $P(x_3 = 0)P(x_2 = 0) = (0.7)(0.8)(0.8) = 0.448$	$P(x_4 = 0, x_3 = 1, x_2 = 0) = P(x_4 = 0|x_3 = 1, x_2 = 0)$ $P(x_3 = 1)P(x_2 = 0) = (0.1)(0.2)(0.8) = 0.016$
$x_4 = 1$		$P(x_4 = 1, x_3 = 0, x_2 = 0) = P(x_4 = 1|x_3 = 0, x_2 = 0)$ $P(x_3 = 0)P(x_2 = 0) = (0.3)(0.8)(0.8) = 0.192$	$P(x_4 = 1, x_3 = 1, x_2 = 0) = P(x_4 = 1|x_3 = 1, x_2 = 0)$ $P(x_3 = 1)P(x_2 = 0) = (0.9)(0.2)(0.8) = 0.144$
		$x_2 = 1$	
		$x_3 = 0$	$x_3 = 1$
$x_4 = 0$		$P(x_4 = 0, x_3 = 0, x_2 = 1) = P(x_4 = 0|x_3 = 0, x_2 = 1)$ $P(x_3 = 0)P(x_2 = 1) = (0.1)(0.8)(0.2) = 0.016$	$P(x_4 = 0, x_3 = 1, x_2 = 1) = P(x_4 = 0|x_3 = 1, x_2 = 1)$ $P(x_3 = 1)P(x_2 = 1) = (0.1)(0.2)(0.2) = 0.004$
$x_4 = 1$		$P(x_4 = 1, x_3 = 0, x_2 = 1) = P(x_4 = 1|x_3 = 0, x_2 = 1)$ $P(x_3 = 0)P(x_2 = 1) = (0.9)(0.8)(0.2) = 0.144$	$P(x_4 = 1, x_3 = 1, x_2 = 1) = P(x_4 = 1|x_3 = 1, x_2 = 1)$ $P(x_3 = 1)P(x_2 = 1) = (0.9)(0.2)(0.2) = 0.036$

By marginalizing x_3 and x_2 out of $P(x_4, x_3, x_2)$, we obtain $P(x_4)$:

$$P(x_4 = 0) = P(x_4 = 0, x_3 = 0, x_2 = 0) + P(x_4 = 0, x_3 = 1, x_2 = 0)$$
$$+ P(x_4 = 0, x_3 = 0, x_2 = 1) + P(x_4 = 0, x_3 = 1, x_2 = 1)$$
$$= 0.448 + 0.016 + 0.016 + 0.004 = 0.484$$

Bayesian Network

$$P(x_4 = 1) = P(x_4 = 1, x_3 = 0, x_2 = 0) + P(x_4 = 1, x_3 = 1, x_2 = 0)$$
$$+ P(x_4 = 1, x_3 = 0, x_2 = 1) + P(x_4 = 1, x_3 = 1, x_2 = 1)$$
$$= 0.192 + 0.144 + 0.144 + 0.036 = 0.516.$$

Now we use $P(x_6)$ to compute $P(x_4, x_3, x_2 | x_6 = 1)$:

$$P(x_4, x_3, x_2 | x_6 = 1) = P(x_4 | x_3, x_2) P(x_3 | x_6 = 1) P(x_2)$$
$$= P(x_4 | x_3, x_2) \frac{P(x_6 = 1 | x_3) P(x_3)}{P(x_6 = 1)} P(x_2):$$

	$x_2 = 0$	
	$x_3 = 0$	$x_3 = 1$
$x_4 = 0$	$P(x_4 = 0 \mid x_3 = 0, x_2 = 0)$	$P(x_4 = 0 \mid x_3 = 1, x_2 = 0)$
	$\dfrac{P(x_6 = 1\mid x_3 = 0) P(x_3 = 0)}{P(x_6 = 1)}$	$\dfrac{P(x_6 = 1\mid x_3 = 1) P(x_3 = 1)}{P(x_6 = 1)}$
	$P(x_2 = 0)$	$P(x_2 = 0)$
	$= (0.7) \dfrac{(0.3)(0.8)}{0.42} (0.8)$	$= (0.1) \dfrac{(0.9)(0.2)}{0.42} (0.8)$
	$= 0.32$	$= 0.034$
$x_4 = 1$	$P(x_4 = 1 \mid x_3 = 0, x_2 = 0)$	$P(x_4 = 1 \mid x_3 = 1, x_2 = 0)$
	$\dfrac{P(x_6 = 1\mid x_3 = 0) P(x_3 = 0)}{P(x_6 = 1)}$	$\dfrac{P(x_6 = 1\mid x_3 = 1) P(x_3 = 1)}{P(x_6 = 1)}$
	$P(x_2 = 0)$	$P(x_2 = 0)$
	$= (0.3) \dfrac{(0.3)(0.8)}{0.42} (0.8)$	$= (0.9) \dfrac{(0.9)(0.2)}{0.42} (0.8)$
	$= 0.137$	$= 0.309$
	$x_2 = 1$	
	$x_3 = 0$	$x_3 = 1$
$x_4 = 0$	$P(x_4 = 0 \mid x_3 = 0, x_2 = 1)$	$(x_4 = 0 \mid x_3 = 1, x_2 = 1)$
	$\dfrac{P(x_6 = 1\mid x_3 = 0) P(x_3 = 0)}{P(x_6 = 1)}$	$\dfrac{P(x_6 = 1\mid x_3 = 1) P(x_3 = 1)}{P(x_6 = 1)}$
	$P(x_2 = 1)$	$P(x_2 = 1)$
	$= (0.1) \dfrac{(0.3)(0.8)}{0.42} (0.2)$	$= (0.1) \dfrac{(0.9)(0.2)}{0.42} (0.2)$
	$= 0.011$	$= 0.009$

(continued)

$x_4 = 1$	$P(x_4 = 1 \mid x_3 = 0, x_2 = 1)$	$(x_4 = 1 \mid x_3 = 1, x_2 = 1)$
	$\dfrac{P(x_6 = 1 \mid x_3 = 0)P(x_3 = 0)}{P(x_6 = 1)}$	$\dfrac{P(x_6 = 1 \mid x_3 = 1)P(x_3 = 1)}{P(x_6 = 1)}$
	$P(x_2 = 1)$	$P(x_2 = 1)$
	$= (0.9)\dfrac{(0.3)(0.8)}{0.42}(0.2)$	$= (0.9)\dfrac{(0.9)(0.2)}{0.42}(0.2)$
	$= 0.103$	$= 0.077$

We obtain $P(x_4 = 1 \mid x_6 = 1)$ by marginalizing x_3 and x_2 out of $P(x_4, x_3, x_2 \mid x_6 = 1)$:

$$P(x_4 = 1 \mid x_6 = 1) = P(x_4 = 1, x_3 = 0, x_2 = 0 \mid x_6 = 1) + P(x_4 = 1, x_3 = 1, x_2 = 0 \mid x_6 = 1)$$
$$+ P(x_4 = 1, x_3 = 0, x_2 = 1 \mid x_6 = 1) + P(x_4 = 1, x_3 = 1, x_2 = 1 \mid x_6 = 1)$$
$$= 0.137 + 0.309 + 0.103 + 0.077 = 0.626.$$

In comparison with $P(x_4 = 1) = 0.516$ that we computed earlier on, the evidence $x_6 = 1$ changes the probability of $x_4 = 1$ to 0.626.

We obtain $P(x_2 = 1 \mid x_6 = 1)$ by marginalizing x_4 and x_3 out of $P(x_4, x_3, x_2 \mid x_6 = 1)$:

$$P(x_2 = 1 \mid x_6 = 1) = P(x_4 = 0, x_3 = 0, x_2 = 1 \mid x_6 = 1) + P(x_4 = 1, x_3 = 0, x_2 = 1 \mid x_6 = 1)$$
$$+ P(x_4 = 0, x_3 = 1, x_2 = 1 \mid x_6 = 1) + P(x_4 = 1, x_3 = 1, x_2 = 1 \mid x_6 = 1)$$
$$= 0.011 + 0.103 + 0.009 + 0.077 = 0.2.$$

The evidence on $x_6 = 1$ does not change the probability of $x_2 = 1$ from its prior probability of 0.2 because x_6 is affected by x_3 only. The evidence on $x_6 = 1$ brings the need to update the posterior probability of x_3, which in turn brings the need to update the posterior probability of x_4 since x_3 affects x_4.

Generally, we conduct the probability inference about a query variable by first obtaining the joint probability distribution that contains the query variable and then marginalizing nonquery variables out of the joint probability distribution to obtain the probability of the query variable. Regardless of what new evidence about a specific value of a variable is obtained, the conditional probability distribution for each node with parent(s), P(child|parent(s)), which is given in a Bayesian network, does not change. However, all other probabilities, including the conditional probabilities P(parent|child) and the probabilities of other variables than the evidence variable, may change, depending on whether or not those

Bayesian Network

probabilities are affected by the evidence variable. All the probabilities that are affected by the evidence variable need to be updated, and the updated probabilities should be used for the probabilistic inference when a new evidence is obtained. For example, if we continue from Example 13.2 and obtain a new evidence of $x_4 = 1$ after updating the probabilities for the evidence of $x_6 = 1$ in Example 13.2, all the updated probabilities from Example 13.2 should be used to conduct the probabilistic inference for the new evidence of $x_4 = 1$, for example the probabilistic inference to determine $P(x_3 = 1 | x_4 = 1)$ and $P(x_2 = 1 | x_4 = 1)$.

Example 13.3

Continuing with all the updated posterior probabilities for the evidence of $x_6 = 1$ from Example 13.2, we now obtain a new evidence of $x_4 = 1$. What are the posterior probabilities of $x_2 = 1$ and $x_3 = 1$? In other words, starting with all the updated probabilities from Example 13.2, what are $P(x_3 = 1 | x_4 = 1)$ and $P(x_2 = 1 | x_4 = 1)$?

The probabilistic inference is presented next:

$$P(x_3, x_2 | x_4 = 1) = \frac{P(x_4 = 1 | x_3, x_2) P(x_3 | x_6 = 1) P(x_2 | x_6 = 1)}{P(x_4 = 1 | x_6 = 1)}$$

$$P(x_3 = 0, x_2 = 0 | x_4 = 1) = \frac{P(x_4 = 1 | x_3 = 0, x_2 = 0) P(x_3 = 0 | x_6 = 1) P(x_2 = 0 | x_6 = 1)}{P(x_4 = 1 | x_6 = 1)}$$

$$= \frac{(0.3)(1 - 0.429)(1 - 0.2)}{(0.626)} = 0.219$$

$$P(x_3 = 0, x_2 = 1 | x_4 = 1) = \frac{P(x_4 = 1 | x_3 = 0, x_2 = 1) P(x_3 = 0 | x_6 = 1) P(x_2 = 1 | x_6 = 1)}{P(x_4 = 1 | x_6 = 1)}$$

$$= \frac{(0.9)(1 - 0.429)(0.2)}{(0.626)} = 0.164$$

$$P(x_3 = 1, x_2 = 0 | x_4 = 1) = \frac{P(x_4 = 1 | x_3 = 1, x_2 = 0) P(x_3 = 1 | x_6 = 1) P(x_2 = 0 | x_6 = 1)}{P(x_4 = 1 | x_6 = 1)}$$

$$= \frac{(0.9)(0.429)(1 - 0.2)}{(0.626)} = 0.494$$

$$P(x_3 = 1, x_2 = 1 | x_4 = 1) = \frac{P(x_4 = 1 | x_3 = 1, x_2 = 1) P(x_3 = 1 | x_6 = 1) P(x_2 = 1 | x_6 = 1)}{P(x_4 = 1 | x_6 = 1)}$$

$$= \frac{(0.9)(0.429)(0.2)}{(0.626)} = 0.123$$

We obtain $P(x_3 = 1|x_4 = 1)$ by marginalizing x_2 out of $P(x_3, x_2|x_4 = 1)$:

$$P(x_3 = 1|x_4 = 1) = P(x_3 = 1, x_2 = 0|x_4 = 1) + P(x_3 = 1, x_2 = 1|x_4 = 1)$$

$$= 0.494 + 0.123 = 0.617$$

Since x_3 affects both x_6 and x_4, we raise the probability of $x_3 = 1$ from 0.2 to 0.429 when we have the evidence of $x_6 = 1$, and then we raise the probability of $x_3 = 1$ again from 0.429 to 0.617 when we have the evidence of $x_4 = 1$.

We obtain $P(x_2 = 1|x_4 = 1)$ by marginalizing x_3 out of $P(x_3, x_2|x_4 = 1)$:

$$P(x_2 = 1|x_4 = 1) = P(x_3 = 0, x_2 = 1|x_4 = 1) + P(x_3 = 1, x_2 = 1|x_4 = 1)$$

$$= 0.164 + 0.123 = 0.287.$$

Since x_2 affects x_4 but not x_6, the probability of $x_2 = 1$ remains the same at 0.2 when we have the evidence on $x_6 = 1$, and then we raise the probability of $x_2 = 1$ from 0.2 to 0.287 when we have the evidence on $x_4 = 1$. It is not a big increase since $x_3 = 1$ may also produce the evidence on $x_4 = 1$.

Algorithms that are used to make the probability inference need to search for a path from the evidence variable to the query variable and to update and infer the probabilities along the path, as we did manually in Examples 13.2 and 13.3. The search and the probabilistic inference require large amounts of computation, as seen from Examples 13.2 and 13.3. Hence, it is crucial to develop computational efficient algorithms for conducting the probabilistic inference in a Bayesian network, for example those in HUGIN (www.hugin.com), which is a software package for Bayesian network.

13.3 Learning of a Bayesian Network

Learning the structure of a Bayesian network and conditional probabilities and prior probabilities in a Bayesian network from training data is a topic under extensive research. In general, we would like to construct the structure of a Bayesian network based on our domain knowledge. However, when we do not have adequate knowledge about the domain but only data of some observable variables in the domain, we need to uncover associations between variables using data mining techniques

Bayesian Network

such as association rules in Chapter 12 and statistical techniques such as tests on the independence of variables.

When all the variables in a Bayesian network are observable to obtain data records of the variables, the conditional probability tables of nodes with parent(s) and the prior probabilities of nodes without parent(s) can be estimated using the following formulas as those in Equations 3.6 and 3.7:

$$P(x=a) = \frac{N_{x=a}}{N} \tag{13.7}$$

$$P(x=a|z=b) = \frac{N_{x=a \& z=b}}{N_{z=b}}, \tag{13.8}$$

where
  N is the number of data points in the data set
  $N_{x=a}$ is the number of data points with $x = a$
  $N_{z=b}$ is the number of data points with $z = b$
  $N_{x=a \& z=b}$ is the number of data points with $x = a$ and $z = b$

Russell et al. (1995) developed the gradient ascent method, which is similar to the gradient decent method for artificial neural network, to learn an entry in a conditional probability table in a Bayesian network when the entry cannot be learned from the training data. Let $w_{ij} = P(x_i|z_j)$ be such an entry in a conditional probability table for the node x taking its ith value with parent(s) z taking the jth value in a Bayesian network. Let h denote a hypothesis about the value of w_{ij}. Given the training data set, we want to find the maximum likelihood hypothesis h that maximizes $P(D|h)$.

$$h = \arg\max_h P(D|h) = \arg\max_h \ln P(D|h).$$

The following gradient ascent is performed to update w_{ij}:

$$w_{ij}(t+1) = w_{ij}(t+1) + \alpha \frac{\partial \ln P(D|h)}{\partial w_{ij}}, \tag{13.9}$$

where α is the learning rate. Denoting $P(D|h)$ by $P_h(D)$ and using $\partial \ln f(x)/\partial x = [1/f(x)][\partial f(x)/\partial x]$, we have

$$\frac{\partial \ln P(D|h)}{\partial w_{ij}} = \frac{\partial \ln P_h(D)}{\partial w_{ij}} = \frac{\partial \ln \prod_{d \in D} P_h(d)}{\partial w_{ij}}$$

$$= \sum_{d \in D} \frac{1}{P_h(d)} \frac{\partial P_h(d)}{\partial w_{ij}} = \sum_{d \in D} \frac{1}{P_h(d)} \frac{\partial \sum_{i',j'} P_h(d|x_{i'}, z_{j'}) P_h(x_{i'}, z_{j'})}{\partial w_{ij}}$$

$$= \sum_{d \in D} \frac{1}{P_h(d)} \frac{\partial \sum_{i',j'} P_h(d|x_{i'}, z_{j'}) P_h(x_{i'}|z_{j'}) P_h(z_{j'})}{\partial w_{ij}}$$

$$= \sum_{d \in D} \frac{1}{P_h(d)} \frac{\partial \sum_{i',j'} P_h(d|x_{i'}, z_{j'}) w_{i'j'} P_h(z_{j'})}{\partial w_{ij}}$$

$$= \sum_{d \in D} \frac{1}{P_h(d)} P_h(d|x_i, z_j) P_h(z_j) = \sum_{d \in D} \frac{1}{P_h(d)} \frac{P_h(x_i, z_j|d) P_h(d)}{P_h(x_i, z_j)} P_h(z_j)$$

$$= \sum_{d \in D} \frac{P_h(x_i, z_j|d)}{P_h(x_i, z_j)} P_h(z_j) = \sum_{d \in D} \frac{P_h(x_i, z_j|d)}{P_h(x_i|z_j)} = \sum_{d \in D} \frac{P_h(x_i, z_j|d)}{w_{ij}}.$$

(13.10)

Plugging Equation 13.10 into 13.9, we obtain:

$$w_{ij}(t+1) = w_{ij}(t+1) + \alpha \frac{\partial \ln P(D|h)}{\partial w_{ij}} = w_{ij}(t+1) + \alpha \sum_{d \in D} \frac{P_h(x_i, z_j|d)}{w_{ij}(t)}, \quad (13.11)$$

where $P_h(x_i, z_j|d)$ can be obtained using the probabilistic inference described in Section 13.2. After using Equation 13.11 to update w_{ij}, we need to ensure

$$\sum_i w_{ij}(t+1) = 1 \quad (13.12)$$

by performing the normalization

$$w_{ij}(t+1) = \frac{w_{ij}(t+1)}{\sum_i w_{ij}(t+1)}. \quad (13.13)$$

13.4 Software and Applications

Bayes server (www.bayesserver.com) and HUGIN (www.hugin.com) are two software packages that support Bayesian network. Some applications of Bayesian network in bioinformatics and some other fields can be found in Davis (2003), Diez et al. (1997), Jiang and Cooper (2010), Pourret et al. (2008).

Exercises

13.1 Consider the Bayesian network in Figure 13.2 and the probability distributions in Tables 13.2 through 13.13. Given $x_6 = 1$, what is the probability of $x_7 = 1$? In other words, what is $P(x_1 = 1 | x_6 = 1)$?

13.2 Continuing with all the updated posterior probabilities for the evidence of $x_6 = 1$ from Example 13.2 and Exercise 13.1, we now obtain a new evidence of $x_4 = 1$. What is the posterior probability of $x_7 = 1$? In other words, what is $P(x_1 = 1 | x_4 = 1)$?

13.3 Repeat Exercise 13.1 to determine $P(x_1 = 1 | x_6 = 1)$.

13.4 Repeat Exercise 13.2 to determine $P(x_1 = 1 | x_4 = 1)$.

13.5 Repeat Exercise 13.1 to determine $P(y = 1 | x_6 = 1)$.

13.6 Repeat Exercise 13.2 to determine $P(y = 1 | x_4 = 1)$.

Part IV

Algorithms for Mining Data Reduction Patterns

14

Principal Component Analysis

Principal component analysis (PCA) is a statistical technique of representing high-dimensional data in a low-dimensional space. PCA is usually used to reduce the dimensionality of data so that the data can be further visualized or analyzed in a low-dimensional space. For example, we may use PCA to represent data records with 100 attribute variables by data records with only 2 or 3 variables. In this chapter, a review of multivariate statistics and matrix algebra is first given to lay the mathematical foundation of PCA. Then, PCA is described and illustrated. A list of software packages that support PCA is provided. Some applications of PCA are given with references.

14.1 Review of Multivariate Statistics

If x_i is a continuous random variable with continuous values and probability density function $f_i(x_i)$, the mean and variance of the random variable, u_i and σ_i^2, are defined as follows:

$$u_i = E(x_i) = \int_{-\infty}^{\infty} x_i f_i(x_i) dx_i \qquad (14.1)$$

$$\sigma_i^2 = \int_{-\infty}^{\infty} (x_i - u_i)^2 f_i(x_i) dx_i. \qquad (14.2)$$

If x_i is a discrete random variable with discrete values and probability function $P(x_i)$,

$$u_i = E(x_i) = \sum_{\text{all values of } x_i} x_i P(x_i) \qquad (14.3)$$

$$\sigma_i^2 = \sum_{\text{all values of } x_i} (x_i - u_i)^2 P(x_i). \qquad (14.4)$$

If x_i and x_j are continuous random variables with the joint probability density function $f_{ij}(x_i, x_j)$, the covariance of two random variables, x_i and x_j, is defined as follows:

$$\sigma_{ij} = E(x_i - \mu_i)(x_j - \mu_j) = \int_{-\infty}^{\infty}\int_{-\infty}^{\infty} (x_i - u_i)(x_j - u_j) f_{ij}(x_i, x_j) dx_i dx_j. \qquad (14.5)$$

If x_i and x_j are discrete random variables with the joint probability density function $P(x_i, x_j)$,

$$\sigma_{ij} = E(x_i - \mu_i)(x_j - \mu_j) = \sum_{\text{all values of } x_i} \sum_{\text{all values of } x_j} (x_i - \mu_i)(x_j - \mu_j) P(x_i, x_j). \quad (14.6)$$

The correlation coefficient is

$$\rho_{ij} = \frac{\sigma_{ij}}{\sqrt{\sigma_i}\sqrt{\sigma_j}}. \quad (14.7)$$

For a vector of random variables, $x = (x_1, x_2, \ldots, x_p)$, the mean vector is:

$$E(x) = \begin{bmatrix} E(x_1) \\ E(x_2) \\ \vdots \\ E(x_p) \end{bmatrix} = \begin{bmatrix} \mu_1 \\ \mu_2 \\ \vdots \\ \mu_p \end{bmatrix} = \mu, \quad (14.8)$$

and the variance–covariance matrix is

$$\Sigma = E(x-\mu)(x-\mu)' = E\left(\begin{bmatrix} x_1-\mu_1 \\ x_2-\mu_2 \\ \vdots \\ x_p-\mu_p \end{bmatrix} \begin{bmatrix} x_1-\mu_1 & x_2-\mu_2 & \cdots & x_p-\mu_p \end{bmatrix}\right)$$

$$= E\begin{pmatrix} (x_1-\mu_1)^2 & (x_1-\mu_1)(x_2-\mu_2) & \cdots & (x_1-\mu_1)(x_p-\mu_p) \\ (x_2-\mu_2)(x_1-\mu_1) & (x_2-\mu_2)^2 & \cdots & (x_1-\mu_1)(x_2-\mu_2) \\ \vdots & \vdots & \ddots & \vdots \\ (x_p-\mu_p)(x_1-\mu_1) & (x_p-\mu_p)(x_2-\mu_2) & \cdots & (x_p-\mu_p)^2 \end{pmatrix}$$

$$= \begin{pmatrix} E(x_1-\mu_1)^2 & E(x_1-\mu_1)(x_2-\mu_2) & \cdots & E(x_1-\mu_1)(x_p-\mu_p) \\ E(x_2-\mu_2)(x_1-\mu_1) & E(x_2-\mu_2)^2 & \cdots & E(x_2-\mu_2)(x_p-\mu_p) \\ \vdots & \vdots & \ddots & \vdots \\ E(x_p-\mu_p)(x_1-\mu_1) & E(x_p-\mu_p)(x_2-\mu_2) & \cdots & E(x_p-\mu_p)^2 \end{pmatrix}$$

$$= \begin{bmatrix} \sigma_1 & \sigma_{12} & \cdots & \sigma_{1p} \\ \sigma_{21} & \sigma_2 & \cdots & \sigma_{2p} \\ \vdots & \vdots & \ddots & \vdots \\ \sigma_{p1} & \sigma_{p2} & \cdots & \sigma_p \end{bmatrix}. \quad (14.9)$$

Principal Component Analysis

Example 14.1

Compute the mean vector and variance–covariance matrix of two variables in Table 14.1.

The data set in Table 14.1 is a part of the data set for the manufacturing system in Table 1.4 and includes two attribute variables, x_7 and x_8, for nine cases of single-machine faults. Table 14.2 shows the joint and marginal probabilities of these two variables.

The mean and variance of x_7 are

$$u_7 = E(x_7) = \sum_{\substack{\text{all values} \\ \text{of } x_7}} x_7 P(x_7) = 0 \times \frac{4}{9} + 1 \times \frac{5}{9} = \frac{5}{9}$$

$$\sigma_7^2 = \sum_{\substack{\text{all values} \\ \text{of } x_7}} (x_7 - u_7)^2 P(x_7) = \left(0 - \frac{5}{9}\right)^2 \times \frac{4}{9} + \left(1 - \frac{5}{9}\right)^2 \times \frac{5}{9} = 0.2469.$$

TABLE 14.1

Data Set for System Fault Detection with Two Quality Variables

Instance (Faulty Machine)	x_7	x_8
1 (M1)	1	0
2 (M2)	0	1
3 (M3)	1	1
4 (M4)	0	1
5 (M5)	1	0
6 (M6)	1	0
7 (M7)	1	0
8 (M8)	0	1
9 (M9)	0	0

TABLE 14.2

Joint and Marginal Probabilities of Two Quality Variables

$P(x_7, x_8)$		x_8	$P(x_7)$
x_7	0	1	
0	$\frac{1}{9}$	$\frac{3}{9}$	$\frac{1}{9} + \frac{3}{9} = \frac{4}{9}$
1	$\frac{4}{9}$	$\frac{1}{9}$	$\frac{4}{9} + \frac{1}{9} = \frac{5}{9}$
$P(x_8)$	$\frac{1}{9} + \frac{4}{9} = \frac{5}{9}$	$\frac{3}{9} + \frac{1}{9} = \frac{4}{9}$	1

The mean and variance of x_8 are

$$u_8 = E(x_8) = \sum_{\substack{\text{all values} \\ \text{of } x_8}} x_8 P(x_8) = 0 \times \frac{5}{9} + 1 \times \frac{4}{9} = \frac{4}{9}$$

$$\sigma_8^2 = \sum_{\substack{\text{all values} \\ \text{of } x_8}} (x_8 - u_8)^2 P(x_8) = \left(0 - \frac{4}{9}\right)^2 \times \frac{5}{9} + \left(1 - \frac{4}{9}\right)^2 \times \frac{4}{9} = 0.2469.$$

The covariance of x_7 and x_8 are

$$\sigma_{78} = \sum_{\substack{\text{all values} \\ \text{of } x_7}} \sum_{\substack{\text{all values} \\ \text{of } x_8}} (x_7 - \mu_7)(x_8 - \mu_8) P(x_7, x_8)$$

$$= \left(0 - \frac{5}{9}\right)\left(0 - \frac{4}{9}\right) \times \frac{1}{9} + \left(0 - \frac{5}{9}\right)\left(1 - \frac{4}{9}\right) \times \frac{3}{9} + \left(1 - \frac{5}{9}\right)\left(0 - \frac{4}{9}\right) \times \frac{4}{9}$$

$$+ \left(1 - \frac{5}{9}\right)\left(1 - \frac{4}{9}\right) \times \frac{1}{9} = -0.1358.$$

The mean vector of $x = (x_7, x_8)$ is

$$\mu = \begin{bmatrix} \mu_7 \\ \mu_8 \end{bmatrix} = \begin{bmatrix} \frac{5}{9} \\ \frac{4}{9} \end{bmatrix}$$

$$\Sigma = \begin{bmatrix} \sigma_{77} & \sigma_{78} \\ \sigma_{87} & \sigma_{88} \end{bmatrix} = \begin{bmatrix} 0.2469 & -0.1358 \\ -0.1358 & 0.2469 \end{bmatrix}.$$

14.2 Review of Matrix Algebra

Given a vector of p variables:

$$x = \begin{bmatrix} x_1 \\ x_2 \\ \vdots \\ x_p \end{bmatrix}, \quad x' = \begin{bmatrix} x_1 & x_2 & \cdots & x_p \end{bmatrix}, \quad (14.10)$$

Principal Component Analysis

$x_1, x_2, \ldots, x_p$ are linearly dependent if there exists a set of constants, $c_1, c_2, \ldots, c_p$, not all zero, which makes the following equation hold:

$$c_1 x_1 + c_2 x_2 + \cdots + c_p x_p = 0. \tag{14.11}$$

Similarly, $x_1, x_2, \ldots, x_p$ are linearly independent if there exists only one set of constants, $c_1 = c_2 = \ldots = c_i = 0$, which makes the following equation hold:

$$c_1 x_1 + c_2 x_2 + \cdots + c_p x_p = 0. \tag{14.12}$$

The length of the vector, x, is computed as follows:

$$L_x = \sqrt{x_1^2 + x_2^2 + \cdots x_p^2} = \sqrt{x'x}. \tag{14.13}$$

Figure 14.1 plots a two-dimensional vector, $x' = (x_1, x_2)$, and shows the computation of the length of the vector.

Figure 14.2 shows the angle θ between two vectors, $x' = (x_1, x_2)$, $y' = (y_1, y_2)$, which is computed as follows:

$$\cos(\theta_1) = \frac{x_1}{L_x} \tag{14.14}$$

$$\sin(\theta_1) = \frac{x_2}{L_x} \tag{14.15}$$

$$\cos(\theta_2) = \frac{y_1}{L_y} \tag{14.16}$$

$$\sin(\theta_2) = \frac{y_2}{L_y} \tag{14.17}$$

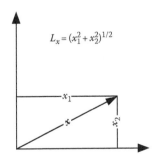

FIGURE 14.1
Computation of the length of a vector.

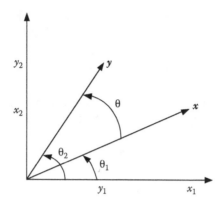

FIGURE 14.2
Computation of the angle between two vectors.

$$\cos(\theta) = \cos(\theta_2 - \theta_1) = \cos(\theta_2)\cos(\theta_1) + \sin(\theta_2)\sin(\theta_1)$$

$$= \left(\frac{y_1}{L_y}\right)\left(\frac{x_1}{L_x}\right) + \left(\frac{y_2}{L_y}\right)\left(\frac{x_2}{L_x}\right) = \frac{x_1 y_1 + x_2 y_x}{L_x L_y} = \frac{x'y}{L_x L_y}. \quad (14.18)$$

Based on the computation of the angle between two vectors, x' and y', the two vectors are orthogonal, that is, $\theta = 90°$ or $270°$, or $\cos(\theta) = 0$, only if $x'y = 0$.

A $p \times p$ square matrix, A, is symmetric if $A = A'$, that is, $a_{ij} = a_{ji}$, for $i = 1, ..., p$, and $j = 1, ..., p$. An identity matrix is the following:

$$I = \begin{bmatrix} 1 & 0 & \cdots & 0 \\ 0 & 1 & \cdots & 0 \\ \vdots & \vdots & \ddots & \vdots \\ 0 & 0 & \cdots & 1 \end{bmatrix},$$

and we have

$$AI = IA = A. \quad (14.19)$$

The inverse of the matrix A is denoted as A^{-1}, and we have

$$AA^{-1} = A^{-1}A = I. \quad (14.20)$$

The inverse of A exists if the k columns of A, $a_1, a_2, ..., a_p$, are linearly independent.

Principal Component Analysis

Let $|A|$ denote the determinant of a square $p \times p$ matrix A. $|A|$ is computed as follows:

$$|A| = a_{11} \quad \text{if } p = 1 \tag{14.21}$$

$$|A| = \sum_{j=1}^{p} a_{1j} |A_{1j}| (-1)^{1+j} = \sum_{j=1}^{p} a_{ij} |A_{ij}| (-1)^{i+j} \quad \text{if } p > 1, \tag{14.22}$$

where
A_{1j} is the $(p-1) \times (p-1)$ matrix obtained by removing the first row and the jth column of A
A_{ij} is the $(p-1) \times (p-1)$ matrix obtained by removing the ith row and the jth column of A. For a 2×2 matrix:

$$A = \begin{bmatrix} a_{11} & a_{12} \\ a_{21} & a_{22} \end{bmatrix},$$

the determinant of A is

$$|A| = \begin{vmatrix} a_{11} & a_{12} \\ a_{21} & a_{22} \end{vmatrix} = \sum_{j=1}^{2} a_{1j} |A_{1j}| (-1)^{1+j}$$

$$= a_{11} |A_{11}| (-1)^{1+1} + a_{12} |A_{12}| (-1)^{1+2} = a_{11} a_{22} - a_{12} a_{21}. \tag{14.23}$$

For the identity matrix I,

$$|I| = 1. \tag{14.24}$$

The calculation of the determinant of a matrix A is illustrated next using the variance–covariance matrix of x_7 and x_8 from Table 14.1:

$$A = \begin{bmatrix} 0.2469 & -0.1358 \\ -0.1358 & 0.2469 \end{bmatrix} = 0.2469 \times 0.2469 - (-0.1358)(-0.1358) = 0.0425.$$

Let A be a $p \times p$ square matrix and I be the $p \times p$ identity matrix. The values $\lambda_1, \ldots, \lambda_p$ are called eigenvalues of the matrix A if they satisfy the following equation:

$$|A - \lambda I| = 0. \tag{14.25}$$

Example 14.2

Compute the eigenvalues of the following matrix A, which is obtained from Example 14.1:

$$A = \begin{bmatrix} 0.2469 & -0.1358 \\ -0.1358 & 0.2469 \end{bmatrix}$$

$$|A - \lambda I| = \left\| \begin{bmatrix} 0.2469 & -0.1358 \\ -0.1358 & 0.2469 \end{bmatrix} - \lambda \begin{bmatrix} 1 & 0 \\ 0 & 1 \end{bmatrix} \right\| = \begin{vmatrix} 0.2469 - \lambda & -0.1358 \\ -0.1358 & 0.2469 - \lambda \end{vmatrix} = 0$$

$$(0.2469 - \lambda)(0.2469 - \lambda) - 0.0184 = 0$$

$$\lambda^2 - 0.4938\lambda + 0.0426 = 0$$

$$\lambda_1 = 0.3824 \quad \lambda_2 = 0.1115.$$

Let A be a $p \times p$ square matrix and λ be an eigenvalue of A. The vector x is the eigenvector of A associated with the eigenvalue λ, if x is a nonzero vector and satisfies the following equation:

$$Ax = \lambda x. \tag{14.26}$$

The normalized eigenvector with the unit length, e, is computed as follows:

$$e = \frac{x}{\sqrt{x'x}}. \tag{14.27}$$

Example 14.3

Compute the eigenvectors associated with the eigenvalues in Example 14.2. The eigenvectors associated with the eigenvalues $\lambda_1 = 0.3824$ and $\lambda_2 = 0.1115$ of the following square matrix A in Example 14.2 are computed next:

$$A = \begin{bmatrix} 0.2469 & -0.1358 \\ -0.1358 & 0.2469 \end{bmatrix}$$

$$Ax = \lambda_1 x$$

$$\begin{bmatrix} 0.2469 & -0.1358 \\ -0.1358 & 0.2469 \end{bmatrix} \begin{bmatrix} x_1 \\ x_2 \end{bmatrix} = 0.3824 \begin{bmatrix} x_1 \\ x_2 \end{bmatrix}$$

$$\begin{cases} 0.2469x_1 - 0.1358x_2 = 0.3824x_1 \\ -0.1358x_1 + 0.2469x_2 = 0.3824x_2 \end{cases}$$

$$\begin{cases} 0.1355x_1 + 0.1358x_2 = 0 \\ 0.1358x_1 + 0.1355x_2 = 0. \end{cases}$$

The two equations are identical. Hence, there are many solutions. Setting $x_1 = 1$ and $x_2 = -1$, we have

$$x = \begin{bmatrix} 1 \\ -1 \end{bmatrix} \quad e = \begin{bmatrix} \dfrac{1}{\sqrt{2}} \\ \dfrac{-1}{\sqrt{2}} \end{bmatrix}$$

$$Ax = \lambda_2 x$$

$$\begin{bmatrix} 0.2469 & -0.1358 \\ -0.1358 & 0.2469 \end{bmatrix} \begin{bmatrix} x_1 \\ x_2 \end{bmatrix} = 0.1115 \begin{bmatrix} x_1 \\ x_2 \end{bmatrix}$$

$$\begin{cases} 0.2469x_1 - 0.1358x_2 = 0.1115x_1 \\ -0.1358x_1 + 0.2469x_2 = 0.1115x_2 \end{cases}$$

$$\begin{cases} 0.1354x_1 - 0.1358x_2 = 0 \\ 0.1358x_1 - 0.1354x_2 = 0. \end{cases}$$

The aforementioned two equations are identical and thus have many solutions. Setting $x_1 = 1$ and $x_2 = 1$, we have

$$x = \begin{bmatrix} 1 \\ 1 \end{bmatrix} \quad e = \begin{bmatrix} \dfrac{1}{\sqrt{2}} \\ \dfrac{1}{\sqrt{2}} \end{bmatrix}.$$

In this example, the two eigenvectors associated with the two eigenvalues are chosen such that they are orthogonal.

Let A be a $p \times p$ symmetric matrix and (λ_i, e_i), $i = 1, \ldots, p$ be p pairs of eigenvalues and eigenvectors of A, with e_i, $i = 1, \ldots, p$, being chosen to be mutually orthogonal. A spectral decomposition of A is given next:

$$A = \sum_{i=1}^{p} \lambda_i e_i e_i'. \tag{14.28}$$

Example 14.4

Compute the spectral decomposition of the matrix in Examples 14.2 and 14.3.

The spectral decomposition of the following symmetric matrix in Examples 14.2 and 14.3 is illustrated next:

$$A = \begin{bmatrix} 0.2469 & -0.1358 \\ -0.1358 & 0.2469 \end{bmatrix}$$

$$\lambda_1 = 0.3824 \quad \lambda_2 = 0.1115$$

$$e_1 = \begin{bmatrix} \dfrac{1}{\sqrt{2}} \\ \dfrac{-1}{\sqrt{2}} \end{bmatrix}$$

$$e_2 = \begin{bmatrix} \dfrac{1}{\sqrt{2}} \\ \dfrac{1}{\sqrt{2}} \end{bmatrix}$$

$$\begin{bmatrix} 0.2469 & -0.1358 \\ -0.1358 & 0.2469 \end{bmatrix} = 0.3824 \begin{bmatrix} \dfrac{1}{\sqrt{2}} \\ \dfrac{-1}{\sqrt{2}} \end{bmatrix} \begin{bmatrix} \dfrac{1}{\sqrt{2}} & \dfrac{-1}{\sqrt{2}} \end{bmatrix} + 0.1115 \begin{bmatrix} \dfrac{1}{\sqrt{2}} \\ \dfrac{1}{\sqrt{2}} \end{bmatrix} \begin{bmatrix} \dfrac{1}{\sqrt{2}} & \dfrac{1}{\sqrt{2}} \end{bmatrix}$$

$$= \begin{bmatrix} 0.1912 & -0.1912 \\ -0.1912 & 0.1912 \end{bmatrix} + \begin{bmatrix} 0.0558 & 0.0558 \\ 0.0558 & 0.0558 \end{bmatrix}$$

$$A = \lambda_1 e_1 e_1' + \lambda_2 e_2 e_2'.$$

Principal Component Analysis

A $p \times p$ symmetric matrix A is called a positive definite matrix if it satisfies the following for any nonzero vector $= \begin{bmatrix} x_1 \\ x_2 \\ \vdots \\ x_p \end{bmatrix} \neq \begin{bmatrix} 0 \\ 0 \\ \vdots \\ 0 \end{bmatrix}$:

$$x'Ax > 0.$$

A $p \times p$ symmetric matrix A is a positive definite matrix if and only if every eigenvalue of A is greater than or equal to zero (Johnson and Wichern, 1998). For example, the following matrix 2×2 A is a positive definite matrix with two positive eigenvalues:

$$A = \begin{bmatrix} 0.2469 & -0.1358 \\ -0.1358 & 0.2469 \end{bmatrix}$$

$$\lambda_1 = 0.3824 \quad \lambda_2 = 0.1115.$$

Let A be a $p \times p$ positive definite matrix with the eigenvalues sorted in the order of $\lambda_1 \geq \lambda_2 \geq \cdots \geq \lambda_p \geq 0$ and associated normalized eigenvectors $e_1, e_2, \ldots, e_p$, which are orthogonal. The quadratic form, $(x'Ax)/(x'x)$, is maximized to λ_1 when $x = e_1$, and this quadratic form is minimized to λ_p when $x = e_p$ (Johnson and Wichern, 1998). That is, we have the following:

$$\max_{x \neq 0} \frac{x'Ax}{x'x} = \lambda_1 \quad \text{attained by } x = e_1 \quad \text{or}$$

$$e_1'Ae_1 = e_1' \left(\sum_{i=1}^{p} \lambda_i e_i e_i' \right) e_1 = \lambda_1 = \max_{x \neq 0} \frac{x'Ax}{x'x} \tag{14.29}$$

$$\min_{x \neq 0} \frac{x'Ax}{x'x} = \lambda_p \quad \text{attained by } x = e_p \quad \text{or}$$

$$e_p'Ae_p = e_p' \left(\sum_{i=1}^{p} \lambda_i e_i e_i' \right) e_p = \lambda_p = \min_{x \neq 0} \frac{x'Ax}{x'x} \tag{14.30}$$

and

$$\max_{x \perp e_1, \ldots, e_i} \frac{x'Ax}{x'x} = \lambda_{i+1} \text{ attained by } x = e_{i+1}, \quad i = 1, \ldots, p-1. \tag{14.31}$$

14.3 Principal Component Analysis

Principal component analysis explains the variance–covariance matrix of variables. Given a vector of variables $x' = [x_1, ..., x_p]$ with the variance–covariance matrix Σ, the following is a linear combination of these variables:

$$y_i = a_i'x = a_{i1}x_1 + a_{i2}x_2 + \cdots + a_{ip}x_p. \tag{14.32}$$

The variance and covariance of y_i can be computed as follows:

$$\text{var}(y_i) = a_i' \Sigma a_i \tag{14.33}$$

$$\text{cov}(y_i, y_j) = a_i' \Sigma a_j. \tag{14.34}$$

The principal components $y' = [y_1, y_2, ..., y_p]$ are chosen to be linear combinations of x' that satisfy the following:

$$y_1 = a_1'x = a_{11}x_1 + a_{12}x_2 + \cdots + a_{1p}x_p,$$

$$a_1'a_1 = 1, \; a_1 \text{ is chosen to maximize var}(y_1) \tag{14.35}$$

$$y_2 = a_2'x = a_{21}x_1 + a_{22}x_2 + \cdots + a_{2p}x_p,$$

$$a_2'a_2 = 1, \text{cov}(y_2, y_1) = 0, \; a_2 \text{ is chosen to maximize var}(y_2)$$

$$\vdots$$

$$y_i = a_i'x = a_{i1}x_1 + a_{i2}x_2 + \cdots + a_{ip}x_p,$$

$$a_i'a_i = 1, \text{cov}(y_i, y_j) = 0 \quad \text{for } j < i, \; a_i \text{ is chosen to maximize var}(y_i).$$

Let (λ_i, e_i), $i = 1, ..., p$, be eigenvalues and orthogonal eigenvectors of Σ, $e_i'e_i = 1$, and $\lambda_1 \geq \lambda_2 \geq \cdots \geq \lambda_p \geq 0$. Setting $a_1 = e_1, ..., a_p = e_p$, we have

$$y_i = e_i'x \quad i = 1, ..., p \tag{14.36}$$

$$e_i'e_i = 1$$

$$\text{var}(y_i) = e_i' \Sigma e_i = \lambda_i$$

$$\text{cov}(y_i, y_j) = e_i' \Sigma e_j = 0 \quad \text{for } j < i.$$

Principal Component Analysis

Based on Equations 14.29 through 14.31, y_i, $i = 1, \ldots, p$, set by Equation 14.36, satisfy the requirement of the principal components in Equation 14.35. Hence, the principal components are determined using Equation 14.36.

Let $x_1, \ldots, x_p$ have variances of $\sigma_1, \ldots, \sigma_p$, respectively. The sum of variances of $x_1, \ldots, x_p$ is equal to the sum of variances of $y_1, \ldots, y_p$ (Johnson and Wichern, 1998):

$$\sum_{i=1}^{p} \text{var}(x_i) = \sigma_1 + \cdots + \sigma_p = \sum_{i=1}^{p} \text{var}(y_i) = \lambda_1 + \cdots + \lambda_p. \tag{14.37}$$

Example 14.5

Determine the principal components of the two variables in Example 14.1. For the two variables $x' = [x_7, x_8]$ in Table 14.1 and Example 14.1, the variance–covariance matrix Σ is

$$\Sigma = \begin{bmatrix} 0.2469 & -0.1358 \\ -0.1358 & 0.2469 \end{bmatrix},$$

with eigenvalues and eigenvectors determined in Examples 14.2 and 14.3:

$$\lambda_1 = 0.3824 \quad \lambda_2 = 0.1115$$

$$e_1 = \begin{bmatrix} \frac{1}{\sqrt{2}} \\ \frac{-1}{\sqrt{2}} \end{bmatrix}$$

$$e_2 = \begin{bmatrix} \frac{1}{\sqrt{2}} \\ \frac{1}{\sqrt{2}} \end{bmatrix}.$$

The principal components are

$$y_1 = e_1' x = \frac{1}{\sqrt{2}} x_7 - \frac{1}{\sqrt{2}} x_8$$

$$y_2 = e_2' x = \frac{1}{\sqrt{2}} x_7 + \frac{1}{\sqrt{2}} x_8.$$

The variances of y_1 and y_2 are

$$\text{var}(y_1) = \text{var}\left(\frac{1}{\sqrt{2}}x_7 - \frac{1}{\sqrt{2}}x_8\right)$$

$$= \left(\frac{1}{\sqrt{2}}\right)^2 \text{var}(x_7) + \left(\frac{-1}{\sqrt{2}}\right)^2 \text{var}(x_8) + 2\left(\frac{1}{\sqrt{2}}\right)\left(\frac{-1}{\sqrt{2}}\right)\text{cov}(x_7, x_8)$$

$$= \frac{1}{2}(0.2469) + \frac{1}{2}(0.2469) - (-0.1358) = 0.3827 = \lambda_1$$

$$\text{var}(y_2) = \text{var}\left(\frac{1}{\sqrt{2}}x_7 + \frac{1}{\sqrt{2}}x_8\right)$$

$$= \left(\frac{1}{\sqrt{2}}\right)^2 \text{var}(x_7) + \left(\frac{1}{\sqrt{2}}\right)^2 \text{var}(x_8) + 2\left(\frac{1}{\sqrt{2}}\right)\left(\frac{1}{\sqrt{2}}\right)\text{cov}(x_7, x_8)$$

$$= \frac{1}{2}(0.2469) + \frac{1}{2}(0.2469) + (-0.1358) = 0.1111 = \lambda_2.$$

We also have

$$\text{var}(x_7) + \text{var}(x_8) = 0.2469 + 0.2469 = \text{var}(y_1) + \text{var}(y_2) = 0.3827 + 0.1111.$$

The proportion of the total variances accounted for by the first principal component y_1 is $0.3824/0.4939 = 0.7742$ or 77%. Since most of the total variances in $x' = [x_7, x_8]$ is accounted by y_1, we may use y_1 to replace and represent originally the two variables x_7, x_8 without loss of much variances. This is the basis of applying PCA for reducing the dimensions of data by using a few principal components to represent a large number of variables in the original data while still accounting for much of variances in the data. Using a few principal components to represent the data, the data can be further visualized in a one-, two-, or three-dimensional space of the principal components to observe data patterns, or can be mined or analyzed to uncover data patterns of principal components. Note that the mathematical meaning of each principal component as the linear combination of the original data variable does not necessarily have a meaningful interpretation in the problem domain. Ye (1997, 1998) shows examples of interpreting data that are not represented in their original problem domain.

14.4 Software and Applications

PCA is supported by many statistical software packages, including SAS (www.sas.com), SPSS (www.spss.com), and Statistica (www.statistica.com). Some applications of PCA in the manufacturing fields are described in Ye (2003, Chapter 8).

Principal Component Analysis

Exercises

14.1 Determine the nine principal components of $x_1, \ldots, x_9$ in Table 8.1 and identify the principal components that can be used to account for 90% of the total variances of data.

14.2 Determine the principal components of x_1 and x_2 in Table 3.2.

14.3 Repeat Exercise 14.2 for $x_1, \ldots, x_9$, and identify the principal components that can be used to account for 90% of the total variances of data.

15
Multidimensional Scaling

Multidimensional scaling (MDS) aims at representing high-dimensional data in a low-dimensional space so that data can be visualized, analyzed, and interpreted in the low-dimensional space to uncover useful data patterns. This chapter describes MDS, software packages supporting MDS, and some applications of MDS with references.

15.1 Algorithm of MDS

We are given n data items in the p-dimensional space, $x_i = (x_{i1}, \ldots, x_{ip})$, $i = 1, \ldots, n$, along with the dissimilarity δ_{ij} of each pair of n data items, x_i and x_j, and the rank order of these dissimilarities from the least similar pair to the most similar pair:

$$\delta_{i_1 j_1} \leq \delta_{i_2 j_2} \leq \cdots \leq \delta_{i_M j_M}, \tag{15.1}$$

where M denotes the total number of different data pairs, and $M = n(n-1)/2$ for n data items. MDS (Young and Hamer, 1987) is to find coordinates of the n data items in a q-dimensional space, $z_i = (z_{i1}, \ldots, x_{iq})$, $i = 1, \ldots, n$, with q being much smaller than p, while preserving the dissimilarities of n data items given in Equation 15.1. MDS is nonmetric if only the rank order of the dissimilarities in Equation 15.1 is preserved. Metric MDS goes further to preserve the magnitudes of the dissimilarities. This chapter describes nonmetric MDS.

Table 15.1 gives the steps of the MDS algorithm to find coordinates of the n data items in the q-dimensional space, while preserving the dissimilarities of n data points given in Equation 15.1. In Step 1 of the MDS algorithm, the initial configuration for coordinates of n data points in the q-dimensional space is generated using random values so that no two data points are the same.

In Step 2 of the MDS algorithm, the following is used to normalize $x_i = (x_{i1}, \ldots, x_{iq})$, $i = 1, \ldots, n$:

$$\text{normalized } x_{ij} = \frac{x_{ij}}{\sqrt{x_{i1}^2 + \cdots + x_{iq}^2}}. \tag{15.2}$$

TABLE 15.1

MDS Algorithm

Step	Description
1	Generate an initial configuration for the coordinates of n data points in the q-dimensional space, $(x_{11}, \ldots, x_{1q}, \ldots, x_{n1}, \ldots, x_{nq})$, such that no two points are the same
2	Normalize $x_i = (x_{i1}, \ldots, x_{iq})$, $i = 1, \ldots, n$, such that the vector for each data point has the unit length using Equation 15.2
3	Compute S as the stress of the configuration using Equation 15.3
4	REPEAT UNTIL a stopping criterion based on S is satisfied
5	Update the configuration using the gradient decent method and Equations 15.14 through 15.18
6	Normalize $x_i = (x_{i1}, \ldots, x_{iq})$, $i = 1, \ldots, n$, in the configuration using Equation 15.2
7	Compute S of the updated configuration using Equation 15.3

In Step 3 of the MDS algorithm, the following is used to compute the stress of the configuration that measures how well the configuration preserves the dissimilarities of n data points given in Equation 15.1 (Kruskal, 1964a,b):

$$S = \sqrt{\frac{\sum_{ij}(d_{ij} - \hat{d}_{ij})^2}{\sum_{ij} d_{ij}^2}}, \quad (15.3)$$

where d_{ij} measures the dissimilarity of x_i and x_j using their q-dimensional coordinates, and $\hat{d}_{ij}$ gives the desired dissimilarity of x_i and x_j that preserves the dissimilarity order of δ_{ij}s in Equation 15.1 such that

$$\hat{d}_{ij} < \hat{d}_{i'j'} \quad \text{if} \quad \hat{\delta}_{ij} < \hat{\delta}_{i'j'}. \quad (15.4)$$

Note that there are $n(n-1)/2$ different pairs of i and j in Equations 15.3 and 15.4.

The Euclidean distance shown in Equation 15.5, the more general Minkowski r-metric distance shown in Equation 15.6, or some other dissimilarity measure can be used to compute d_{ij}:

$$d_{ij} = \sqrt{\sum_{k=1}^{q}(d_{ik} - d_{jk})^2} \quad (15.5)$$

$$d_{ij} = \left[\sum_{k=1}^{q}(d_{ik} - d_{jk})^r\right]^{\frac{1}{r}}. \quad (15.6)$$

Multidimensional Scaling

$\hat{d}_{ij}$s are predicted from δ_{ij}s by using a monotone regression algorithm described in Kruskal (1964a,b) to produce

$$\hat{d}_{i_1 j_1} \leq \hat{d}_{i_2 j_2} \leq \cdots \leq \hat{d}_{i_M j_M}, \tag{15.7}$$

given Equation 15.1

$$\delta_{i_1 j_1} \leq \delta_{i_2 j_2} \leq \cdots \leq \delta_{i_M j_M}.$$

Table 15.2 describes the steps of the monotone regression algorithm, assuming that there are no ties (equal values) among δ_{ij}s. In Step 2 of the monotone regression algorithm, $\hat{d}_{B_m}$ for the block B_m is computed using the average of d_{ij}s in B_m:

$$\hat{d}_{B_m} = \sum_{d_{ij} \in B_m} \frac{d_{ij}}{N_m}, \tag{15.8}$$

where N_m is the number of d_{ij}s in B_m. If B_m has only one d_{ij}, $\hat{d}_{i_m j_m} = d_{ij}$.

TABLE 15.2

Monotone Regression Algorithm

Step	Description
1	Arrange $\delta_{i_m j_m}$, $m = 1, \ldots, M$, in the order from the smallest to the largest
2	Generate the initial M blocks in the same order in Step 1, $B_1, \ldots, B_M$, such that each block, B_m, has only one dissimilarity value, $d_{i_m j_m}$, and compute $\hat{d}_B$ using Equation 15.8
3	Make the lowest block the active block, and also make it up-active; denote B as the active block, B_- as the next lower block of B, B_+ as the next higher block of B
4	WHILE the active block B is not the highest block
5	IF $\hat{d}_{B_-} < \hat{d}_B < \hat{d}_{B_+}$ /* B is both down-satisfied and up-satisfied, note that the lowest clock is already down-satisfied and the highest block is already up-satisfied */
6	Make the next higher block of B the active block, and make it up-active
7	ELSE
8	IF B is up-active
9	IF $\hat{d}_B < \hat{d}_{B_+}$ /* B is up-satisfied */
10	Make B down-active
11	ELSE
12	Merge B and B_+ to form a new larger block which replaces B and B_+
13	Make the new block as the active block and it is down-active
14	ELSE /* B is down-active */
15	IF $\hat{d}_{B_-} < \hat{d}_B$ /* B is down-satisfied */
16	Make B up-active
17	ELSE
18	Merge B_- and B to form a new larger block which replaces B_- and B
19	Make the new block as the active block and it is up-active
20	$\hat{d}_{ij} = \hat{d}_B$, for each $d_{ij} \in B$ and for each block B in the final sequence of the blocks

In Step 1 of the monotone regression algorithm, if there are ties among δ_{ij}s, these δ_{ij}s with the equal value are arranged in the increasing order of their corresponding d_{ij}s in the q-dimensional space (Kruskal, 1964a,b). Another method of handling ties among δ_{ij}s is to let these δ_{ij}s with the equal value form one single block with their corresponding d_{ij}s in this block.

After using the monotone regression method to obtain $\hat{d}_{ij}$s, we use Equation 15.3 to compute the stress of the configuration in Step 3 of the MDS algorithm. The smaller the S value is, the better the configuration preserves the dissimilarity order in Equation 15.1. Kruskal (1964a,b) considers the S value of 20% indicating a poor fit of the configuration to the dissimilarity order in Equation 15.1, the S value of 10% indicating a fair fit, the S value of 5% indicating a good fit, the S value of 2.5% indicating an excellent fit, and the S value of 0% indicating the best fit. Step 4 of the MDS algorithm evaluates the goodness-of-fit using the S value of the configuration. If the S value of the configuration is not acceptable, Step 5 of the MDS algorithm changes the configuration to improve the goodness-of-fit using the gradient descent method. Step 6 of the MDS algorithm normalizes the vector of each data point in the updated configuration. Step 7 of the MDS algorithm computes the S value of the updated configuration.

In Step 4 of the MDS algorithm, a threshold of goodness-of-fit can be set and used such that the configuration is considered acceptable if S of the configuration is less than or equal to the threshold of goodness-of-fit. Hence, a stopping criterion in Step 4 of the MDS algorithm is having S less than or equal to the threshold of goodness-of-fit. If there is little change in S, that is, S levels off after iterations of updating the configuration, the procedure of updating the configuration can be stopped too. Hence, the change of S, which is smaller than a threshold, is another stopping criterion that can be used in Step 4 of the MDS algorithm.

The gradient descent method of updating the configuration in Step 5 of the MDS algorithm is similar to the gradient descent method used for updating connection weights in the back-propagation learning of artificial neural networks in Chapter 5. The objective of updating the configuration, $(x_{11}, \ldots, x_{1q}, \ldots, x_{n1}, \ldots, x_{nq})$, is to minimize the stress of the configuration in Equation 15.3, which is shown next:

$$S = \sqrt{\frac{\sum_{ij}(d_{ij} - \hat{d}_{ij})^2}{\sum_{ij} d_{ij}^2}} = \sqrt{\frac{S^*}{T^*}}, \qquad (15.9)$$

where

$$S^* = \sum_{ij}(d_{ij} - \hat{d}_{ij})^2 \qquad (15.10)$$

Multidimensional Scaling

$$T^* = \sum_{ij} d_{ij}^2. \tag{15.11}$$

Using the gradient descent method, we update each x_{kl}, $k = 1, \ldots, n$, $l = 1, \ldots, q$, in the configuration as follows (Kruskal, 1964a,b):

$$x_{kl}(t+1) = x_{kl}(t) + \alpha \Delta x_{kl} = x_{kl}(t) + \alpha(g_{kl}) \bigg/ \left(\frac{\sqrt{\sum_{k,l} g_{kl}^2}}{\sqrt{\sum_{k,l} x_{kl}^2}} \right), \tag{15.12}$$

where

$$g_{kl} = -\frac{\partial S}{\partial x_{kl}}, \tag{15.13}$$

and α is the learning rate. For a normalized x, Formula 15.12 becomes

$$x_{kl}(t+1) = x_{kl}(t) + \alpha \Delta x_{kl} = x_{kl}(t) + \alpha \frac{g_{kl}}{\sqrt{\dfrac{\sum_{k,l} g_{kl}^2}{n}}}. \tag{15.14}$$

Kruskal (1964a,b) gives the following formula to compute g_{kl} if d_{ij} is computed using the Minkowski r-metric distance:

$$g_{kl} = -\frac{\partial S}{\partial x_{kl}} = S \sum_{i,j} \left[(\rho^{ki} - \rho^{kj}) \left(\frac{d_{ij} - \hat{d}_{ij}}{S^*} - \frac{d_{ij}}{T^*} \right) \left(\frac{|x_{il} - x_{jl}|^{r-1}}{d_{ij}^{r-1}} \right) \operatorname{sign}(x_{il} - x_{jl}) \right], \tag{15.15}$$

where

$$\rho^{ki} = \begin{cases} 1 & \text{if } k = i \\ 0 & \text{if } k \neq i \end{cases} \tag{15.16}$$

$$\operatorname{sign}(x_{il} - x_{jl}) = \begin{cases} 1 & \text{if } x_{il} - x_{jl} > 0 \\ -1 & \text{if } x_{il} - x_{jl} > 0 \\ 0 & \text{if } x_{il} - x_{jl} = 0 \end{cases}. \tag{15.17}$$

If $r = 2$ in Formula 15.13, that is, the Euclidean distance is used to computer d_{ij},

$$g_{kl} = S \sum_{i,j} \left[(\rho^{ki} - \rho^{kj}) \left(\frac{d_{ij} - \hat{d}_{ij}}{S^*} - \frac{d_{ij}}{T^*} \right) \left(\frac{x_{il} - x_{jl}}{d_{ij}} \right) \right]. \tag{15.18}$$

Example 15.1

Table 15.3 gives three data records of nine quality variables, which is a part of Table 8.1. Table 15.4 gives the Euclidean distance for each pair of the three data points in the nine-dimensional space. This Euclidean distance for a pair of data point, x_i and x_j, is taken as δ_{ij}. Perform the MDS of this data set with only one iteration of the configuration update for $q = 2$, the stopping criterion of $S \leq 5\%$, and $\alpha = 0.2$.

This data set has three data points, $n = 3$, in a nine-dimensional space. We have $\delta_{12} = 2.65$, $\delta_{13} = 2.65$, and $\delta_{23} = 2$. In Step 1 of the MDS algorithm described in Table 15.1, we generate an initial configuration of the three data points in the two-dimensional space:

$$x_1 = (1, 1) \quad x_2 = (0, 1) \quad x_3 = (1, 0.5).$$

In Step 2 of the MDS algorithm, we normalize each data point so that it has the unit length, using Formula 15.2:

$$x_1 = \left(\frac{x_{11}}{\sqrt{x_{11}^2 + x_{12}^2}}, \frac{x_{12}}{\sqrt{x_{11}^2 + x_{12}^2}} \right) = \left(\frac{1}{\sqrt{1^2 + 1^2}}, \frac{1}{\sqrt{1^2 + 1^2}} \right) = (0.71, 0.71)$$

TABLE 15.3

Data Set for System Fault Detection with Three Cases of Single-Machine Faults

Instance (Faulty Machine)	Attribute Variables about Quality of Parts								
	x_1	x_2	x_3	x_4	x_5	x_6	x_7	x_8	x_9
1 (M1)	1	0	0	0	1	0	1	0	1
2 (M2)	0	1	0	1	0	0	0	1	0
3 (M3)	0	0	1	1	0	1	1	1	0

TABLE 15.4

Euclidean Distance for Each Pair of Data Points

	$C_1 = \{x_1\}$	$C_2 = \{x_2\}$	$C_3 = \{x_3\}$
$C_1 = \{x_1\}$		2.65	2.65
$C_2 = \{x_2\}$			2
$C_3 = \{x_3\}$			

Multidimensional Scaling

$$x_2 = \left(\frac{x_{21}}{\sqrt{x_{21}^2+x_{22}^2}}, \frac{x_{22}}{\sqrt{x_{21}^2+x_{22}^2}}\right) = \left(\frac{0}{\sqrt{0^2+1^2}}, \frac{1}{\sqrt{0^2+1^2}}\right) = (0,1)$$

$$x_3 = \left(\frac{x_{31}}{\sqrt{x_{31}^2+x_{32}^2}}, \frac{x_{32}}{\sqrt{x_{31}^2+x_{32}^2}}\right) = \left(\frac{1}{\sqrt{1^2+0.5^2}}, \frac{0.5}{\sqrt{1^2+0.5^2}}\right) = (0.89, 0.45).$$

The distance between each pair of the three data points in the two-dimensional space is computed using their initial coordinates:

$$d_{12} = \sqrt{(x_{11}-x_{21})^2+(x_{12}-x_{22})^2} = \sqrt{(0.71-0)^2+(0.71-1)^2} = 0.77$$

$$d_{13} = \sqrt{(x_{11}-x_{31})^2+(x_{12}-x_{32})^2} = \sqrt{(0.71-0.89)^2+(0.71-0.45)^2} = 0.32$$

$$d_{23} = \sqrt{(x_{21}-x_{31})^2+(x_{22}-x_{32})^2} = \sqrt{(0-0.89)^2+(1-0.45)^2} = 1.05.$$

Before we compute the stress of the initial configuration using Formula 15.3, we need to use the monotone regression algorithm in Table 15.2 to compute $\hat{d}_{ij}$. In Step 1 of the monotone regression algorithm, we arrange $\delta_{i_m j_m}$, $m = 1, \ldots, M$, in the order from the smallest to the largest, where $M = 3$:

$$\delta_{23} < \delta_{12} = \delta_{13}.$$

Since there is a tie between δ_{12} and δ_{13}, δ_{12} and δ_{13} are arranged in the increasing order of $d_{13} = 0.32$ and $d_{12} = 0.77$:

$$\delta_{23} < \delta_{13} < \delta_{12}.$$

In Step 2 of the monotone regression algorithm, we generate the initial M blocks in the same order in Step 1, $B_1, \ldots, B_M$, such that each block, B_m, has only one dissimilarity value, $d_{i_m j_m}$:

$$B_1 = \{d_{23}\} \quad B_2 = \{d_{13}\} \quad B_3 = \{d_{12}\}.$$

We compute $\hat{d}_B$ using Formula 15.8:

$$\hat{d}_{B_1} = \sum_{d_{ij} \in B_1} \frac{d_{ij}}{n_1} = \frac{d_{23}}{1} = 1.05$$

$$\hat{d}_{B_2} = \sum_{d_{ij} \in B_2} \frac{d_{ij}}{n_2} = \frac{d_{13}}{1} = 0.32$$

$$\hat{d}_{B_3} = \sum_{d_{ij} \in B_3} \frac{d_{ij}}{n_3} = \frac{d_{12}}{1} = 0.77.$$

In Step 3 of the monotone regression algorithm, we make the lowest block B_1 the active block:

$$B = B_1 \quad B_- = \emptyset \quad B_+ = B_2,$$

and make B up-active. In Step 4 of the monotone regression algorithm, we check that the active block B_1 is not the highest block. In Step 5 of the monotone regression algorithm, we check that $\hat{d}_B > \hat{d}_{B_+}$ and thus B is not up-satisfied. We go to Step 8 of the monotone regression algorithm and check that B is up-active. In Step 9 of the monotone regression algorithm, we check that $\hat{d}_B > \hat{d}_{B_+}$ and thus B is not up-satisfied. We go to Step 12 and merge B and B_+ to form a new larger block B_{12} to replace B_1 and B_2:

$$B_{12} = \{d_{23}, d_{13}\}$$

$$\hat{d}_{B_{12}} = \sum_{d_{ij} \in B_{12}} \frac{d_{ij}}{n_{12}} = \frac{d_{23} + d_{13}}{2} = \frac{1.05 + 0.32}{2} = 0.69$$

$$B_{12} = \{d_{23}, d_{13}\} \quad B_3 = \{d_{12}\}$$

$$\hat{d}_{B_3} = \sum_{d_{ij} \in B_3} \frac{d_{ij}}{n_3} = \frac{d_{12}}{1} = 0.77.$$

In Step 13 of the monotone regression algorithm, we make the new block B_{12} the active block and also make it down-active:

$$B = B_{12} \quad B_- = \emptyset \quad B_+ = B_3.$$

Going back to Step 4, we check that the active block B_{12} is not the highest block. In Step 5, we check that B is both up-satisfied with $\hat{d}_{B_{12}} < \hat{d}_{B_3}$ and down-satisfied. Therefore, we execute Step 6 to make B_3 the active block and make it up-active:

$$B_{12} = \{d_{23}, d_{13}\} \quad B_3 = \{d_{12}\}$$

$$\hat{d}_{B_{12}} = \sum_{d_{ij} \in B_{12}} \frac{d_{ij}}{n_{12}} = \frac{d_{23} + d_{13}}{2} = \frac{1.05 + 0.32}{2} = 0.69$$

$$\hat{d}_{B_3} = \sum_{d_{ij} \in B_3} \frac{d_{ij}}{n_3} = \frac{d_{12}}{1} = 0.77$$

$$B = B_3 \quad B_- = B_{12} \quad B_+ = \emptyset.$$

Going back to Step 4 again, we check that the active block B is the highest block, get out of the WHILE loop, execute Step 20—the last step of the monotone regression algorithm, and assign the following values of $\hat{d}_{ij}$s:

$$\hat{d}_{12} = \hat{d}_{B_3} = 0.77$$

$$\hat{d}_{13} = \hat{d}_{B_{12}} = 0.69$$

$$\hat{d}_{23} = \hat{d}_{B_{12}} = 0.69.$$

With those $\hat{d}_{ij}$ values and the d_{ij} values:

$$d_{12} = 0.77$$

$$d_{13} = 0.32$$

$$d_{23} = 1.05,$$

we now execute Step 3 of the MDS algorithm to compute the stress of the initial configuration using Equations 15.9 through 15.11:

$$S^* = \sum_{ij}(d_{ij} - \hat{d}_{ij})^2 = (0.77 - 0.77)^2 + (0.32 - 0.69)^2 + (1.05 - 0.69)^2 = 0.27$$

$$T^* = \sum_{ij} d_{ij}^2 = 0.77^2 + 0.32^2 + 1.05^2 = 0.61$$

$$S = \sqrt{\frac{S^*}{T^*}} = \sqrt{\frac{0.27}{0.61}} = 0.67.$$

This stress level indicates a poor goodness-of-fit. In Step 4 of the MDS algorithm, we check that S does not satisfy the stopping criterion of the REPEAT loop. In Step 5 of the MDS algorithm, we update the configuration using Equations 15.14, 15.16 and 15.18 with $k = 1, 2, 3$ and $l = 1, 2$:

$$g_{kl} = g_{11} = S \sum_{i,j} \left[(\rho^{ki} - \rho^{kj}) \left(\frac{d_{ij} - \hat{d}_{ij}}{S^*} - \frac{d_{ij}}{T^*} \right) \left(\frac{x_{il} - x_{jl}}{d_{ij}} \right) \right]$$

$$= (0.67) \sum_{i,j} \left[(\rho^{1i} - \rho^{1j}) \left(\frac{d_{ij} - \hat{d}_{ij}}{S^*} - \frac{d_{ij}}{T^*} \right) \left(\frac{x_{i1} - x_{j1}}{d_{ij}} \right) \right]$$

$$= (0.67) \left[(\rho^{11} - \rho^{12}) \left(\frac{d_{12} - \hat{d}_{12}}{S^*} - \frac{d_{12}}{T^*} \right) \left(\frac{x_{11} - x_{21}}{d_{12}} \right) \right.$$

$$+ (\rho^{11} - \rho^{13}) \left(\frac{d_{13} - \hat{d}_{13}}{S^*} - \frac{d_{13}}{T^*} \right) \left(\frac{x_{11} - x_{31}}{d_{13}} \right)$$

$$\left. + (\rho^{12} - \rho^{13}) \left(\frac{d_{23} - \hat{d}_{23}}{S^*} - \frac{d_{23}}{T^*} \right) \left(\frac{x_{21} - x_{31}}{d_{23}} \right) \right]$$

$$= (0.67) \left[(1-0) \left(\frac{0.77 - 0.77}{0.27} - \frac{0.77}{0.61} \right) \left(\frac{0.71 - 0}{0.77} \right) \right.$$

$$+ (1-0) \left(\frac{0.32 - 0.69}{0.27} - \frac{0.32}{0.61} \right) \left(\frac{0.71 - 0.89}{0.32} \right)$$

$$\left. + (0-0) \left(\frac{1.05 - 0.69}{0.27} - \frac{1.05}{0.61} \right) \left(\frac{0 - 0.89}{1.05} \right) \right]$$

$$= -0.13$$

$$g_{kl} = g_{12} = S \sum_{i,j} \left[(\rho^{ki} - \rho^{kj}) \left(\frac{d_{ij} - \hat{d}_{ij}}{S^*} - \frac{d_{ij}}{T^*} \right) \left(\frac{x_{il} - x_{jl}}{d_{ij}} \right) \right]$$

$$= (0.67) \sum_{i,j} \left[(\rho^{1i} - \rho^{1j}) \left(\frac{d_{ij} - \hat{d}_{ij}}{S^*} - \frac{d_{ij}}{T^*} \right) \left(\frac{x_{i2} - x_{j2}}{d_{ij}} \right) \right]$$

$$= (0.67) \left[(\rho^{11} - \rho^{12}) \left(\frac{d_{12} - \hat{d}_{12}}{S^*} - \frac{d_{12}}{T^*} \right) \left(\frac{x_{12} - x_{22}}{d_{12}} \right) \right.$$

$$+ (\rho^{11} - \rho^{13}) \left(\frac{d_{13} - \hat{d}_{13}}{S^*} - \frac{d_{13}}{T^*} \right) \left(\frac{x_{12} - x_{32}}{d_{13}} \right)$$

$$\left. + (\rho^{12} - \rho^{13}) \left(\frac{d_{23} - \hat{d}_{23}}{S^*} - \frac{d_{23}}{T^*} \right) \left(\frac{x_{22} - x_{32}}{d_{23}} \right) \right]$$

$$= (0.67) \left[(1-0) \left(\frac{0.77 - 0.77}{0.27} - \frac{0.77}{0.61} \right) \left(\frac{0.71 - 1}{0.77} \right) \right.$$

$$+ (1-0) \left(\frac{0.32 - 0.69}{0.27} - \frac{0.32}{0.61} \right) \left(\frac{0.71 - 0.45}{0.32} \right)$$

$$\left. + (0-0) \left(\frac{1.05 - 0.69}{0.27} - \frac{1.05}{0.61} \right) \left(\frac{1 - 0.45}{1.05} \right) \right]$$

$$= -0.71$$

Multidimensional Scaling

$$g_{kl} = g_{21} = S \sum_{i,j} \left[(\rho^{ki} - \rho^{kj}) \left(\frac{d_{ij} - \hat{d}_{ij}}{S^*} - \frac{d_{ij}}{T^*} \right) \left(\frac{x_{il} - x_{jl}}{d_{ij}} \right) \right]$$

$$= (0.67) \sum_{i,j} \left[(\rho^{2i} - \rho^{2j}) \left(\frac{d_{ij} - \hat{d}_{ij}}{S^*} - \frac{d_{ij}}{T^*} \right) \left(\frac{x_{i1} - x_{j1}}{d_{ij}} \right) \right]$$

$$= (0.67) \left[(\rho^{21} - \rho^{22}) \left(\frac{d_{12} - \hat{d}_{12}}{S^*} - \frac{d_{12}}{T^*} \right) \left(\frac{x_{11} - x_{21}}{d_{12}} \right) \right.$$

$$+ (\rho^{21} - \rho^{23}) \left(\frac{d_{13} - \hat{d}_{13}}{S^*} - \frac{d_{13}}{T^*} \right) \left(\frac{x_{11} - x_{31}}{d_{13}} \right)$$

$$\left. + (\rho^{22} - \rho^{23}) \left(\frac{d_{23} - \hat{d}_{23}}{S^*} - \frac{d_{23}}{T^*} \right) \left(\frac{x_{21} - x_{31}}{d_{23}} \right) \right]$$

$$= (0.67) \left[(0 - 1) \left(\frac{0.77 - 0.77}{0.27} - \frac{0.77}{0.61} \right) \left(\frac{0.71 - 0}{0.77} \right) \right.$$

$$+ (0 - 0) \left(\frac{0.32 - 0.69}{0.27} - \frac{0.32}{0.61} \right) \left(\frac{0.71 - 0.89}{0.32} \right)$$

$$\left. + (1 - 0) \left(\frac{1.05 - 0.69}{0.27} - \frac{1.05}{0.61} \right) \left(\frac{0 - 0.89}{1.05} \right) \right]$$

$$= 1.07$$

$$g_{kl} = g_{22} = S \sum_{i,j} \left[(\rho^{ki} - \rho^{kj}) \left(\frac{d_{ij} - \hat{d}_{ij}}{S^*} - \frac{d_{ij}}{T^*} \right) \left(\frac{x_{il} - x_{jl}}{d_{ij}} \right) \right]$$

$$= (0.67) \sum_{i,j} \left[(\rho^{2i} - \rho^{2j}) \left(\frac{d_{ij} - \hat{d}_{ij}}{S^*} - \frac{d_{ij}}{T^*} \right) \left(\frac{x_{i2} - x_{j2}}{d_{ij}} \right) \right]$$

$$= (0.67) \left[(\rho^{21} - \rho^{22}) \left(\frac{d_{12} - \hat{d}_{12}}{S^*} - \frac{d_{12}}{T^*} \right) \left(\frac{x_{12} - x_{22}}{d_{12}} \right) \right.$$

$$+ (\rho^{21} - \rho^{23}) \left(\frac{d_{13} - \hat{d}_{13}}{S^*} - \frac{d_{13}}{T^*} \right) \left(\frac{x_{12} - x_{32}}{d_{13}} \right)$$

$$\left. + (\rho^{22} - \rho^{23}) \left(\frac{d_{23} - \hat{d}_{23}}{S^*} - \frac{d_{23}}{T^*} \right) \left(\frac{x_{22} - x_{32}}{d_{23}} \right) \right]$$

$$= (0.67) \left[(0 - 1) \left(\frac{0.77 - 0.77}{0.27} - \frac{0.77}{0.61} \right) \left(\frac{0.71 - 1}{0.77} \right) \right.$$

$$+ (0 - 0) \left(\frac{0.32 - 0.69}{0.27} - \frac{0.32}{0.61} \right) \left(\frac{0.71 - 0.45}{0.32} \right)$$

$$\left. + (1 - 0) \left(\frac{1.05 - 0.69}{0.27} - \frac{1.05}{0.61} \right) \left(\frac{1 - 0.45}{1.05} \right) \right]$$

$$= -0.45$$

$$g_{kl} = g_{31} = S \sum_{i,j} \left[(\rho^{ki} - \rho^{kj}) \left(\frac{d_{ij} - \hat{d}_{ij}}{S^*} - \frac{d_{ij}}{T^*} \right) \left(\frac{x_{il} - x_{jl}}{d_{ij}} \right) \right]$$

$$= (0.67) \sum_{i,j} \left[(\rho^{3i} - \rho^{3j}) \left(\frac{d_{ij} - \hat{d}_{ij}}{S^*} - \frac{d_{ij}}{T^*} \right) \left(\frac{x_{i1} - x_{j1}}{d_{ij}} \right) \right]$$

$$= (0.67) \left[(\rho^{31} - \rho^{32}) \left(\frac{d_{12} - \hat{d}_{12}}{S^*} - \frac{d_{12}}{T^*} \right) \left(\frac{x_{11} - x_{21}}{d_{12}} \right) \right.$$

$$+ (\rho^{31} - \rho^{33}) \left(\frac{d_{13} - \hat{d}_{13}}{S^*} - \frac{d_{13}}{T^*} \right) \left(\frac{x_{11} - x_{31}}{d_{13}} \right)$$

$$\left. + (\rho^{32} - \rho^{33}) \left(\frac{d_{23} - \hat{d}_{23}}{S^*} - \frac{d_{23}}{T^*} \right) \left(\frac{x_{21} - x_{31}}{d_{23}} \right) \right]$$

$$= (0.67) \left[(0 - 0) \left(\frac{0.77 - 0.77}{0.27} - \frac{0.77}{0.61} \right) \left(\frac{0.71 - 0}{0.77} \right) \right.$$

$$+ (0 - 1) \left(\frac{0.32 - 0.69}{0.27} - \frac{0.32}{0.61} \right) \left(\frac{0.71 - 0.89}{0.32} \right)$$

$$\left. + (0 - 1) \left(\frac{1.05 - 0.69}{0.27} - \frac{1.05}{0.61} \right) \left(\frac{0 - 0.89}{1.05} \right) \right]$$

$$= 0.90$$

$$g_{kl} = g_{32} = S \sum_{i,j} \left[(\rho^{ki} - \rho^{kj}) \left(\frac{d_{ij} - \hat{d}_{ij}}{S^*} - \frac{d_{ij}}{T^*} \right) \left(\frac{x_{il} - x_{jl}}{d_{ij}} \right) \right]$$

$$= (0.67) \sum_{i,j} \left[(\rho^{3i} - \rho^{3j}) \left(\frac{d_{ij} - \hat{d}_{ij}}{S^*} - \frac{d_{ij}}{T^*} \right) \left(\frac{x_{i2} - x_{j2}}{d_{ij}} \right) \right]$$

$$= (0.67) \left[(\rho^{31} - \rho^{32}) \left(\frac{d_{12} - \hat{d}_{12}}{S^*} - \frac{d_{12}}{T^*} \right) \left(\frac{x_{12} - x_{22}}{d_{12}} \right) \right.$$

$$+ (\rho^{31} - \rho^{33}) \left(\frac{d_{13} - \hat{d}_{13}}{S^*} - \frac{d_{13}}{T^*} \right) \left(\frac{x_{12} - x_{32}}{d_{13}} \right)$$

$$\left. + (\rho^{32} - \rho^{33}) \left(\frac{d_{23} - \hat{d}_{23}}{S^*} - \frac{d_{23}}{T^*} \right) \left(\frac{x_{22} - x_{32}}{d_{23}} \right) \right]$$

$$= (0.67) \left[(0 - 0) \left(\frac{0.77 - 0.77}{0.27} - \frac{0.77}{0.61} \right) \left(\frac{0.71 - 1}{0.77} \right) \right.$$

$$+ (0 - 1) \left(\frac{0.32 - 0.69}{0.27} - \frac{0.32}{0.61} \right) \left(\frac{0.71 - 0.45}{0.32} \right)$$

$$\left. + (0 - 1) \left(\frac{1.05 - 0.69}{0.27} - \frac{1.05}{0.61} \right) \left(\frac{1 - 0.45}{1.05} \right) \right]$$

$$= 0.77$$

Multidimensional Scaling

$$x_{kl}(t+1) = x_{kl}(t) + \alpha \Delta x_{kl} = x_{kl}(t) + \alpha \frac{g_{kl}}{\sqrt{\dfrac{\sum_{k,l} g_{kl}^2}{n}}}$$

$$x_{11}(1) = x_{11}(0) + 0.2 \frac{g_{11}}{\sqrt{\dfrac{g_{11}^2 + g_{12}^2 + g_{21}^2 + g_{22}^2 + g_{31}^2 + g_{32}^2}{3}}}$$

$$= 0.71 + 0.2 \frac{-0.13}{\sqrt{\dfrac{(-0.13)^2 + (-0.71)^2 + 1.07^2 + (-0.45)^2 + 0.90^2 + 0.77^2}{3}}} = 0.70$$

$$x_{12}(1) = x_{12}(0) + 0.2 \frac{g_{12}}{\sqrt{\dfrac{g_{11}^2 + g_{12}^2 + g_{21}^2 + g_{22}^2 + g_{31}^2 + g_{32}^2}{3}}}$$

$$= 0.71 + 0.2 \frac{-0.71}{\sqrt{\dfrac{(-0.13)^2 + (-0.71)^2 + 1.07^2 + (-0.45)^2 + 0.90^2 + 0.77^2}{3}}} = 0.63$$

$$x_{21}(1) = x_{21}(0) + 0.2 \frac{g_{21}}{\sqrt{\dfrac{g_{11}^2 + g_{12}^2 + g_{21}^2 + g_{22}^2 + g_{31}^2 + g_{32}^2}{3}}}$$

$$= 0 + 0.2 \frac{1.07}{\sqrt{\dfrac{(-0.13)^2 + (-0.71)^2 + 1.07^2 + (-0.45)^2 + 0.90^2 + 0.77^2}{3}}} = 0.12$$

$$x_{22}(1) = x_{22}(0) + 0.2 \frac{g_{22}}{\sqrt{\dfrac{g_{11}^2 + g_{12}^2 + g_{21}^2 + g_{22}^2 + g_{31}^2 + g_{32}^2}{3}}}$$

$$= 1 + 0.2 \frac{-0.45}{\sqrt{\dfrac{(-0.13)^2 + (-0.71)^2 + 1.07^2 + (-0.45)^2 + 0.90^2 + 0.77^2}{3}}} = 0.95$$

$$x_{31}(1) = x_{31}(0) + 0.2 \frac{g_{21}}{\sqrt{\dfrac{g_{11}^2 + g_{12}^2 + g_{21}^2 + g_{22}^2 + g_{31}^2 + g_{32}^2}{3}}}$$

$$= 0.89 + 0.2 \frac{0.90}{\sqrt{\dfrac{(-0.13)^2 + (-0.71)^2 + 1.07^2 + (-0.45)^2 + 0.90^2 + 0.77^2}{3}}} = 0.99$$

$$x_{32}(1) = x_{32}(0) + 0.2 \frac{g_{22}}{\sqrt{\frac{g_{11}^2 + g_{12}^2 + g_{21}^2 + g_{22}^2 + g_{31}^2 + g_{32}^2}{3}}}$$

$$= 0.45 + 0.2 \frac{0.77}{\sqrt{\frac{(-0.13)^2 + (-0.71)^2 + 1.07^2 + (-0.45)^2 + 0.90^2 + 0.77^2}{3}}} = 0.54.$$

Hence, after the update of the initial configuration in Step 5 of the MDS algorithm, we obtain:

$$x_1 = (0.70, 0.63) \quad x_2 = (0.12, 0.95) \quad x_3 = (0.99, 0.54).$$

In Step 6 of the MDS algorithm, we normalize each x_i:

$$x_1 = \left(\frac{0.70}{\sqrt{0.70^2 + 0.63^2}}, \frac{0.63}{\sqrt{0.70^2 + 0.63^2}} \right) = (0.74, 0.67)$$

$$x_2 = \left(\frac{0.12}{\sqrt{0.12^2 + 0.95^2}}, \frac{0.95}{\sqrt{0.12^2 + 0.95^2}} \right) = (0.13, 0.99)$$

$$x_3 = \left(\frac{0.99}{\sqrt{0.99^2 + 0.54^2}}, \frac{0.54}{\sqrt{0.99^2 + 0.54^2}} \right) = (0.88, 0.48).$$

15.2 Number of Dimensions

The MDS algorithm in Section 15.1 starts with the given q—the number of dimensions. Before obtaining the final result of MDS for a data set, it is recommended that several q values are used to obtain the MDS result for each q value, plot the stress of the MDS result versus q, and select q at the elbow of this plot along with the corresponding MDS result. Figure 15.1 slows a plot of stress versus q, and the q value at the elbow of this plot is 2. The q value at the elbow of the stress-q plot is chosen because the stress improves much before the elbow point but levels off after the elbow point. For example, in the study by Ye (1998), the MDS results for q = 1, 2, 3, 4, 5, and 6 are obtained. The stress values of these MDS results show the elbow point at q = 3.

Multidimensional Scaling

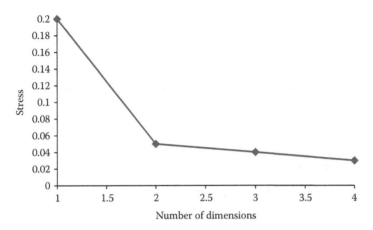

FIGURE 15.1
An example of plotting the stress of a MDS result versus the number of dimensions.

15.3 INDSCALE for Weighted MDS

In the study by Ye (1998), a number of human subjects (including expert programmers and novice programmers) are given a list of C programming concepts and are asked to rate the dissimilarity for each pair of these concepts. Hence, a dissimilarity matrix of the C programming concepts was obtained from each subject. Considering each programming concept as a data point, individual difference scaling (INDSCALE) is used in the study to take the dissimilarity matrices of data points from the subjects as the inputs and produce the outputs including both the configuration of each data point's coordinates in a q-dimensional space for the entire group of the subjects and a weight vector for each subject. The weight vector for a subject contains a weight value for the subject in each dimension. Applying the weight vector of a subject to the group configuration of concept coordinates gives the configuration of concept coordinates by the subject—the organization of the C programming concepts by each subject. Since the different weight vectors of the individual subjects reflect their differences in knowledge organization, the study applies the angular analysis of variance (ANAVA) to the weight vectors of the individual subjects to analyze the angular differences of the weight vectors and evaluate the significance of knowledge organization differences between two skill groups of experts and novices.

In general, INDSCALE or weighted MDS takes object dissimilarity matrices of n objects from m subjects and produces the group configuration of object coordinates:

$$x_i = (x_{i1}, \ldots, x_{iq}), \quad i = 1, \ldots, n,$$

and weight vectors of individual subjects:

$$w_j = (w_{j1}, \ldots, x_{jq}), \quad j = 1, \ldots, m.$$

The weight vector of a subject reflects the relative salience of each dimension in the configuration space to the subject.

15.4 Software and Applications

MDS is supported in many statistical software packages, including SAS MDS and INDSCALE procedures (www.sas.com). An application of MDS and INDSCALE to determining expert-novice differences in knowledge representation is described in Section 15.3 with details in Ye (1998).

Exercises

15.1 Continue Example 15.1 to perform the next iteration of the configuration update.

15.2 Consider the data set consisting of the three data points in instances #4, 5, and 6 in Table 8.1. Use the Euclidean distance for each pair of the three data points in the nine-dimensional space, x_i and x_j, as δ_{ij}. Perform the MDS of this data set with only one iteration of the configuration update for $q = 3$, the stopping criterion of $S \leq 5\%$, and $\alpha = 0.2$.

15.3 Consider the data set in Table 8.1 consisting of nine data points in instances 1–9. Use the Euclidean distance for each pair of the nine data points in the nine-dimensional space, x_i and x_j, as δ_{ij}. Perform the MDS of this data set with only one iteration of the configuration update for $q = 1$, the stopping criterion of $S \leq 5\%$, and $\alpha = 0.2$.

Part V

Algorithms for Mining Outlier and Anomaly Patterns

16

Univariate Control Charts

Outliers and anomalies are data points that deviate largely from the norm where the majority of data points follow. Outliers and anomalies may be caused by a fault of a manufacturing machine and thus an out of control manufacturing process, an attack whose behavior differs largely from normal use activities on computer and network systems, and so on. Detecting outliers and anomalies are important in many fields. For example, detecting an out of control manufacturing process quickly is important for reducing manufacturing costs by not producing defective parts. An early detection of a cyber attack is crucial to protect computer and network systems from being compromised.

Control chart techniques define and detect outliers and anomalies on a statistical basis. This chapter describes univariate control charts that monitor one variable for anomaly detection. Chapter 17 describes multivariate control charts that monitor multiple variables simultaneously for anomaly detection. The univariate control charts described in this chapter include Shewhart control charts, cumulative sum (CUSUM) control charts, exponentially weighted moving average (EWMA) control charts, and cuscore control charts. A list of software packages that support univariate control charts is provided. Some applications of univariate control charts are given with references.

16.1 Shewhart Control Charts

Shewhart control charts include variable control charts, each of which monitors a variable with numeric values (e.g., a diameter of a hole manufactured by a cutting machine), and attribute control charts, each of which monitors an attribute summarizing categorical values (e.g., the fraction of defective or nondefective parts). When samples of data points can be observed, variable control charts, for example, $\bar{x}$ control charts for detecting anomalies concerning the mean of a process, and R and s control charts for detecting anomalies concerning the variance of a process, are applicable. When only individual data points can be observed, variable control charts, for example, individual control charts, are applicable. For a data set with individual data points rather than samples of data points, both CUSUM control charts in

251

TABLE 16.1

Samples of Data Observations

Sample	Data Observations in Each Sample	Sample Mean	Sample Standard Deviation
1	$x_{11}, \ldots, x_{1j}, \ldots, x_{1n}$	$\bar{x}_1$	s_1
...	...	...	...
i	$x_{i1}, \ldots, x_{ij}, \ldots, x_{in}$	$\bar{x}_i$	s_i
...	...	...	...
m	$x_{m1}, \ldots, x_{mj}, \ldots, x_{mn}$	$\bar{x}_m$	s_m

Section 16.2 and EWMA control charts in Section 16.3 have advantages over individual control charts.

We describe the $\bar{x}$ control charts to illustrate how Shewhart control charts work. Consider a variable x that takes m samples of n data observations from a process as shown in Table 16.1. The $\bar{x}$ control chart assumes that x is normally distributed with the mean μ and the standard deviation σ when the process is in control.

$\bar{x}_i$ and s_i, $i = 1, \ldots, m$, in Table 16.1 are computed as follows:

$$\bar{x}_i = \frac{\sum_{j=1}^{n} x_{ij}}{n} \tag{16.1}$$

$$s_i = \sqrt{\frac{\sum_{j=1}^{n} (x_{ij} - \bar{x}_i)^2}{n-1}}. \tag{16.2}$$

The mean μ and the standard deviation σ are estimated using $\bar{\bar{x}}$ and $\bar{s}$:

$$\bar{\bar{x}} = \frac{\sum_{i=1}^{m} \bar{x}_i}{m} \tag{16.3}$$

$$\bar{s} = \frac{\sum_{i=1}^{m} s_i}{m}. \tag{16.4}$$

If the sample size n is large, $\bar{x}_i$ follows a normal distribution according to the central limit theory. The probability that $\bar{x}_i$ falls within the three standard deviations from the mean is approximately 99.7% based on the probability density function of a normal distribution:

$$P(\bar{\bar{x}} - 3\bar{s} \leq \bar{x}_i \leq \bar{\bar{x}} + 3\bar{s}) = 99.7\%. \tag{16.5}$$

Since the probability that $\bar{x}_i$ falls beyond the three standard deviations from the mean is only 0.3%, such $\bar{x}_i$ is considered an outlier or anomaly that may be caused by the process being out of control. Hence, the estimated mean and the 3-sigma control limits are typically used as the centerline and the control limits (UCL for upper control limit and LCL for lower control limit), respectively, for the in-control process mean in the $\bar{x}$ control chart:

$$\text{Centerline} = \bar{\bar{x}} \qquad (16.6)$$

$$UCL = \bar{\bar{x}} + 3\bar{s} \qquad (16.7)$$

$$LCL = \bar{\bar{x}} - 3\bar{s}. \qquad (16.8)$$

The $\bar{x}$ control chart monitors $\bar{x}_i$ from sample i of data observations. If $\bar{x}_i$ falls within [LCL, UCL], the process is considered in control; otherwise, an anomaly is detected and the process is considered out of control.

Using the 3-sigma control limits in the $\bar{x}$ control chart, there is still 0.3% probability that the process is in control but a data observation falls outside the control limit and an out-of-control signal is generated by the $\bar{x}$ control chart. If the process is in control but the control chart gives an out-of-control signal, the signal is a false alarm. The rate of false alarms is the ratio of the number of false alarms to the total number of data samples being monitored. If the process is out of control and the control chart generates an out-of-control signal, we have a hit. The rate of hits is the ratio of the number of hits to the total numbers of data samples. Using the 3-sigma control limits, we should have the hit rate of 99.7% and the false alarm rate of 0.3%.

If the sample size n is not large, the estimation of the standard deviation by $\bar{s}$ may be off to a certain extent, and the coefficient for $\bar{s}$ in Formulas 16.7 and 16.8 may need to be set to a different value than 3 in order to set the appropriate control limits so that a vast majority of the data population falls in the control limits statistically. Montgomery (2001) gives appropriate coefficients to set the control limits for various values of the sample size n.

The $\bar{x}$ control chart shows how statistical control charts such as Shewhart control charts establish the centerline and control limits based on the probability distribution of the variable of interest and the estimation of distribution parameters from data samples. In general, the centerline of a control chart is set to the expected value of the variable, and the control limits are set so that a vast majority of the data population fall in the control limits statistically. Hence, the norm of data and anomalies are defined statistically, depending on the probability distribution of data and estimation of distribution parameters.

Shewhart control charts are sensitive to the assumption that the variable of interest follows a normal distribution. A deviation from this normality assumption may cause a Shewhart control chart such as the $\bar{x}$ control chart to perform poorly, for example, giving an out-of-control signal when the process is truly in control or giving no signal when the process is truly out of control. Because Shewhart control charts monitor and evaluate only one data sample or one individual data observation at a time, Shewhart control charts are not effective at detecting small shifts, e.g., small shifts of a process mean monitored by the $\bar{x}$ control chart. CUSUM control charts in Section 16.2 and EWMA control charts in Section 16.3 are less sensitive to the normality assumption of data and are effective at detecting small shifts. CUSUM control charts and EWMA control charts can be used to monitor both data samples and individual data observations. Hence, CUSUM control charts and EWMA control charts are more practical.

16.2 CUSUM Control Charts

Given a time series of data observations for a variable x, x_1, ..., x_n, the cumulative sum up to the ith observation is (Montgomery, 2001; Ye, 2003, Chapter 3)

$$CS_i = \sum_{j=1}^{i}(x_i - \mu_0), \qquad (16.9)$$

where μ_0 is the target value of the process mean. If the process is in control, data observations are expected to randomly fluctuate around the process mean, and thus CS_i stays around zero. However, if the process is out of control with a shift of x values from the process mean, CS_i keeps increasing for a positive shift (i.e., $x_i - \mu_0 > 0$) or decreasing for a negative shift. Even if there is a small shift, the effect of the small shift keeps accumulating in CS_i and becomes large to be defected. Hence, a CUSUM control chart is more effective than a Shewhart control chart to detect small shifts since a Shewhart control chart examines only one data sample or one data observation. Formula 16.9 is used to monitor individual data observations. If samples of data points can be observed, x_i in Formula 16.9 can be replaced by $\bar{x}_i$ to monitor the sample average.

If we are interested in detecting only a positive shift, a one-side CUSUM chart can be constructed to monitor the CS_i^+ statistic:

$$CS_i^+ = \max\left[0, x_i - (\mu_0 + K) + CS_{i-1}^+\right], \qquad (16.10)$$

where K is called the reference value specifying how much increase from the process mean μ_0 we are interested in detecting. Since we expect $x_i \geq \mu_0 + K$ as a result of the positive shift K from the process mean μ_0, we expect $x_i - (\mu_0 + K)$ to be positive and expect CS_i^+ to keep increasing with i. In case that some x_i makes $x_i - (\mu_0 + K) + CS_{i-1}^+$ have a negative value, CS_i^+ takes the value of 0 according to

Univariate Control Charts

Formula 16.10 since we are interested in only the positive shift. One method of specifying K is to use the standard deviation σ of the process. For example, $K = 0.5\sigma$ indicates that we are interested in detecting a shift of 0.5σ above the target mean. If the process is in control, we expect CS_i^+ to stay around zero. Hence, CS_i^+ is initially set to zero:

$$CS_0^+ = 0. \tag{16.11}$$

When CS_i^+ exceeds the decision threshold H, the process is considered out of control. Typically $H = 5\sigma$ is used as the decision threshold so that a low rate of false alarms can be achieved (Montgomery, 2001). Note that $H = 5\sigma$ is greater than the 3-sigma control limits used for the $\bar{x}$ control chart in Section 16.1 since CS_i^+ accumulates the effects of multiple data observations whereas the $\bar{x}$ control chart examines only one data observation or data sample.

If we are interested in detecting only a negative shift, $-K$, from the process mean, a one-side CUSUM chart can be constructed to monitor the CS_i^- statistic:

$$CS_i^- = \max\left[0, (\mu_0 - K) - x_i + CS_{i-1}^-\right]. \tag{16.12}$$

Since we expect $x_i \leq \mu_0 - K$ as a result of the negative shift, $-K$, from the process mean μ_0, we expect $(\mu_0 - K) - x_i$ to be positive and expect CS_i^- to keep increasing with i. $H = 5\sigma$ is typically used as the decision threshold to achieve a low rate of false alarms (Montgomery, 2001). CS_i^- is initially set to zero since we expect CS_i^- to stay around zero if the process is in control:

$$CS_0^- = 0. \tag{16.13}$$

A two-side CUSUM control chart can be used to monitor both CS_i^+ using the one-side upper CUSUM and CS_i^- using the one-side lower CUSUM for the same x_i. If either CS_i^+ or CS_i^- exceeds the decision threshold H, the process is considered out of control.

Example 16.1

Consider the launch temperature data in Table 1.5 and presented in Table 16.2 as a sequence of data observations over time. Given the following information:

$$\mu_0 = 69$$
$$\sigma = 7$$
$$K = 0.5\sigma = (0.5)(7) = 3.5$$
$$H = 5\sigma = (5)(7) = 35,$$

use a two-side CUSUM control chart to monitor the launch temperature.

TABLE 16.2

Data Observations of the Launch Temperature from the Data Set of O-Rings with Stress along with Statistics for the Two-Side CUSUM Control Chart

Data Observation i	Launch Temperature x_i	CS_i^+	CS_i^-
1	66	0	0
2	70	0	0
3	69	0	0
4	68	0	0
5	67	0	0
6	72	0	0
7	73	0.5	0
8	70	0	0
9	57	0	8.5
10	63	0	11
11	70	0	6.5
12	78	5.5	0
13	67	0	0
14	53	0	12.5
15	67	0	11
16	75	2.5	1.5
17	70	0	0
18	81	8.5	0
19	76	12	0
20	79	18.5	0
21	75	21	0
22	76	24.5	0
23	58	10	7.5

With CS_i^+ and CS_i^- initially set to zero, that is, $CS_0^+ = 0$ and $CS_0^- = 0$, we compute CS_1^+ and CS_1^-:

$$CS_1^+ = \max\left[0, x_1 - (\mu_0 + K) + CS_0^+\right] = \max\left[0, 66 - (69 + 3.5) + 0\right] = \max[0, -6.5] = 0$$

$$CS_1^- = \max\left[0, (\mu_0 - K) - x_1 + CS_0^-\right] = \max\left[0, (69 - 3.5) - 66 + 0\right] = \max[0, -0.5] = 0,$$

and then CS_2^+ and CS_2^-:

$$CS_2^+ = \max\left[0, x_2 - (\mu_0 + K) + CS_1^+\right] = \max\left[0, 70 - (69 + 3.5) + 0\right] = \max[0, -2.5] = 0$$

$$CS_2^- = \max\left[0, (\mu_0 - K) - x_1 + CS_0^-\right] = \max\left[0, (69 - 3.5) - 70 + 0\right] = \max[0, -4.5] = 0.$$

Univariate Control Charts

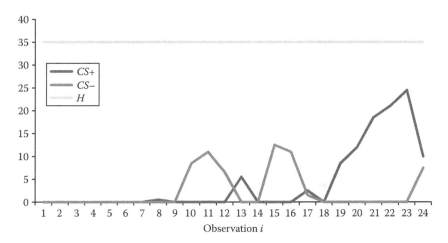

FIGURE 16.1
Two-side CUSUM control chart for the launch temperature in the data set of O-ring with stress.

The values of CS_i^+ and CS_i^- for $i = 3, \ldots, 23$ are shown in Table 16.2. Figure 16.1 shows the two-side CUSUM control chart. The CS_i^+ and CS_i^- values for all the 23 observations do not exceed the decision threshold $H = 35$. Hence, no anomalies of the launch temperature are detected. If the decision threshold is set to $H = 3\sigma = (3)(7) = 21$, the observation $i = 22$ will be signaled as an anomaly because $CS_{22}^+ = 24.5 > H$.

After an out-of-control signal is generated, the CUSUM control chart will reset CS_i^+ and CS_i^- to their initial value of zero and use the initial value of zero to compute CS_i^+ and CS_i^- for the next observation.

16.3 EWMA Control Charts

An EWMA control chart for a variable x with independent data observations, x_i, monitors the following statistic (Montgomery, 2001; Ye, 2003, Chapter 4):

$$z_i = \lambda x_i + (1-\lambda)z_{i-1} \tag{16.14}$$

where λ is a weight in (0, 1]:

$$z_0 = \mu. \tag{16.15}$$

The control limits are (Montgomery, 2001; Ye, 2003, Chapter 3):

$$UCL = \mu + L\sigma\sqrt{\frac{\lambda}{2-\lambda}} \tag{16.16}$$

$$LCL = \mu - L\sigma\sqrt{\frac{\lambda}{2-\lambda}}. \tag{16.17}$$

The weight λ determines the relative impacts of the current data observation x_i and previous data observations as captured through z_{i-1} on z_i. If we express z_i using $x_i, x_{i-1}, \ldots, x_1$,

$$\begin{aligned}
z_i &= \lambda x_i + (1-\lambda) z_{i-1} \\
&= \lambda x_i + (1-\lambda)[\lambda x_{i-1} + (1-\lambda) z_{i-2}] \\
&= \lambda x_i + (1-\lambda)\lambda x_{i-1} + (1-\lambda)^2 z_{i-2} \\
&= \lambda x_i + (1-\lambda)\lambda x_{i-1} + (1-\lambda)^2 [\lambda x_{i-2} + (1-\lambda) z_{i-3}] \\
&= \lambda x_i + (1-\lambda)\lambda x_{i-1} + (1-\lambda)^2 \lambda x_{i-2} + (1-\lambda)^3 z_{i-3} \\
&\cdots \\
&= \lambda x_i + (1-\lambda)\lambda x_{i-1} + (1-\lambda)^2 \lambda x_{i-2} + \cdots + (1-\lambda)^{i-2} \lambda x_2 + (1-\lambda)^{i-1} \lambda x_1 \quad (16.18)
\end{aligned}$$

we can see the weights on $x_i, x_{i-1}, \ldots, x_1$ decreasing exponentially. For example, for $\lambda = 0.3$, we have the weight of 0.3 for x_i, $(0.7)(0.3) = 0.21$ for x_{i-1}, $(0.7)^2(0.3) = 0.147$ for x_{i-2}, $(0.7)^3(0.3) = 0.1029$ for x_{i-3}, ..., as illustrated in Figure 16.2. This gives the term of EWMA. The larger the λ value is, the less impact the past observations and the more impact the current observation have on the current EWMA statistic, z_i.

In Formulas 16.14 through 16.17, setting λ and L in the following ranges usually works well (Montgomery, 2001; Ye, 2003, Chapter 4):

$$0.05 \leq \lambda \leq 0.25$$

$$2.6 \leq L \leq 3.$$

A data sample can be used to compute the sample average and the sample standard deviation as the estimates of μ and σ, respectively.

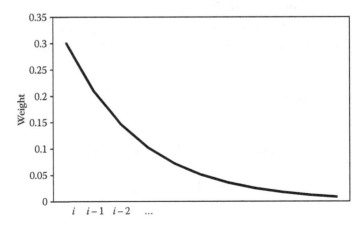

FIGURE 16.2
Exponentially decreasing weights on data observations.

Univariate Control Charts

Example 16.2

Consider the launch temperature data in Table 1.5 and presented in Table 16.3 as a sequence of data observations over time. Given the following:

$$\mu = 69$$
$$\sigma = 7$$
$$\lambda = 0.2$$
$$L = 3,$$

use an EWMA control chart to monitor the launch temperature. We first compute the control limits:

$$UCL = \mu + L\sigma\sqrt{\frac{\lambda}{2-\lambda}} = 69 + (3)(7)\sqrt{\frac{0.3}{2-0.3}} = 77.82$$

TABLE 16.3

Data Observations of the Launch Temperature from the Data Set of O-Rings with Stress along with the EWMA Statistic for the EWMA Control Chart

Data Observation i	Launch Temperature x_i	z_i
1	66	68.4
2	70	68.72
3	69	68.78
4	68	68.62
5	67	68.30
6	72	69.04
7	73	69.83
8	70	69.86
9	57	67.29
10	63	66.43
11	70	67.15
12	78	69.32
13	67	68.85
14	53	65.68
15	67	65.95
16	75	67.76
17	70	68.21
18	81	70.76
19	76	71.81
20	79	73.25
21	75	73.60
22	76	74.08
23	58	70.86

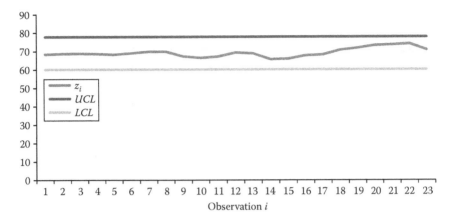

FIGURE 16.3
EWMA control chart to monitor the launch temperature from the data set of O-rings with stress.

$$LCL = \mu - L\sigma\sqrt{\frac{\lambda}{2-\lambda}} = 69 - (3)(7)\sqrt{\frac{0.3}{2-0.3}} = 60.18.$$

Using $z_0 = \mu = 69$, we compute the EWMA statistic:

$$z_1 = \lambda x_1 + (1-\lambda)z_0 = (0.2)(66) + (1-0.2)(69) = 68.4$$

$$z_2 = \lambda x_2 + (1-\lambda)z_1 = (0.2)(70) + (1-0.2)(68.4) = 68.72.$$

The values of the EWMA statistic for other data observations are given in Table 16.3. The EWMA statistic values of all the 23 data observations stay in the control limits $[LCL, UCL] = [60.18, 77.82]$, and no anomalies are detected. Figure 16.3 plots the EWMA control chart with the EWMA statistic and the control limits.

If data observations are autocorrelated (see Chapter 18 for the description of autocorrelation), we can first build a 1-step ahead prediction model of autocorrelated data, compare a data observation with its 1-step predicted value to obtain the error or residual, and use an EWMA control chart to monitor the residual data (Montgomery and Mastrangelo, 1991). The 1-step ahead predicted value for x_i is computed as follows:

$$z_{i-1} = \lambda x_{i-1} + (1-\lambda)z_{i-2}, \qquad (16.19)$$

where $0 < \lambda \le 1$. That is, z_{i-1} is the EWMA of $x_{i-1}, \ldots, x_1$ and is used as the prediction for x_i. The prediction error or residual is then computed:

$$e_i = x_i - z_{i-1}. \qquad (16.20)$$

Univariate Control Charts

In Formula 16.19, λ can be set to minimize the sum of squared prediction errors on the training data set:

$$\lambda = \arg\min_{\lambda} \sum_i e_i^2. \tag{16.21}$$

If the 1-step ahead prediction model represents the autocorrelated data well, e_is should be independent of each other and normally distributed with the mean of zero and the standard deviation of σ_e. An EWMA control chart for monitoring e_i has the centerline at zero and the following control limits:

$$UCL_{e_i} = L\hat{\sigma}_{e_{i-1}} \tag{16.22}$$

$$LCL_{e_i} = -L\hat{\sigma}_{e_{i-1}} \tag{16.23}$$

$$\hat{\sigma}_{e_{i-1}}^2 = \alpha e_{i-1}^2 + (1-\alpha)\hat{\sigma}_{e_{i-2}}^2, \tag{16.24}$$

where L is set to a value such that $2.6 \leq L \leq 3$, $0 < \alpha \leq 1$ and $\hat{\sigma}_{e_{i-1}}^2$ gives the estimate of σ_e for x_i using the exponentially weighted moving average of the prediction errors. Using Equation 16.20, which gives $x_i = e_i + z_{i-1}$, the control limits for monitoring x_i directly instead of e_i are:

$$UCL_{x_i} = z_{i-1} + L\hat{\sigma}_{e_{i-1}} \tag{16.25}$$

$$LCL_{x_i} = z_{i-1} - L\hat{\sigma}_{e_{i-1}} \tag{16.26}$$

Like CUSUM control charts, EWMA control charts are more robust to the normality assumption of data than Shewhart control charts (Montgomery, 2001). Unlike Shewhart control charts, CUSUM control charts and EWMA control charts are effective at detecting anomalies of not only large shifts but also small shifts since CUSUM control charts and EWMA control charts take into account the effects of multiple data observations.

16.4 Cuscore Control Charts

The control charts described in Sections 16.1 through 16.3 detect the out-of-control shifts from the mean or the standard deviation. Cumulative score (cuscore) control charts (Luceno, 1999) are designed to detect the change from any specific form of an in-control data model to any specific form of an out-of-control data model. For example, a cuscore control chart can be constructed to detect a change of the slope in a linear model of in-control data as follows:

In-control data model:

$$y_t = \theta_0 t + \varepsilon_t \qquad (16.27)$$

Out-of-control data model:

$$y_t = \theta t + \varepsilon_t, \quad \theta \neq \theta_0, \qquad (16.28)$$

where ε_t is a random variable with the normal distribution, the mean $\mu = 0$ and the standard deviation σ. For another example, we can have a cuscore control chart to detect the presence of a sine wave in an in-control process with random variations from the mean of T:

In-control data model:

$$y_t = T + \theta_0 \sin\left(\frac{2\pi t}{p}\right) + \varepsilon_t, \quad \theta_0 = 0 \qquad (16.29)$$

Out-of-control data model:

$$y_t = T + \theta \sin\left(\frac{2\pi t}{p}\right) + \varepsilon_t. \qquad (16.30)$$

To construct the cuscore statistic, we consider y_t as a function of x_t and the parameter θ that differentiates an out-of-control process from an in-control process:

$$y_t = f(x_t, \theta) \qquad (16.31)$$

and when the process is in control, we have

$$\theta = \theta_0. \qquad (16.32)$$

In the two examples shown in Equations 16.27 through 16.30, x_t include only t, and $\theta = \theta_0$ when the process is in control.

The residual, ε_t, can be computed by subtracting the predicted value $\hat{y}_t$ from the observed value of y_t:

$$\varepsilon_t = y_t - \hat{y}_t = y_t - f(x_t, \theta) = g(y_t, x_t, \theta). \qquad (16.33)$$

When the process is in-control, we have $\theta = \theta_0$ and expect $\varepsilon_1, \varepsilon_2, \ldots, \varepsilon_n$ to be independent of each other, each of which is a random variable of white noise with independent data observations, a normal distribution, the mean $\mu = 0$

Univariate Control Charts

and the standard deviation σ. That is, the random variables, $\varepsilon_1, \varepsilon_2, \ldots, \varepsilon_n$, have a joint multivariate normal distribution with the following joint probability density function:

$$P(\varepsilon_1, \ldots, \varepsilon_n \mid \theta = \theta_0) = \frac{1}{(2\pi)^{\frac{n}{2}}} e^{-\frac{1}{2}\sum_{t=1}^{n} \frac{\varepsilon_{t0}^2}{\sigma^2}}. \tag{16.34}$$

Taking a natural logarithm of Equation 16.34, we have

$$l(\varepsilon_1, \ldots, \varepsilon_n \mid \theta = \theta_0) = -\frac{2}{n} \ln(2\pi) - \frac{1}{2\sigma^2} \sum_{t=1}^{n} \varepsilon_{t0}^2. \tag{16.35}$$

As seen from Equation 16.33, ε_t is a function of θ, and $P(\varepsilon_1, \ldots, \varepsilon_n)$ in Equation 16.34 reaches the maximum likelihood value if the process is in control with $\theta = \theta_0$ and we have independently, identically normally distributed ε_{t0}, $t = 1, \ldots, n$, plugged into Equation 16.34. If the process is out-of-control and $\theta \neq \theta_0$, Equation 16.34 is not the correct joint probability density function for $\varepsilon_1, \varepsilon_2, \ldots, \varepsilon_n$ and thus does not give the maximum likelihood of $\varepsilon_1, \varepsilon_2, \ldots, \varepsilon_n$. Hence, if the process is in control with $\theta = \theta_0$, we have

$$\frac{\partial l(\varepsilon_1, \ldots, \varepsilon_n \mid \theta = \theta_0)}{\partial \theta} = 0. \tag{16.36}$$

Using Equation 16.35 to substitute $l(\varepsilon_1, \ldots, \varepsilon_n \mid \theta = \theta_0)$ in Equation 16.36 and dropping all the terms not related to θ when taking the derivative, we have

$$\sum_{t=1}^{n} \varepsilon_{t0} \left(-\frac{\partial \varepsilon_{t0}}{\partial \theta} \right) = 0. \tag{16.37}$$

The cuscore statistic for a cuscore control chart to monitor is Q_0:

$$Q_0 = \sum_{t=1}^{n} \varepsilon_{t0} \left(-\frac{\partial \varepsilon_{t0}}{\partial \theta} \right) = \sum_{t=1}^{n} \varepsilon_{t0} d_{t0} \tag{16.38}$$

where

$$d_{t0} = -\frac{\partial \varepsilon_{t0}}{\partial \theta}. \tag{16.39}$$

Based on Equation 16.37, Q_0 is expected to stay near zero if the process is in control with $\theta = \theta_0$. If θ shifts from θ_0, Q_0 departs from zero not randomly but in a consistent manner.

For example, to detect a change of the slope from a linear model of in-control data described in Equations 16.27 and 16.28, a cuscore control chart monitors:

$$Q_0 = \sum_{t=1}^{n} \varepsilon_{t0}\left(-\frac{\partial \varepsilon_{t0}}{\partial \theta}\right) = \sum_{t=1}^{n} \varepsilon_{t0}\left(-\frac{\partial(y_t - \theta t)}{\partial \theta}\right) = \sum_{t=1}^{n}(y_t - \theta_0 t)t. \quad (16.40)$$

If the slope θ of the in-control linear model changes from θ_0, $(y_t - \theta_0 t)$ in Equation 16.40 contains t, which is multiplied by another t to make Q_0 keep increasing (if $y_t - \theta_0 t > 0$) or decreasing (if $y_t - \theta_0 t < 0$) rather than randomly varying around zero. Such a consistent departure of Q_0 from zero causes the slope of the line connecting Q_0 values over time to increase or decrease from zero, which can be used to signal the presence of an anomaly.

To detect a sine wave in an in-control process with the mean of T and random variations described in Equations 16.29 and 16.30, the cuscore statistic for a cuscore control chart is

$$Q_0 = \sum_{t=1}^{n} \varepsilon_{t0}\left(-\frac{\partial \varepsilon_{t0}}{\partial \theta}\right) = \sum_{t=1}^{n}(y_t - T)\left[-\frac{\partial\left(y_t - T - \theta \sin\left(\frac{2\pi t}{p}\right)\right)}{\partial \theta}\right]$$

$$= \sum_{t=1}^{n}(y_t - T)\sin\left(\frac{2\pi t}{p}\right). \quad (16.41)$$

If the sine wave is present in y_t, $(y_t - T)$ in Equation 16.41 contains $\sin(2\pi t/p)$, which is multiplied by another $\sin(2\pi t/p)$ to make Q_0 keep increasing (if $y_t - T > 0$) or decreasing (if $y_t - T < 0$) rather than randomly varying around zero.

To detect a mean shift of K from μ_0 as in a CUSUM control chart described in Equations 16.9, 16.10, and 16.12, we have:

In-control data model:

$$y_t = \mu_0 + \theta_0 K + \varepsilon_t, \quad \theta_0 = 0 \quad (16.42)$$

Out-of-control data model:

$$y_t = \mu_0 + \theta K + \varepsilon_t, \quad \theta \neq \theta_0 \quad (16.43)$$

$$Q_0 = \sum_{t=1}^{n} \varepsilon_{t0}\left(-\frac{\partial \varepsilon_{t0}}{\partial \theta}\right) = \sum_{t=1}^{n}(y_t - \mu_0)\left[-\frac{\partial(y_t - \mu_0 - \theta K)}{\partial \theta}\right] = \sum_{t=1}^{n}(y_t - \mu_0)K. \quad (16.44)$$

If the mean shift of K from μ_0 occurs, $(y_t - \mu_0)$ in Equation 16.44 contains K, which is multiplied by another K to make Q_0 keep increasing (if $y_t - \mu_0 > 0$) or decreasing (if $y_t - \mu_0 < 0$) rather than randomly varying around zero.

Since cuscore control charts allow us to detect a specific form of an anomaly given a specific form of in-control data model, cuscore control charts allow us to monitor and detect a wider range of in-control vs. out-of-control situations than Shewhart control charts, CUSUM control charts, and EWMA control charts.

16.5 Receiver Operating Curve (ROC) for Evaluation and Comparison of Control Charts

Different values of the decision threshold parameters used in various control charts, for example, the 3-sigma in a $\bar{x}$ control chart, H in an CUSUM control chart, and L in an EWMA control chart, produce different rates of false alarms and hits. Suppose that in Example 16.1 any value of $x_i \geq 75$ is truly an anomaly. Hence, seven data observations, observations #12, 16, 18, 19, 20, 21, and 22, have $x_i \geq 75$ and are truly anomalies. If the decision threshold is set to a value greater than or equal to the maximum value of CS_i^+ and CS_i^- for all the 23 data observations, for example, $H = 24.5$, CS_i^+ and CS_i^- for all the 23 data observations do not exceed H, and the two-side CUSUM control chart does not signal any data observation as an anomaly. We have no false alarms and zero hits, that is, we have the false alarm rate of 0% and the hit rate of 0%. If the decision threshold is set to a value smaller than the minimum value of CS_i^+ and CS_i^- for all the 23 data observations, for example, $H = -1$, CS_i^+ and CS_i^- for all the 23 data observations exceed H, and the two-side CUSUM control chart signals every data observation as an anomaly, producing 7 hits on all the true anomalies (observations #12, 16, 18, 19, 20, 21, and 22) and 16 false alarms, that is, the hit rate of 100% and the false alarm rate of 100%. If the decision threshold is set to $H = 0$, the two-side CUSUM control charts signals data observations #7, 9, 10, 11, 12, 14, 15, 16, 18, 19, 20, 21, and 22 as anomalies, producing 7 out-of-control signals on all the 7 true anomalies (the hit rate of 100%) and 7 out-of-control signals on observations #7, 9, 10, 11, 14, 15 and 23 out of 16 in-control data observations (the false alarm rate of 44%). Table 16.4 lists pairs of the false alarm rate and the hit rate for other values of H for the two-side CUSUM control chart in Example 16.1.

A ROC plots pairs of the hit rate and the false alarm rate for various values of a decision threshold. Figure 16.4 plots the ROC for the two-side CUSUM control chart in Example 16.1, given seven true anomalies on observations #12, 16, 18, 19, 20, 21, and 22. Unlike a pair of the false alarm rate and the hit rate for a specific value of a decision threshold, ROC gives a complete picture of performance by an anomaly detection technique. The closer the ROC is to the top-left corner (representing the false alarm rate 0% and the hit rate 100%) of the chart, the better performance the anomaly detection technique produces. Since it is difficult to set the decision thresholds for two different anomaly detection techniques so that their performance can be compared fairly, ROC can be plotted

TABLE 16.4

Pairs of the False Alarm Rate and the Hit Rate for Various Values of the Decision Threshold H for the Two-Side CUSUM Control Chart in Example 16.1

H	False Alarm Rate	Hit Rate
−1	1	1
0	0.44	1
0.5	0.38	1
2.5	0.38	0.86
5.5	0.38	0.71
6.5	0.31	0.71
8.5	0.25	0.57
10	0.19	0.57
11	0.06	0.57
12	0.06	0.43
12.5	0	0.43
18.5	0	0.29
21	0	0.14
24.5	0	0

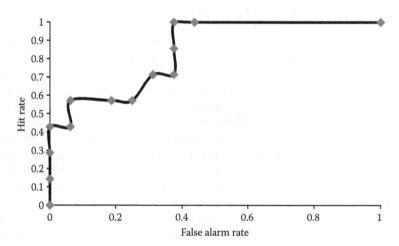

FIGURE 16.4
ROC for the two-side CUSUM control chart in Example 16.1.

for each technique in the same chart to compare ROCs for two techniques and examine which ROC is closer to the top-left corner of the chart to determine which technique produces better detection performance. Ye et al. (2002b) show the use of ROCs for a comparison of cyber attack detection performance by two control chart techniques.

16.6 Software and Applications

Minitab (www.minitab.com) supports statistical process control charts. Applications of univariate control charts to manufacturing quality and cyber intrusion detection can be found in Ye (2003, Chapter 4), Ye (2008), Ye et al. (2002a, 2004), and Ye and Chen (2003).

Exercises

16.1 Consider the launch temperature data and the following information in Example 16.1:

$$\mu_0 = 69$$

$$K = 3.5$$

construct a cuscore control chart using Equation 16.44 to monitor the launch temperature.

16.2 Plot the ROCs for the CUSUM control chart in Example 16.1, the EWMA control chart in Example 16.2, and the cuscore control chart in Exercise 16.1, in the same chart, and compare the performance of these control chart techniques.

16.3 Collect the data of daily temperatures in the last 12 months in your city, consider the temperature data in each month as a data sample, and construct a $\bar{x}$ control chart to monitor the local temperature and detect any anomaly.

16.4 Consider the same data set consisting of 12 monthly average temperatures obtained from Exercise 16.3 and use $\bar{\bar{x}}$ and $\bar{s}$ obtained from Exercise 16.3 to estimate μ_0 and σ. Set $K = 0.5\sigma$, and $H = 5\sigma$. Construct a two-side CUSUM control chart to monitor the data of the monthly average temperatures and detect any anomaly.

16.5 Consider the data set and the μ_0 and K values in Exercise 16.4. Construct a cuscore control chart to monitor the data of the monthly average temperatures and detect any anomaly.

16.6 Consider the data set and the estimate of μ_0 and σ in Exercise 16.4. Set $\lambda = 0.1$ and $L = 3$. Construct an EWMA control chart to monitor the data of the monthly average temperatures.

16.7 Repeat Exercise 16.6 but with $\lambda = 0.3$ and compare the EWMA control charts in Exercises 16.6 and 16.7.

17
Multivariate Control Charts

Multivariate control charts monitor multiple variables simultaneously for anomaly detection. This chapter describes three multivariate statistical control charts: Hotelling's T^2 control charts, multivariate EWMA control charts, and chi-square control charts. Some applications of multivariate control charts are given with references.

17.1 Hotelling's T^2 Control Charts

Let $x_i = (x_{i1}, \ldots, x_{ip})'$ denote an ith observation of random variables, $x_{i1}, \ldots, x_{ip}$, which follow a multivariate normal distribution (see the probability density function of a multivariate normal distribution in Chapter 16) with the mean vector of μ and the variance–covariance matrix Σ (see the definition of the variance–covariance matrix in Chapter 14). Given a data sample with n data observations, the sample mean vector $\bar{x}$ and the sample variance–covariance matrix S:

$$\bar{x} = \begin{bmatrix} \bar{x}_1 \\ \vdots \\ \bar{x}_p \end{bmatrix} \tag{17.1}$$

$$S = \frac{1}{n-1} \sum_{i=1}^{n} (x_i - \bar{x})(x_i - \bar{x})', \tag{17.2}$$

can be used to estimate μ and Σ, respectively. Hotelling's T^2 statistic for an observation x_i is (Chou et al., 1999; Everitt, 1979; Johnson and Wichern, 1998; Mason et al., 1995, 1997a,b; Mason and Young, 1999; Ryan, 1989):

$$T^2 = (x_i - \bar{x})' S^{-1} (x_i - \bar{x}), \tag{17.3}$$

where
  S^{-1} is the inverse of S
  Hotelling's T^2 statistic measures the statistical distance of x_i from $\bar{x}$

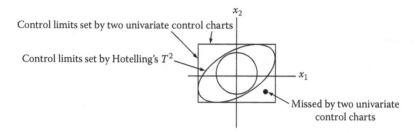

FIGURE 17.1
An illustration of statistical distance measured by Hotelling's T^2 and control limits of Hotelling's T^2 control charts and univariate control charts.

Suppose that we have $\bar{x} = 0$ at the origin of the two-dimensional space of x_1 and x_2 in Figure 17.1. In Figure 17.1, the data points x_is with the same statistical distance from $\bar{x}$ lie at the ellipse by taking into account the variance and covariance of x_1 and x_2, whereas all the data points x_is with the same Euclidean distance lie at the circle. The larger the value of Hotelling's T^2 statistic for an observation x_i, the larger statistical distance x_i is from $\bar{x}$.

A Hotelling's T^2 control chart monitors the Hotelling's T^2 statistic in Equation 17.3. If $x_{i1}, \ldots, x_{ip}$ follow a multivariate normal distribution, a transformed value of the Hotelling's T^2 statistic:

$$\frac{n(n-p)}{p(n+1)(n-1)} T^2$$

follows a F distribution with p and $n - p$ degrees of freedom. Hence, the tabulated F value for a given level of significance, for example, $\alpha = 0.05$, can be used as the signal threshold. If the transformed value of the Hotelling's T^2 statistic for an observation x_i is greater than the signal threshold, a Hotelling's T^2 control chart signals x_i as an anomaly. A Hotelling's T^2 control chart can detect both mean shifts and counter-relationships. Counter-relationships are large deviations from the covariance structure of the variables.

Figure 17.1 illustrates the control limits set by two individual $\bar{x}$ control charts for x_1 and x_2, respectively, and the control limits set by a Hotelling's T^2 control chart based on the statistical distance. Because each of the individual $\bar{x}$ control charts for x_1 and x_2 does not include the covariance structure of x_1 and x_2, a data observation deviating from the covariance structure of x_1 and x_2 is missed by each of the individual $\bar{x}$ control charts but detected by the Hotelling's T^2 control chart, as illustrated in Figure 17.1. It is pointed out in Ryan (1989) that Hotelling's T^2 control charts are more sensitive to counter-relationships than mean shifts. For example, if two variables have a positive correlation and a mean shift occurs with both variables but in the same direction to maintain their correlation, Hotelling's T^2 control charts may not detect the mean shift (Ryan, 1989). Hotelling's T^2 control charts are also sensitive to the multivariate normality assumption.

Example 17.1

The data set of the manufacturing system in Table 14.1, which is copied in Table 17.1, includes two attribute variables, x_7 and x_8, in nine cases of single-machine faults. The sample mean vector and the sample variance–covariance matrix are computed in Chapter 14 and given next. Construct a Hotelling's T^2 control chart to determine if the first data observation $x = (x_7, x_8) = (1, 0)$ is an anomaly.

$$\bar{x} = \begin{bmatrix} \bar{x}_7 \\ \bar{x}_8 \end{bmatrix} = \begin{bmatrix} \frac{5}{9} \\ \frac{4}{9} \end{bmatrix}$$

$$S = \begin{bmatrix} 0.2469 & -0.1358 \\ -0.1358 & 0.2469 \end{bmatrix}.$$

For the first data observation $x = (x_7, x_8) = (1, 0)$, we compute the value of the Hotelling's T^2 statistic:

$$T^2 = (x_i - \bar{x})'S^{-1}(x_i - \bar{x}) = \begin{bmatrix} 1 - \frac{5}{9} & 0 - \frac{4}{9} \end{bmatrix} \begin{bmatrix} 0.2469 & -0.1358 \\ -0.1358 & 0.2469 \end{bmatrix}^{-1} \begin{bmatrix} 1 - \frac{5}{9} \\ 0 - \frac{4}{9} \end{bmatrix}$$

$$= \begin{bmatrix} \frac{4}{9} & -\frac{4}{9} \end{bmatrix} \begin{bmatrix} 5.8070 & 3.1939 \\ 3.1939 & 5.8070 \end{bmatrix} \begin{bmatrix} \frac{4}{9} \\ -\frac{4}{9} \end{bmatrix} = 0.1435.$$

TABLE 17.1

Data Set for System Fault Detection with Two Quality Variables

Instance (Faulty Machine)	x_7	x_8
1 (M1)	1	0
2 (M2)	0	1
3 (M3)	1	1
4 (M4)	0	1
5 (M5)	1	0
6 (M6)	1	0
7 (M7)	1	0
8 (M8)	0	1
9 (M9)	0	0

The transformed T^2 has the value

$$\frac{n(n-p)}{p(n+1)(n-1)}T^2 = \frac{(9)(9-2)}{(2)(9+1)(9-1)}(0.1435) = 0.0502.$$

The tabulated F value for $\alpha = 0.05$ with 2 and 7 degrees of freedom is 4.74, which is used as the signal threshold. Since $0.0502 < 4.74$, the Hotelling's T^2 control chart does not signal $x = (x_7, x_8) = (1, 0)$ as an anomaly.

17.2 Multivariate EWMA Control Charts

Hotelling's T^2 control charts are a multivariate version of univariate $\bar{x}$ control charts in Chapter 16. Multivariate EWMA control charts are a multivariate version of EWMA control charts in Chapter 16. A multivariate EWMA control chart monitors the following statistic (Ye, 2003, Chapter 4):

$$T^2 = z_i' S_z^{-1} z_i, \tag{17.4}$$

where

$$z_i = \lambda x_i + (1-\lambda) z_{i-1}, \tag{17.5}$$

λ is a weight in (0, 1],

$$z_0 = \mu \quad \text{or} \quad \bar{x} \tag{17.6}$$

$$S_z = \frac{\lambda}{2-\lambda}[1-(1-\lambda)^{2i}]S \tag{17.7}$$

and S is the sample variance–covariance matrix of x.

17.3 Chi-Square Control Charts

Since Hotelling's T^2 control charts and multivariate EWMA control charts require computing the inverse of a variance–covariance matrix, these control charts are not scalable to a large number of variables. The presence of linearly correlated variables creates the difficulty of obtaining the inverse of

Multivariate Control Charts

a variance–covariance matrix. To address these problems, chi-square control charts are developed (Ye et al., 2002b, 2006). A chi-square control chart monitors the chi-square statistic for an observation $x_i = (x_{i1}, \ldots, x_{ip})'$ as follows:

$$\chi^2 = \sum_{j=1}^{p} \frac{(x_{ij} - \bar{x}_j)^2}{\bar{x}_j}. \tag{17.8}$$

For example, the data set of the manufacturing system in Table 17.1 includes two attribute variables, x_7 and x_8, in nine cases of single-machine faults. The sample mean vector is computed in Chapter 14 and given next:

$$\bar{x} = \begin{bmatrix} \bar{x}_7 \\ \bar{x}_8 \end{bmatrix} = \begin{bmatrix} \frac{5}{9} \\ \frac{4}{9} \end{bmatrix}.$$

The chi-square statistic for the first data observation in Table 17.1 $x = (x_7, x_8) = (1, 0)$ is

$$\chi^2 = \sum_{j=7}^{8} \frac{(x_{1j} - \bar{x}_j)^2}{\bar{x}_j} = \frac{(x_{17} - \bar{x}_7)^2}{\bar{x}_7} + \frac{(x_{18} - \bar{x}_8)^2}{\bar{x}_8} = \frac{\left(1 - \frac{5}{9}\right)^2}{\frac{5}{9}} + \frac{\left(0 - \frac{4}{9}\right)^2}{\frac{4}{9}} = 0.8.$$

If the p variables are independent and p is large, the chi-square statistic follows a normal distribution based on the central limit theorem. Given a sample of in-control data observations, the sample mean $\overline{\chi^2}$ and the sample variance s_{χ^2} of the chi-square statistic can be computed and used to set the control limits:

$$UCL = \overline{\chi^2} + L s_{\chi^2} \tag{17.9}$$

$$LCL = \overline{\chi^2} - L s_{\chi^2}. \tag{17.10}$$

If we let $L = 3$, we have the 3-sigma control limits. If the value of the chi-square statistic for an observation falls beyond [LCL, UCL], the chi-square control chart signals an anomaly.

In the work by Ye et al. (2006), chi-square control charts are compared with Hotelling's T^2 control charts in their performance of detecting mean shifts and counter-relationships for four types of data: (1) data with correlated and normally distributed variables, (2) data with uncorrelated and normally distributed variables, (3) data with auto-correlated and normally distributed variables, and (4) non-normally distributed variables without correlations or auto-correlations. The testing results show that chi-square control charts perform better or as well as Hotelling's T^2 control charts for data of types 2, 3, and 4. Hotelling's T^2 control charts perform better than chi-square control charts for data of type 1 only. However, for data of type 1, we can use techniques such as principal component analysis in Chapter 14 to obtain principal components. Then a chi-square control chart can be used to monitor principal components that are independent variables.

17.4 Applications

Applications of Hotelling's T^2 control charts and chi-square control charts to cyber attack detection for monitoring computer and network data and detecting cyber attacks as anomalies can be found in the work by Ye and her colleagues (Emran and Ye, 2002; Ye, 2003, Chapter 4; Ye, 2008; Ye and Chen, 2001; Ye et al., 2001, 2003, 2004, 2006). There are also applications of multivariate control charts in manufacturing (Ye, 2003, Chapter 4) and other fields.

Exercises

17.1 Use the data set of x_4, x_5, and x_6 in Table 8.1 to estimate the parameters for a Hotelling's T^2 control chart and construct the Hotelling's T^2 control chart with $\alpha = 0.05$ for the data set of x_4, x_5, and x_6 in Table 4.6 to monitor the data and detect any anomaly.

17.2 Use the data set of x_4, x_5, and x_6 in Table 8.1 to estimate the parameters for a chi-square control chart and construct the chi-square control chart with $L = 3$ for the data set of x_4, x_5, and x_6 in Table 4.6 to monitor the data and detect any anomaly.

17.3 Repeat Example 17.1 for the second observations.

Part VI

Algorithms for Mining Sequential and Temporal Patterns

18
Autocorrelation and Time Series Analysis

Time series data consist of data observations over time. If data observations are correlated over time, time series data are autocorrelated. Time series analysis was introduced by Box and Jenkins (1976) to model and analyze time series data with autocorrelation. Time series analysis has been applied to real-world data in many fields, including stock prices (e.g., S&P 500 index), airline fares, labor force size, unemployment data, and natural gas price (Yaffee and McGee, 2000). There are stationary and nonstationary time series data that require different statistical inference procedures. In this chapter, autocorrelation is defined. Several types of stationarity and nonstationarity time series are explained. Autoregressive and moving average (ARMA) models of stationary series data are described. Transformations of nonstationary series data into stationary series data are presented, along with autoregressive, integrated, moving average (ARIMA) models. A list of software packages that support time series analysis is provided. Some applications of time series analysis are given with references.

18.1 Autocorrelation

Equation 14.7 in Chapter 14 gives the correlation coefficient of two variables x_i and x_j:

$$\rho_{ij} = \frac{\sigma_{ij}}{\sqrt{\sigma_{ii}}\sqrt{\sigma_{jj}}},$$

where Equations 14.4 and 14.6 give

$$\sigma_i^2 = \sum_{\substack{\text{all values} \\ \text{of } x_i}} (x_i - u_i)^2 p_i(x_i)$$

$$\sigma_{ij} = \sum_{\substack{\text{all values} \\ \text{of } x_i}} \sum_{\substack{\text{all values} \\ \text{of } x_j}} (x_i - \mu_i)(x_j - \mu_j) p(x_i, x_j).$$

Given a variable x and a sample of its time series data x_t, $t = 1, \ldots, n$, we obtain the lag-k autocorrelation function (ACF) coefficient by replacing the variables x_i and x_j in the aforementioned equations with x_t and x_{t-k}, which are two data observations with time lags of k:

$$\text{ACF}(k) = \rho_k = \frac{\sum_{t=k+1}^{n}(x_t - \bar{x})(x_{t-k} - \bar{x})/(n-k)}{\sum_{t=1}^{n}(x_t - \bar{x})^2/n}, \qquad (18.1)$$

where $\bar{x}$ is the sample average. If time series data are statistically independent at lag-k, ρ_k is zero. If x_t and x_{t-k} change from $\bar{x}$ in the same way (e.g., both increasing from $\bar{x}$), ρ_k is positive. If x_t and x_{t-k} change from $\bar{x}$ in the opposite way (e.g., one increasing and another decreasing from $\bar{x}$), ρ_k is negative.

The lag-k partial autocorrelation function (PACF) coefficient measures the autocorrelation of lag-k, which is not accounted for by the autocorrelation of lags 1 to $k-1$. PACF for lag-1 and lag-2 are given next (Yaffee and McGee, 2000):

$$\text{PACF}(1) = \rho_1 \qquad (18.2)$$

$$\text{PACF}(2) = \frac{\rho_2 - \rho_1^2}{1 - \rho_1^2}. \qquad (18.3)$$

18.2 Stationarity and Nonstationarity

Stationarity usually refers to weak stationarity that requires the mean and variance of time series data not changing over time. A time series is strictly stationary if the autocovariance $\sigma_{t,t-k}$ does not change over time t but depends only on the number of time lags k in addition to the fixed mean and the constant variance. For example, a Gaussian time series that has a multivariate normal distribution is a strict stationary series because the mean, variance, and autocovariance of the series do not change over time. ARMA models are used to model stationary time series.

Nonstationarity may be caused by

- Outliers (see the description in Chapter 16)
- Random walk in which each observation randomly deviates from the previous observation without reversion to the mean
- Deterministic trend (e.g., a linear trend that has values changing over time at a constant rate)

- Changing variance
- Cycles with a data pattern that repeats periodically, including seasonable cycles with annual periodicity
- Others that make the mean or variance of a time series changes over time

A nonstationary series must be transformed into a stationary series in order to build an ARMA model.

18.3 ARMA Models of Stationary Series Data

ARMA models apply to time series data with weak stationarity. An autoregressive (AR) model of order p, AR(p), describes a time series in which the current observation of a variable x is a function of its previous p observation(s) and a random error:

$$x_t = \phi_1 x_{t-1} + \cdots + \phi_p x_{t-p} + e_t. \tag{18.4}$$

For example, time series data for the approval of president's job performance based on the Gallup poll is modeled as AR(1) (Yaffee and McGee, 2000):

$$x_t = \phi_1 x_{t-1} + e_t. \tag{18.5}$$

Table 18.1 gives a time series of an AR(1) model with $\phi_1 = 0.9$, $x_0 = 3$, and a white noise process for e_t with the mean of 0 and the standard deviation of 1.

TABLE 18.1

Time Series of an AR(1) Model with $\phi_1 = 0.9$, $x_0 = 3$, and a White Noise Process for e_t

t	e_t	x_t
1	0.166	2.866
2	−0.422	2.157
3	−1.589	0.353
4	0.424	0.741
5	0.295	0.962
6	−0.287	0.579
7	−0.140	0.381
8	0.985	1.328
9	−0.370	0.825
10	−0.665	0.078

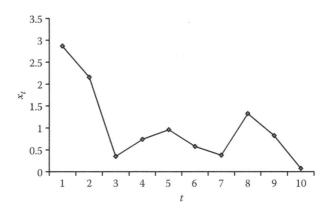

FIGURE 18.1
Time series data generated using an AR(1) model with $\phi_1 = 0.9$ and a white noise process for e_t.

Figure 18.1 plots this AR(1) time series. As seen in Figure 18.1, the effect of the initial x value, $x_0 = 3$, diminishes quickly.

A moving average (MA) model of order q, MA(q), describes a time series in which the current observation of a variable is an effect of a random error at the current time and random errors at previous q time points:

$$x_t = e_t - \theta_1 e_{t-1} - \cdots - \theta_q e_{t-q}. \tag{18.6}$$

For example, time series data from the epidemiological tracking on the proportion of the total population reported to have a disease (e.g., AIDS) is modeled as MV(1) (Yaffee and McGee, 2000)

$$x_t = e_t - \theta_1 e_{t-1}. \tag{18.7}$$

Table 18.2 gives a time series of an MA(1) model with $\theta_1 = 0.9$ and a white noise process for e_t with the mean of 0 and the standard deviation of 1. Figure 18.2 plots this MA(1) time series. As seen in Figure 18.2, $-0.9e_{t-1}$ in Formula 18.7 tends to bring x_t in the opposite direction of x_{t-1}, making x_ts oscillating.

An ARMA model, ARMA(p, q), describes a time series with both autoregressive and moving average characteristics:

$$x_t = \phi_1 x_{t-1} + \cdots + \phi_p x_{t-p} + e_t - \theta_1 x_{t-1} - \cdots - \theta_q x_{t-q}. \tag{18.8}$$

ARMA(p, 0) denotes an AR(p) model, and ARMA(0, q) denotes an MA(q) model. Generally, a smooth time series has high AR coefficients and low MA coefficients, and a time series affected dominantly by random errors has high MA coefficients and low AR coefficients.

TABLE 18.2
Time Series of an MA(1) Model with $\theta_1 = 0.9$ and a White Noise Process for e_t

t	e_t	x_t
0	0.649	
1	0.166	−0.418
2	−0.422	−0.046
3	−1.589	−1.548
4	0.424	1.817
5	0.295	−1.340
6	−0.287	0.919
7	−0.140	−0.967
8	0.985	1.856
9	−0.370	−2.040
10	−0.665	1.171

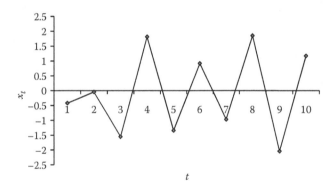

FIGURE 18.2
Time series data generated using an MA(1) model with $\theta_1 = 0.9$ and a white noise process for e_t.

18.4 ACF and PACF Characteristics of ARMA Models

ACF and PACF described in Section 18.1 provide analytical tools to reveal and identify the autoregressive order or the moving average order in an ARMA model for a time series. The characteristics of ACF and PACF for time series data generated by AR, MA and ARMA models are described next.

For an AR(1) time series:

$$x_t = \phi_1 x_{t-1} + e_t,$$

ACF(k) is (Yeffee and McGee, 2000)

$$\text{ACF}(k) = \phi_1^k. \tag{18.9}$$

If $\phi_1 < 1$, AR(1) is stationary with the exponential decline in the absolute value of ACF over time since ACF(k) decreases with k and eventually diminishes. If $\phi_1 > 0$, ACF(k) is positive. If $\phi_1 < 0$, ACF(k) is oscillating in that it is negative for $k = 1$, positive for $k = 2$, negative for $k = 3$, positive for $k = 4$, and so on. If $\phi_1 \geq 1$, AR(1) is nonstationary. For a stationary AR(2) time series:

$$x_t = \phi_1 x_{t-1} + \phi_2 x_{t-2} + e_t,$$

ACF(k) is positive with the exponential decline in the absolute value of ACF over time if $\phi_1 > 0$ and $\phi_2 > 0$, and ACF(k) is oscillating with the exponential decline in the absolute value of ACF over time if $\phi_1 < 0$ and $\phi_2 > 0$.

PACF(k) for an autoregressive series AR(p) carries through lag p and become zero after lag p. For AR(1), PACF(1) is positive if $\phi_1 > 0$ or negative if $\phi_1 < 0$, and PACF(k) for $k \geq 2$ is zero. For AR(2), PACF(1) and PACF(2) are positive if $\phi_1 > 0$ and $\phi_2 > 0$, PACF(1) is negative and PACF(2) is positive if $\phi_1 < 0$ and $\phi_2 > 0$, and PACF(k) for $k \geq 3$ is zero. Hence, PACF identifies the order of an autoregressive time series.

For an MA(1) time series,

$$x_t = e_t - \theta_1 e_{t-1},$$

ACF(1) is not zero as follows (Yeffee and McGee, 2000):

$$\text{ACF}(1) = \frac{-\theta_1}{1+\theta_1^2}, \tag{18.10}$$

and ACF(k) is zero for $k > 1$. Similarly, for an MA(2) time series, ACF(1) and ACF(2) are negative, and ACF(q) is zero for $q > 2$. For an MA(q), we have (Yeffee and McGee, 2000)

$$\text{ACF}(k) \neq 0 \quad \text{if } k \leq q$$
$$\text{ACF}(k) = 0 \quad \text{if } k > q.$$

Unlike an autoregressive time series whose ACF declines exponentially over time, a moving average time series has a finite memory since the autocorrelation of MA(q) carries only through lag q. Hence, ACF identifies the order of a moving average time series. A moving average time series has PACF whose magnitude exponentially declines over time. For MA(1), PACF(k) is negative if $\theta_1 > 0$, and PACF(k) is oscillating between positive and negative values with the exponential decline in the magnitude of PACF(k) over time. For MA(2), PACF(k) is negative with the exponential decline in the magnitude of PACF over time if $\theta_1 > 0$ and $\theta_2 > 0$, and ACF(k) is oscillating with the exponential decline in the absolute value of ACF over time if $\theta_1 < 0$ and $\theta_2 < 0$.

The aforementioned characteristics of autoregressive and moving average time series are combined in mixed time series with ARMA(p, q) models where $p > 0$ and $q > 0$. For example, for an ARMA(1,1) with $\phi_1 > 0$ and $\theta_1 < 0$, ACF declines exponentially overtime and PACF is oscillating with the exponential decline over time.

The parameters in an ARMA model can be estimated from a sample of time series data using the unconditional least-squares method, the conditional least-squares method, or the maximum likelihood method (Yeffee and McGee, 2000), which are supported in statistical software such as SAS (www.sas.com) and SPSS (www.ibm.com/software/analytics/spss/).

18.5 Transformations of Nonstationary Series Data and ARIMA Models

For nonstationary series caused by outliers, random walk, deterministic trend, changing variance, and cycles and seasonality, which are described in Section 18.2, methods of transforming those nonstationary series into stationary series are described next.

When outliers are detected in a time series, they can be removed and replaced using the average of the series. A random walk has each observation randomly deviating from the previous observation without reversion to the mean. Drunken drivers and birth rates have the behavior of a random walk (Yeffee and McGee, 2000). Differencing is applied to a random walk series as follows:

$$e_t = x_t - x_{t-1} \tag{18.11}$$

to obtain a stationary series of residual e_t, which is then modeled as an ARMA model. A deterministic trend such as the following linear trend:

$$x_t = a + bt + e_t, \tag{18.12}$$

can be removed by de-trending. The de-trending includes first building a regression model to capture the trend (e.g., a linear model for a linear trend or a polynomial model for a higher-order trend) and then obtaining a stationary series of residual e_t through differencing between the observed value and the predicted value from the regression model. For a changing variance with the variance of a time series expanding, contracting, or fluctuating over time, the natural log transformation or a power transformation (e.g., square and square root) can be considered to stabilize the variance (Yeffee and McGee, 2000). The natural log and power transformations belong to the family of Box–Cox transformations, which are defined as (Yaffee and McGee, 2000):

$$y_t = \frac{(x_t + c)^\lambda - 1}{\lambda} \quad \text{if } 0 < \lambda \leq 1$$

$$y_t = \ln x_t + c \quad \text{if } \lambda = 0$$

(18.13)

where
x_t is the original time series
y_t is the transformed time series
c is a constant
λ is a shape parameter

For a time series consisting of cycles, some of which are seasonable with annual periodicity, cyclic or seasonal differencing can be performed as follows:

$$e_t = x_t - x_{t-d},$$ (18.14)

where d is the number of time lags that a cycle spans.

The regular differencing and the cyclic/seasonal differencing can be added to an ARMA model to become an ARIMA model where I stands for integrated:

$$x_t - x_{t-d} = \phi_1 x_{t-1} + \cdots + \phi_p x_{t-p} + e_t - \theta_1 x_{t-1} - \cdots - \theta_q x_{t-q}.$$ (18.15)

18.6 Software and Applications

SAS (www.sas.com), SPSS (www.ibm.com/software/analytics/spss/), and MATLAB (www.mathworks.com) support time series analysis. In the work by Ye and her colleagues (Ye, 2008, Chapters 10 and 17), time series analysis is applied to uncovering and identifying autocorrelation characteristics of normal use and cyber attack activities using computer and network data. Time series models are built based on these characteristics and used in cuscore control charts as described in Chapter 16 to detect the presence of cyber attacks. The applications of time series analysis for forecasting can be found in Yaffee and McGee (2000).

Exercises

18.1 Construct time series data following an ARMA(1,1) model.

18.2 For the time series data in Table 18.1, compute ACF(1), ACF(2), ACF(3), PACF(1), and PACF(2).

18.3 For the time series data in Table 18.2, compute ACF(1), ACF(2), ACF(3), PACF(1), and PACF(2).

19
Markov Chain Models and Hidden Markov Models

Markov chain models and hidden Markov models have been widely used to build models and make inferences of sequential data patterns. In this chapter, Markov chain models and hidden Markov models are described. A list of data mining software packages that support the learning and inference of Markov chain models and hidden Markov models is provided. Some applications of Markov chain models and hidden Markov models are given with references.

19.1 Markov Chain Models

A Markov chain model describes the first-order discrete-time stochastic process of a system with the Markov property that the probability the system state at time n does not depend on previous system states leading to the system state at time $n-1$ but only the system state at $n-1$:

$$P(s_n|s_{n-1},\ldots,s_1) = P(s_n|s_{n-1}) \quad \text{for all } n, \tag{19.1}$$

where s_n is the system state at time n. A stationary Markov chain model has an additional property that the probability of a state transition from time $n-1$ to n is independent of time n:

$$P(s_n = j|s_{n-1} = i) = P(j|i), \tag{19.2}$$

where $p(j|i)$ is the probability that the system is in state j at one time given the system is in state i at the previous time. A stationary Markov model is simply called a Markov model in the following text.

If the system has a finite number of states, $1, \ldots, S$, a Markov chain model is defined by the state transition probabilities, $P(j|i)$, $i = 1, \ldots, S$, $j = 1, \ldots, S$,

$$\sum_{j=1}^{S} P(j|i) = 1, \tag{19.3}$$

and the initial state probabilities, $P(i)$, $i = 1, \ldots, S$,

$$\sum_{i=1}^{S} P(i) = 1, \tag{19.4}$$

where $P(i)$ is the probability that the system is in state i at time 1. The joint probability of a given sequence of system states $s_{n-K+1}, \ldots, s_n$ in a time window of size K including discrete times $n - (K-1), \ldots, n$ is computed as follows:

$$P(s_{n-K+1}, \ldots, s_n) = P(s_{n-K+1}) \prod_{k=K-1}^{1} P(s_{n-k+1} | s_{n-k}). \tag{19.5}$$

The state transition probabilities and the initial state probabilities can be learned from a training data set containing one or more state sequences as follows:

$$P(j|i) = \frac{N_{ji}}{N_{.i}} \tag{19.6}$$

$$P(i) = \frac{N_i}{N}, \tag{19.7}$$

where
 N_{ji} is the frequency that the state transition from state i to state j appears in the training data
 $N_{.i}$ is the frequency that the state transition from state i to any of the states, $1, \ldots, S$, appears in the training data
 N_i is the frequency that state i appears in the training data
 N is the total number of the states in the training data

Markov chain models can be used to learn and classify sequential data patterns. For each target class, sequential data with the target class can be used to build a Markov chain model by learning the state transition probability matrix and the initial probability distribution from the training data according to Equations 19.6 and 19.7. That is, we obtain a Markov chain model for each target class. If we have target classes, $1, \ldots, c$, we build Markov chain models, $M_1, \ldots, M_c$, for these target classes. Given a test sequence, the joint probability of this sequence is computed using Equation 19.5 under each Markov chain model. The test sequence is classified into the target class of the Markov chain model which gives the highest value for the joint probability of the test sequence.

In the applications of Markov chain models to cyber attack detection (Ye et al., 2002c, 2004), computer audit data under the normal use condition and under various attack conditions on computers are collected. There are totally 284 types of audit events in the audit data. Each audit event is considered as one of 284 system states. Each of the conditions (normal use and various attacks) is considered as a target class. The Markov chain model for a target class is learned from the training data under the condition of the target class. For each test sequence of audit events in an observation window, the joint probability of the test sequence is computed under each Markov chain model. The test sequence is classified into one of the conditions (normal or one of the attack types) to determine if an attack is present.

Example 19.1

A system has two states: misuse (m) and regular use (r). A sequence of system states is observed for training a Markov chain model: *mmmrrrrrrmrrmrrmrmmr*. Build a Markov chain model using the observed sequence of system states and compute the probability that the sequence of system states *mmrmrr* is generated by the Markov chain model.

Figure 19.1 shows the states and the state transitions in the observed training sequence of systems states.

Using Equation 19.6 and the training sequence of system states *mmmrrrrrrmrrmrrmrmmr*, we learn the following state transition probabilities:

$$P(m|m) = \frac{N_{mm}}{N_{.m}} = \frac{3}{8},$$

because state transitions 1, 2, and 18 are the state transition of $m \rightarrow m$, and state transitions 1, 2, 3, 10, 13, 16, 18, and 19 are the state transition of $m \rightarrow$ any state:

$$P(r|m) = \frac{N_{rm}}{N_{.m}} = \frac{5}{8},$$

because state transitions 3, 10, 13, 16, and 19 are the state transition of $m \rightarrow m$, and state transitions 1, 2, 3, 10, 13, 16, 18, and 19 are the state transition of $m \rightarrow$ any state:

$$P(m|r) = \frac{N_{mr}}{N_{.r}} = \frac{4}{11},$$

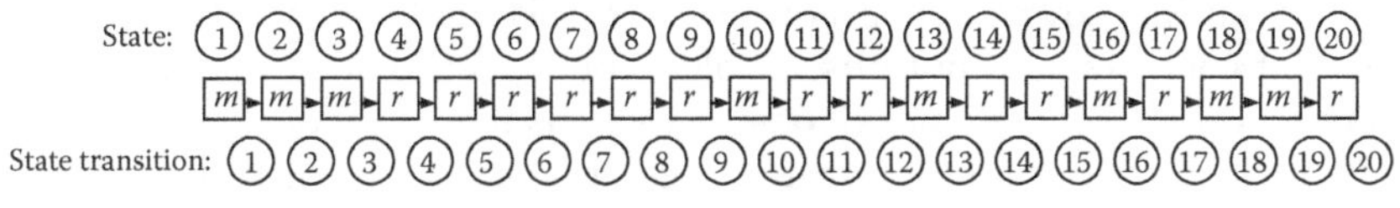

FIGURE 19.1
States and state transitions in Example 19.1.

because state transitions 9, 12, 15, and 17 are the state transition of $r \to m$, and state transitions 4, 5, 6, 7, 8, 9, 11, 12, 14, 15, and 17 are the state transition of $r \to$ any state:

$$P(r|r) = \frac{N_{rr}}{N_{.r}} = \frac{7}{11},$$

because state transitions 4, 5, 6, 7, 8, 11, and 14 are the state transition of $m \to m$, and state transitions 4, 5, 6, 7, 8, 9, 11, 12, 14, 15, and 17 are the state transition of $m \to$ any state.

Using Equation 19.7 and the training sequence of states *mmmrrrrrrmrrmrrmrmmr*, we learn the following initial state probabilities:

$$P(m) = \frac{N_m}{N} = \frac{8}{20},$$

because states 1, 2, 3, 10, 13, 16, 18, and 19 are state m, and there are 20 states in the sequence of states:

$$P(r) = \frac{N_r}{N} = \frac{12}{20},$$

because states 4, 5, 6, 7, 8, 9, 11, 12, 14, 15, 17, 20 are state r, and there are 20 states in the sequence of states.

After learning all the parameters of the Markov chain model, we compute the probability that the model generates the sequence of states: *mmrmrr*.

$$P(mmrmrr) = P(s_1)P(s_2|s_1)P(s_3|s_2)P(s_4|s_3)P(s_5|s_4)P(s_6|s_5)$$
$$= P(m)P(m|m)P(r|m)P(m|r)P(r|m)P(r|r)$$
$$= \left(\frac{8}{20}\right)\left(\frac{3}{8}\right)\left(\frac{5}{8}\right)\left(\frac{4}{11}\right)\left(\frac{5}{8}\right)\left(\frac{7}{11}\right) = 0.014.$$

19.2 Hidden Markov Models

In a hidden Markov model, an observation x is made at each stage, but the state s at each stage is not observable. Although the state at each stage is not observable, the sequence of observations is the result of state transitions and emissions of observation from states upon arrival at each state. In addition to the initial state probabilities and the state transition probabilities, the probability of emitting x from each state s, $P(x|s)$, is also defined as the emission probability in a hidden Markov model:

$$\sum_x P(x|s) = 1. \tag{19.8}$$

It is assumed that observations are independent of each other, and that the emission probability of x from each state s does not depend on other states.

A hidden Markov model is used to determine the probability that a given sequence of observations, $x_1, \ldots, x_N$, at stages $1, \ldots, N$, is generated by the hidden Markov model. Using any path method (Theodoridis and Koutroumbas, 1999), this probability is computed as follows:

$$\sum_{i=1}^{S^N} P(x_1, \ldots, x_N | s_{1_i}, \ldots, s_{N_i}) P(s_{1_i}, \ldots, s_{N_i})$$

$$= \sum_{i=1}^{S^N} P(s_{1_i}) P(x_1 | s_{1_i}) \prod_{n=2}^{N} P(s_{n_i} | s_{n-1_i}) P(x_n | s_{n_i}), \tag{19.9}$$

Where
 i is the index for a possible state sequence, $s_{1_i}, \ldots, s_{N_i}$, there are totally S^N possible state sequences
 $P(s_{1_i})$ is the initial state probability, $P(s_{n_i} | s_{n-1_i})$ is the state transition probability
 $P(x_n | s_{n_i})$ is the emission probability

Figure 19.2 shows stages $1, \ldots, N$, states $1, \ldots, S$, and observations at stages, $x_1, \ldots, x_N$, involved in computing Equation 19.9. To perform the computation in Equation 19.9, we define $\rho(s_n)$ as the probability that (1) state s_n is reached at

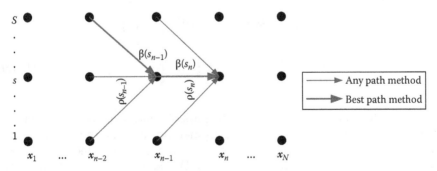

FIGURE 19.2
Any path method and the best path method for a hidden Markov model.

stage n, (2) observations $x_1, \ldots, x_{n-1}$ have been emitted at stages 1 to $n-1$, and (3) observation x_n is emitted from state s_{i_n} at stage n. $\rho(s_n)$ can be computed recursively as follows:

$$\rho(s_n) = \sum_{s_{n-1}=1}^{S} \rho(s_{n-1}) P(s_n | s_{n-1}) P(x_n | s_n), \quad (19.10)$$

$$\rho(s_1) = P(s_1) P(x_1 | s_1). \quad (19.11)$$

That is, $\rho(s_n)$ is the sum of the probabilities that starting from all possible state $s_n = 1, \ldots, S$ at stage $n-1$ with $x_1, \ldots, x_{n-1}$ already emitted, we transition to state s_n at stage n which emits x_n, as illustrated in Figure 19.2. Using Equations 19.10 and 19.11, Equation 19.9 can be computed as follows:

$$\sum_{i=1}^{S^N} P(x_1, \ldots, x_N | s_{1_i}, \ldots, s_{N_i}) P(s_{1_i}, \ldots, s_{N_i}) = \sum_{s_N=1}^{S} \rho(s_N). \quad (19.12)$$

Hence, in any path method, Equations 19.10 through 19.12 are used to compute the probability of a hidden Markov model generating a sequence of observations $x_1, \ldots, x_N$. Any path method starts by computing all $\rho(s_1)$ for $s_1 = 1, \ldots, S$ using Equation 19.11, then uses $\rho(s_1)$ to compute all $\rho(s_2)$, $s_2 = 1, \ldots, S$ using Equation 19.10, and continues all the way to obtain all $\rho(s_N)$ for $s_N = 1, \ldots, S$, which are finally used in Equation 19.12 to complete the computation.

The computational cost of any path method is high because all S^N possible state sequences/paths from stage 1 to stage N are involved in the computation. Instead of using Equation 19.9, the best path method uses Equation 19.13 to compute the probability that a given sequence of observations, $x_1, \ldots, x_N$, at stages $1, \ldots, N$, is generated by the hidden Markov model:

$$\max_{i=1}^{S^N} P(x_1, \ldots, x_N | s_{i_1}, \ldots, s_{i_N}) P(s_{i_1}, \ldots, s_{i_N})$$

$$= \max_{i=1}^{S^N} P(s_{i_1}) P(x_1 | s_{i_1}) \prod_{n=2}^{N} P(s_{i_n} | s_{i_{n-1}}) P(x_n | s_{i_n}). \quad (19.13)$$

That is, instead of summing over all the possible state sequences in Equation 19.9 for any path method, the best path method uses the maximum probability that the sequence of observations, $x_1, \ldots, x_N$, is generated by any possible state sequence from stage 1 to stage N. We define $\beta(s_n)$ as the probability that (1) state s_n is reached at stage n through the best path, (2) observations $x_1, \ldots, x_{n-1}$ have been emitted at stages 1 to $n-1$, and (3) observation x_n is emitted from state s_n at stage n. $\beta(s_n)$ can be computed recursively as follows using Bellman's principle (Theodoridis and Koutroumbas, 1999):

$$\beta(s_n) = \max_{s_{n-1}=1}^{S}\left[\beta(s_{n-1})P(s_n|s_{n-1})P(x_n|s_n)\right] \quad (19.14)$$

$$\beta(s_1) = P(s_1)P(x_1|s_1). \quad (19.15)$$

Equation 19.13 is computed using Equation 19.16:

$$\max_{i=1}^{S^N} P(x_1,\ldots,x_N|s_{i_1},\ldots,s_{i_N})P(s_{i_1},\ldots,s_{i_N}) = \max_{s_N=1}^{S}\beta(s_N). \quad (19.16)$$

The Viterbi algorithm (Viterbi, 1967) is widely used to compute the logarithm transformation of Equations 19.13 through 19.16.

The best path method requires less computational cost of storing and computing the probabilities than any path method because the computation at each stage n involves only the S best paths. However, in comparison with any path method, the best path method is an alternative suboptimal method for computing the probability that a given sequence of observations, $x_1, \ldots, x_N$, at stages 1, ..., N, is generated by the hidden Markov model, because only the best path instead of all possible paths is used to determine the probability of observing $x_1, \ldots, x_N$, given all possible paths in the hidden Markov model that can possibly generate the observation sequence.

Hidden Markov models have been widely used in speed recognition, handwritten character recognition, natural language processing, DNA sequence recognition, and so on. In the application of hidden Markov models to handwritten digit recognition (Bishop, 2006) for recognizing handwritten digits, 0, 1, ..., 9, a hidden Markov model is built for each digit. Each digit is considered to have a sequence of line trajectories, $x_1, \ldots, x_N$, at stages 1, ..., N. Each hidden Markov model has 16 latent states, each of which can emit a line segment of a fixed length with one of 16 possible angles. Hence, the emission distribution can be specified by a 16 × 16 matrix with the probability of emitting each of 16 angles from each of 16 states. The hidden Markov model for each digit is trained to establish the initial probability distribution, the transition probability matrix, and the emission probabilities using 45 handwritten examples of the digit. Given a handwritten digit to recognize, the probability that the handwritten digit is generated by the hidden Markov model for each digit is computed. The handwritten digit is classified as the digit whose hidden Markov model produces the highest probability of generating the handwritten digit.

Hence, to apply hidden Markov models to a classification problem, a hidden Markov model is built for each target class. Given a sequence of observations, the probability of generating this observation sequence by each hidden

Markov model is computed using any path method or the best path method. The given observation sequence is classified into the target class whose hidden Markov model produces the highest probability of generating the observation sequence.

19.3 Learning Hidden Markov Models

The set of model parameters for a hidden Markov model, A, includes the state transition probabilities, $P(j|i)$, the initial state probabilities, $P(i)$, and the emission probabilities, $P(x|i)$:

$$A = \{P(j|i), P(i), P(x|i)\}. \tag{19.17}$$

The model parameters need to be learned from a training data set containing a sequence of N observations, $X = x_1, \ldots, x_N$. Since the states cannot be directly observed, Equations 19.6 and 19.7 cannot be used to learn the model parameters such as the state transition probabilities and the initial state probabilities. Instead, the expectation maximization (EM) method is used to estimate the model parameters, which maximize the probability of obtaining the observation sequence from the model with the estimated model parameters, $P(X|A)$. The EM method has the following steps:

1. Assign initial values of the model parameters, A, and use these values to compute $P(X|A)$.
2. Reestimate the model parameters to obtain $\hat{A}$, and compute $P(X|\hat{A})$.
3. If $P(X|\hat{A}) - P(X|A) > \epsilon$, let $A = \hat{A}$ because $\hat{A}$ improves the probability of obtaining the observation sequence from $\hat{A}$ than A, and go to Step 2; otherwise, stop because $P(\hat{A})$ is worse than or similar to $P(A)$, and take A as the final set of the model parameters.

In Step 3, ϵ is a preset threshold of improvement in the probability of generating the observation sequence X from the model parameters.

$P(X|A)$ and $P(X|\hat{A})$ in the aforementioned EM method are computed using Equation 19.12 for any path method and Equation 19.16 for the best path method. If an observation is discrete and thus an observation sequence is a member of a finite set of observation sequences, the Baum–Welch reestimation method is used to reestimate the model parameters in Step 2 of the aforementioned EM method. Theodoridis and Koutroumbas (1999) describe the Baum–Welch reestimation method as follows. Let $\theta_n(i, j, X|A)$

be the probability that (1) the path goes through state i at stage n, (2) the path goes through state j at the next stage $n+1$, and (3) the model generates the observation sequence X using the model parameters A. Let $\varphi_n(i, X|A)$ be the probability that (1) the path goes through state i at stage n, and (2) the model generates the observation sequence X using the model parameters A. Let $\omega_n(i)$ be the probability of having the observations $x_{n+1}, \ldots, x_N$ at stages $n+1, \ldots, N$, given that the path goes through state i at stage n. For any path method, $\omega_n(i)$ can be computed recursively for $n = N-1, \ldots, 1$ as follows:

$$\omega_n(i) = P(x_{n+1}, \ldots, x_N | s_n = i, A) = \sum_{s_{n+1}=1}^{S} \omega_{n+1}(s_{n+1}) P(s_{n+1} | s_n = i) P(x_{n+1} | s_{n+1})$$

(19.18)

$$\omega_N(i) = 1, \quad i = 1, \ldots, S.$$

(19.19)

For the best path method, $\omega_n(i)$ can be computed recursively for $n = N-1, \ldots, 1$ as follows:

$$\omega_n(i) = P(x_{n+1}, \ldots, x_N | s_n = i, A) = \max_{s_{n+1}=1}^{S} \omega_{n+1}(s_{n+1}) P(s_{n+1} | s_n = i) P(x_{n+1} | s_{n+1})$$

(19.20)

$$\omega_N(i) = 1, \quad i = 1, \ldots, S.$$

(19.21)

We also have

$$\varphi_n(i, X|A) = \rho_n(i)\omega_n(i),$$

(19.22)

where $\rho_n(i)$ denotes $\rho(s_n = i)$, which is computed using Equations 19.10 and 19.11. The model parameter $P(i)$ is the expected number of times that state i occurs at stage 1, given the observation sequence X and the model parameters A, that is, $P(i|X, A)$. The model parameter $P(j|i)$ is the expected number of times of transitions from state i to state j occur, given the observation sequence X and the model parameters A, that is, $P(i, j|X, A)/P(i|X, A)$. The model parameters are reestimated as follows:

$$\hat{P}(i) = P(i|X, A) = \frac{\varphi_1(i, X|A)}{P(X|A)} = \frac{\rho_1(i)\omega_1(i)}{P(X|A)}$$

(19.23)

$$\hat{P}(j|i) = \frac{P(i,j|X,A)}{P(i|X,A)} = \frac{\sum_{n=1}^{N-1}\theta_n(i,j,X|A)/P(X|A)}{\sum_{n=1}^{N-1}\varphi_n(i,X|A)/P(X|A)}$$

$$= \frac{\sum_{n=1}^{N-1}\rho_n(i)P(j|i)P(x_{n+1}|j)\omega_{n+1}(j)/P(X|A)}{\sum_{n=1}^{N-1}\rho_n(i)\omega_n(i)/(X|A)}$$

$$= \frac{\sum_{n=1}^{N-1}\rho_n(i)P(j|i)P(x_{n+1}|j)\omega_{n+1}(j)}{\sum_{n=1}^{N-1}\rho_n(i)\omega_n(i)} \quad (19.24)$$

$$\hat{P}(x=v|i) = \frac{\sum_{n=1}^{N}\varphi_{n\&x=v}(i)/P(X|A)}{\sum_{n=1}^{N}\varphi_n(i)/P(X|A)} = \frac{\sum_{n=1}^{N}\varphi_{n\&x_n=v}(i)}{\sum_{n=1}^{N}\varphi_n(i)}$$

$$= \frac{\sum_{n=1}^{N}\rho_{n\&x=v}(i)\omega_{n\&x_n=v}(i)}{\sum_{n=1}^{N}\rho_n(i)\omega_n(i)} \quad (19.25)$$

where

$$\varphi_{n\&x_n=v}(i) = \begin{cases}\varphi_n(i) & \text{if } x_n = v \\ 0 & \text{if } x_n \neq v\end{cases} \quad (19.26)$$

$$\rho_{n\&x_n=v}(i) = \begin{cases}\rho_n(i) & \text{if } x_n = v \\ 0 & \text{if } x_n \neq v\end{cases} \quad (19.27)$$

$$\omega_{n\&x_n=v}(i) = \begin{cases}\omega_n(i) & \text{if } x_n = v \\ 0 & \text{if } x_n \neq v\end{cases} \quad (19.28)$$

and v is one of the discrete value vectors that x may take.

Example 19.2

A system has two states: misuse (m) and regular use (r), each of which can produce one of three events: F, G, and H. A sequence of five events is observed: FFFHG. Using any path method, perform one iteration of the model parameters reestimation in the EM method of learning a hidden Markov model from the observed sequence of events.

In Step 1 of the EM method, the following arbitrary values are assigned to the model parameters initially:

$$P(m) = 0.4 \quad P(r) = 0.6$$

$$P(m|m) = 0.375 \quad P(r|m) = 0.625 \quad P(m|r) = 0.364 \quad P(r|r) = 0.636$$

$$P(F|m) = 0.7 \quad P(G|m) = 0.1 \quad P(H|m) = 0.2$$

$$P(F|r) = 0.3 \quad P(G|r) = 0.4 \quad P(H|r) = 0.4.$$

Using these model parameters, we compute $P(X = FFFHG|A)$ using Equations 19.10, 19.11 and 19.12 for any path method:

$$\rho_1(m) = \rho(s_1 = m) = P(s_1 = m)P(x_1 = F|s_1 = m) = (0.4)(0.7) = 0.28$$

$$\rho_1(r) = \rho(s_1 = r) = P(s_1 = r)P(x_1 = F|s_1 = r) = (0.6)(0.2) = 0.12$$

$$\rho_2(m) = \rho(s_2 = m) = \sum_{s_1=1}^{2} \rho(s_1)P(s_2|s_1)P(x_2|s_2)$$

$$= \rho(s_1 = m)P(s_2 = m|s_1 = m)P(x_2 = F|s_2 = m)$$
$$+ \rho(s_1 = r)P(s_2 = m|s_1 = r)P(x_2 = F|s_2 = m)$$
$$= (0.28)(0.375)(0.7) + (0.12)(0.364)(0.7) = 0.1060$$

$$\rho_2(r) = \rho(s_2 = r) = \sum_{s_1=1}^{2} \rho(s_1)P(s_2|s_1)P(x_2|s_2)$$

$$= \rho(s_1 = m)P(s_2 = r|s_1 = m)P(x_2 = F|s_2 = r)$$
$$+ \rho(s_1 = r)P(s_2 = r|s_1 = r)P(x_2 = F|s_2 = r)$$
$$= (0.28)(0.625)(0.3) + (0.12)(0.636)(0.3) = 0.0754$$

$$\rho_3(m) = \rho(s_3 = m) = \sum_{s_2=1}^{2} \rho(s_2)P(s_3|s_2)P(x_3|s_3)$$

$$= \rho(s_2 = m)P(s_3 = m|s_2 = m)P(x_3 = F|s_3 = m)$$
$$+ \rho(s_2 = r)P(s_3 = m|s_2 = r)P(x_3 = F|s_3 = m)$$
$$= (0.1060)(0.375)(0.7) + (0.0754)(0.364)(0.7) = 0.0470$$

$$\rho_3(r) = \rho(s_3 = r) = \sum_{s_2=1}^{2} \rho(s_2) P(s_3|s_2) P(x_3|s_3)$$

$$= \rho(s_2 = m) P(s_3 = r|s_2 = m) P(x_3 = F|s_3 = r)$$

$$+ \rho(s_2 = r) P(s_3 = r|s_2 = r) P(x_3 = F|s_3 = r)$$

$$= (0.1060)(0.625)(0.2) + (0.0754)(0.636)(0.2) = 0.0228$$

$$\rho_4(m) = \rho(s_4 = m) = \sum_{s_3=1}^{2} \rho(s_3) P(s_4|s_3) P(x_4|s_4)$$

$$= \rho(s_3 = m) P(s_4 = m|s_3 = m) P(x_4 = H|s_4 = m)$$

$$+ \rho(s_3 = r) P(s_4 = m|s_3 = r) P(x_4 = H|s_4 = m)$$

$$= (0.0470)(0.375)(0.2) + (0.0228)(0.364)(0.2) = 0.0052$$

$$\rho_4(r) = \rho(s_4 = r) = \sum_{s_3=1}^{2} \rho(s_3) P(s_4|s_3) P(x_4|s_4)$$

$$= \rho(s_3 = m) P(s_4 = r|s_3 = m) P(x_4 = H|s_4 = r)$$

$$+ \rho(s_3 = r) P(s_4 = r|s_3 = r) P(x_4 = H|s_4 = r)$$

$$= (0.0470)(0.625)(0.4) + (0.0228)(0.636)(0.4) = 0.0176$$

$$\rho_5(m) = \rho(s_5 = m) = \sum_{s_4=1}^{2} \rho(s_4) P(s_5|s_4) P(x_5|s_5)$$

$$= \rho(s_4 = m) P(s_5 = m|s_4 = m) P(x_5 = G|s_5 = m)$$

$$+ \rho(s_4 = r) P(s_5 = m|s_4 = r) P(x_5 = G|s_5 = m)$$

$$= (0.0052)(0.375)(0.1) + (0.0176)(0.364)(0.1) = 0.0008$$

$$\rho_5(r) = \rho(s_5 = r) = \sum_{s_4=1}^{2} \rho(s_4) P(s_5|s_4) P(x_5|s_5)$$

$$= \rho(s_4 = m) P(s_5 = r|s_4 = m) P(x_5 = G|s_5 = r)$$

$$+ \rho(s_4 = r) P(s_5 = r|s_4 = r) P(x_5 = G|s_5 = r)$$

$$= (0.0052)(0.625)(0.4) + (0.0176)(0.636)(0.4) = 0.0058$$

$$P(X = FFFHG|A) = \sum_{s_5=1}^{2} \rho(s_5) = \rho(s_5 = m) + \rho(s_5 = r) = 0.0008 + 0.0058$$

$$= 0.0066.$$

In Step 2 of the EM method, we use Equations 19.23 through 19.25 to reestimate the model parameters. We first need to use Equations 19.18 and 19.19 to compute $\omega_n(i)$, $n = 5, 4, 3, 2$, and 1, which are used in Equations 19.23 through 19.25:

$$\omega_5(m) = 1 \quad \omega_5(r) = 1$$

$$\omega_4(m) = P(x_5 = G|s_4 = m, A) = \sum_{s_5=1}^{2} \omega_5(s_5) P(s_5|s_4 = m) P(x_5 = G|s_5)$$

$$= \omega_5(m) P(s_5 = m|s_4 = m) P(x_5 = G|s_5 = m)$$
$$+ \omega_5(r) P(s_5 = r|s_4 = m) P(x_5 = G|s_5 = r)$$
$$= (1)(0.375)(0.1) + (1)(0.625)(0.4) = 0.2875$$

$$\omega_4(r) = P(x_5 = G|s_4 = r, A) = \sum_{s_5=1}^{2} \omega_5(s_5) P(s_5|s_4 = r) P(x_5 = G|s_5)$$

$$= \omega_5(m) P(s_5 = m|s_4 = r) P(x_5 = G|s_5 = m)$$
$$+ \omega_5(r) P(s_5 = r|s_4 = r) P(x_5 = G|s_5 = r)$$
$$= (1)(0.364)(0.1) + (1)(0.636)(0.4) = 0.2908$$

$$\omega_3(m) = P(x_4 = H, x_5 = G|s_3 = m, A) = \sum_{s_4=1}^{2} \omega_4(s_4) P(s_4|s_3 = m) P(x_4 = H|s_4)$$

$$= \omega_4(m) P(s_4 = m|s_3 = m) P(x_4 = H|s_4 = m)$$
$$+ \omega_4(r) P(s_4 = r|s_3 = m) P(x_4 = H|s_4 = r)$$
$$= (0.2875)(0.375)(0.2) + (0.2908)(0.625)(0.4)$$
$$= 0.0943$$

$$\omega_3(r) = P(x_4 = H, x_5 = G|s_3 = r, A) = \sum_{s_4=1}^{2} \omega_4(s_4) P(s_4|s_3 = r) P(x_4 = H|s_4)$$

$$= \omega_4(m) P(s_4 = m|s_3 = r) P(x_4 = H|s_4 = m)$$
$$+ \omega_4(r) P(s_4 = r|s_3 = r) P(x_4 = H|s_4 = r)$$
$$= (0.2875)(0.364)(0.2) + (0.2908)(0.636)(0.4)$$
$$= 0.0949$$

$$\omega_2(m) = P(x_3 = F, x_4 = H, x_5 = G|s_2 = m, A) = \sum_{s_3=1}^{2} \omega_3(s_3)P(s_3|s_2 = m)P(x_3 = F|s_3)$$

$$= \omega_3(m)P(s_3 = m|s_2 = m)P(x_3 = F|s_3 = m)$$
$$+ \omega_3(r)P(s_3 = r|s_2 = m)P(x_3 = F|s_3 = r)$$
$$= (0.0943)(0.375)(0.7) + (0.0949)(0.625)(0.2)$$
$$= 0.0366$$

$$\omega_2(r) = P(x_3 = F, x_4 = H, x_5 = G|s_2 = r, A) = \sum_{s_3=1}^{2} \omega_3(s_3)P(s_3|s_2 = r)P(x_3 = F|s_3)$$

$$= \omega_3(m)P(s_3 = m|s_2 = r)P(x_3 = F|s_3 = m)$$
$$+ \omega_3(r)P(s_3 = r|s_2 = r)P(x_3 = F|s_3 = r)$$
$$= (0.0943)(0.364)(0.7) + (0.0949)(0.636)(0.2)$$
$$= 0.0361$$

$$\omega_1(m) = P(x_2 = F, x_3 = F, x_4 = H, x_5 = G|s_1 = m, A) = \sum_{s_2=1}^{2} \omega_2(s_2)P(s_2|s_1 = m)P(x_s = F|s_2)$$

$$= \omega_2(m)P(s_2 = m|s_1 = m)P(x_2 = F|s_2 = m)$$
$$+ \omega_2(r)P(s_2 = r|s_1 = m)P(x_2 = F|s_2 = r)$$
$$= (0.0366)(0.375)(0.7) + (0.0361)(0.625)(0.2)$$
$$= 0.0141$$

$$\omega_1(r) = P(x_2 = F, x_3 = F, x_4 = H, x_5 = G|s_1 = r, A) = \sum_{s_2=1}^{2} \omega_2(s_2)P(s_2|s_1 = r)P(x_s = F|s_2)$$

$$= \omega_2(m)P(s_2 = m|s_1 = r)P(x_2 = F|s_2 = m)$$
$$+ \omega_2(r)P(s_2 = r|s_1 = r)P(x_2 = F|s_2 = r)$$
$$= (0.0366)(0.364)(0.7) + (0.0361)(0.636)(0.2)$$
$$= 0.0139.$$

Now we use Equations 19.23 through 19.25 to reestimate the model parameters:

$$\hat{P}(m) = \frac{p_1(m)\omega_1(m)}{P(X = FFFHG|A)} = \frac{(0.28)(0.0141)}{0.0066} = 0.5982$$

$$\hat{P}(r) = \frac{\rho_1(r)\omega_1(r)}{P(X = FFFHG|A)} = \frac{(0.12)(0.0139)}{0.0066} = 0.2527$$

$$\hat{P}(m|m) = \frac{\sum_{n=1}^{4}\rho_n(m)P(m|m)P(x_{n+1}|m)\omega_{n+1}(m)}{\sum_{n=1}^{4}\rho_n(m)\omega_n(m)}$$

$$= \frac{\begin{bmatrix} \rho_1(m)P(m|m)P(x_2 = F|m)\omega_2(m) \\ +\rho_2(m)P(m|m)P(x_3 = F|m)\omega_3(m) \\ +\rho_3(m)P(m|m)P(x_4 = H|m)\omega_4(m) \\ +\rho_4(m)P(m|m)P(x_5 = G|m)\omega_5(m) \end{bmatrix}}{\begin{bmatrix} \rho_1(m)\omega_1(m) \\ +\rho_2(m)\omega_2(m) \\ +\rho_3(m)\omega_3(m) \\ +\rho_4(m)\omega_4(m) \end{bmatrix}}$$

$$= \frac{\begin{bmatrix} (0.28)(0.375)(0.7)(0.0366) + (0.1060)(0.375)(0.7)(0.0943) \\ +(0.0470)(0.375)(0.2)(0.2875) + (0.0052)(0.375)(0.1)(1) \end{bmatrix}}{[(0.28)(0.0141) + (0.1060)(0.0366) + (0.0470)(0.0943) + (0.0052)(0.2875)]}$$

$$= 0.4742$$

$$\hat{P}(r|m) = \frac{\sum_{n=1}^{4}\rho_n(m)P(r|m)P(x_{n+1}|r)\omega_{n+1}(r)}{\sum_{n=1}^{4}\rho_n(m)\omega_n(m)}$$

$$= \frac{\begin{bmatrix} \rho_1(m)P(r|m)P(x_2 = F|r)\omega_2(r) \\ +\rho_2(m)P(r|m)P(x_3 = F|r)\omega_3(r) \\ +\rho_3(m)P(r|m)P(x_4 = H|r)\omega_4(r) \\ +\rho_4(m)P(r|m)P(x_5 = G|r)\omega_5(r) \end{bmatrix}}{\begin{bmatrix} \rho_1(m)\omega_1(m) \\ +\rho_2(m)\omega_2(m) \\ +\rho_3(m)\omega_3(m) \\ +\rho_4(m)\omega_4(m) \end{bmatrix}}$$

$$= \frac{\begin{bmatrix} (0.28)(0.625)(0.2)(0.0361) + (0.1060)(0.625)(0.2)(0.0949) \\ +(0.0470)(0.625)(0.4)(0.2908) + (0.0052)(0.625)(0.4)(1) \end{bmatrix}}{[(0.28)(0.0141) + (0.1060)(0.0366) + (0.0470)(0.0943) + (0.0052)(0.2875)]}$$

$$= 0.5262$$

$$\hat{P}(m|r) = \frac{\sum_{n=1}^{4} \rho_n(r)P(m|r)P(x_{n+1}|m)\omega_{n+1}(m)}{\sum_{n=1}^{4} \rho_n(r)\omega_n(r)}$$

$$= \frac{\begin{bmatrix} \rho_1(r)P(m|r)P(x_2=F|m)\omega_2(m) \\ +\rho_2(r)P(m|r)P(x_3=F|m)\omega_3(m) \\ +\rho_3(r)P(m|r)P(x_4=H|m)\omega_4(m) \\ +\rho_4(r)P(m|r)P(x_5=G|m)\omega_5(m) \end{bmatrix}}{\begin{bmatrix} \rho_1(r)\omega_1(r) \\ +\rho_2(r)\omega_2(r) \\ +\rho_3(r)\omega_3(r) \\ +\rho_4(r)\omega_4(r) \end{bmatrix}}$$

$$= \frac{\begin{bmatrix} (0.12)(0.364)(0.7)(0.0366) + (0.0754)(0.364)(0.7)(0.0943) \\ +(0.0228)(0.364)(0.2)(0.2875) + (0.0176)(0.364)(0.1)(1) \end{bmatrix}}{[(0.12)(0.0139) + (0.0754)(0.0361) + (0.0228)(0.0949) + (0.0176)(0.2908)]}$$

$$= 0.3469$$

$$\hat{P}(r|r) = \frac{\sum_{n=1}^{4} \rho_n(r)P(r|r)P(x_{n+1}|r)\omega_{n+1}(r)}{\sum_{n=1}^{4} \rho_n(r)\omega_n(r)}$$

$$= \frac{\begin{bmatrix} \rho_1(r)P(r|r)P(x_2=F|r)\omega_2(r) \\ +\rho_2(r)P(r|r)P(x_3=F|r)\omega_3(r) \\ +\rho_3(r)P(r|r)P(x_4=H|r)\omega_4(r) \\ +\rho_4(r)P(r|r)P(x_5=G|r)\omega_5(r) \end{bmatrix}}{\begin{bmatrix} \rho_1(r)\omega_1(r) \\ +\rho_2(r)\omega_2(r) \\ +\rho_3(r)\omega_3(r) \\ +\rho_4(r)\omega_4(r) \end{bmatrix}}$$

$$= \frac{\begin{bmatrix} (0.12)(0.636)(0.2)(0.0361) + (0.0754)(0.636)(0.2)(0.0949) \\ +(0.0228)(0.636)(0.4)(0.2908) + (0.0176)(0.636)(0.4)(1) \end{bmatrix}}{[(0.12)(0.0139) + (0.0754)(0.0361) + (0.0228)(0.0949) + (0.0176)(0.2908)]}$$

$$= 0.6533$$

$$\hat{P}(x=F|m) = \frac{\sum_{n=1}^{5} \rho_{n\&x_n=F}(m)\omega_{n\&x_n=F}(m)}{\sum_{n=1}^{5} \rho_n(i)\omega_n(i)}$$

$$= \frac{\rho_{1\&x_1=F}(m)\omega_{1\&x_1=F}(m) + \rho_{2\&x_2=F}(m)\omega_{2\&x_2=F}(m) + \rho_{3\&x_3=F}(m)\omega_{3\&x_3=F}(m)}{\rho_1(m)\omega_1(m) + \rho_2(m)\omega_2(m) + \rho_3(m)\omega_3(m) + \rho_4(m)\omega_4(m) + \rho_5(m)\omega_5(m)}$$

$$= \frac{(0.28)(0.0141) + (0.1060)(0.0366) + (0.0470)(0.0943) + (0)(0) + (0)(0)}{(0.28)(0.0141) + (0.1060)(0.0366) + (0.0470)(0.0943) + (0.0052)(0.2875) + (0.0058)(1)}$$

$$= 0.6269$$

$$\hat{P}(x=G|m) = \frac{\sum_{n=1}^{5} \rho_{n\&x_n=G}(m)\omega_{n\&x_n=G}(m)}{\sum_{n=1}^{5} \rho_n(m)\omega_n(m)}$$

$$= \frac{\rho_{1\&x_1=G}(m)\omega_{1\&x_1=G}(m) + \rho_{2\&x_2=G}(m)\omega_{2\&x_2=G}(m) + \rho_{3\&x_3=G}(m)\omega_{3\&x_3=G}(m)}{\rho_1(m)\omega_1(m) + \rho_2(m)\omega_2(m) + \rho_3(m)\omega_3(m) + \rho_4(m)\omega_4(m) + \rho_5(m)\omega_5(m)}$$

$$= \frac{(0)(0) + (0)(0) + (0)(0) + (0)(0) + (0.0008)(1)}{(0.28)(0.0141) + (0.1060)(0.0366) + (0.0470)(0.0943) + (0.0052)(0.2875) + (0.0008)(1)}$$

$$= 0.0550$$

$$\hat{P}(x=H|m) = \frac{\sum_{n=1}^{5} \rho_{n\&x_n=H}(m)\omega_{n\&x_n=H}(m)}{\sum_{n=1}^{5} \rho_n(m)\omega_n(m)}$$

$$= \frac{\rho_{1\&x_1=H}(m)\omega_{1\&x_1=H}(m) + \rho_{2\&x_2=H}(m)\omega_{2\&x_2=H}(m) + \rho_{3\&x_3=H}(m)\omega_{3\&x_3=H}(m)}{\rho_1(m)\omega_1(m) + \rho_2(m)\omega_2(m) + \rho_3(m)\omega_3(m) + \rho_4(m)\omega_4(m) + \rho_5(m)\omega_5(m)}$$

$$= \frac{(0)(0) + (0)(0) + (0)(0) + (0.0052)(0.2875) + (0)(0)}{(0.28)(0.0141) + (0.1060)(0.0366) + (0.0470)(0.0943) + (0.0052)(0.2875) + (0.0008)(1)}$$

$$= 0.1027$$

$$\hat{P}(x=F|r) = \frac{\sum_{n=1}^{5} \rho_{n \& x_n =F}(r)\omega_{n \& x_n =F}(r)}{\sum_{n=1}^{5} \rho_n(r)\omega_n(r)}$$

$$= \frac{\rho_{1 \& x_1 =F}(r)\omega_{1 \& x_1 =F}(r) + \rho_{2 \& x_2 =F}(r)\omega_{2 \& x_2 =F}(r) + \rho_{3 \& x_3 =F}(r)\omega_{3 \& x_3 =F}(r)}{\rho_1(r)\omega_1(r) + \rho_2(r)\omega_2(r) + \rho_3(r)\omega_3(r) + \rho_4(r)\omega_4(r) + \rho_5(r)\omega_5(r)}$$

$$= \frac{(0.12)(0.0139) + (0.0754)(0.0361) + (0.0228)(0.0949) + (0)(0) + (0)(0)}{(0.12)(0.0139) + (0.0754)(0.0361) + (0.0228)(0.0949) + (0.0176)(0.2908) + (0.0058)(1)}$$

$$= 0.3751$$

$$\hat{P}(x=G|r) = \frac{\sum_{n=1}^{5} \rho_{n \& x_n =G}(r)\omega_{n \& x_n =G}(r)}{\sum_{n=1}^{5} \rho_n(r)\omega_n(r)}$$

$$= \frac{\rho_{1 \& x_1 =G}(r)\omega_{1 \& x_1 =G}(r) + \rho_{2 \& x_2 =G}(r)\omega_{2 \& x_2 =G}(r) + \rho_{3 \& x_3 =G}(r)\omega_{3 \& x_3 =G}(r)}{\rho_1(r)\omega_1(r) + \rho_2(r)\omega_2(r) + \rho_3(r)\omega_3(r) + \rho_4(r)\omega_4(r) + \rho_5(r)\omega_5(r)}$$

$$= \frac{(0)(0) + (0)(0) + (0)(0) + (0)(0) + (0.0058)(1)}{(0.12)(0.0139) + (0.0754)(0.0361) + (0.0228)(0.0949) + (0.0176)(0.2908) + (0.0058)(1)}$$

$$= 0.3320$$

$$\hat{P}(x=H|r) = \frac{\sum_{n=1}^{5} \rho_{n \& x_n =G}(r)\omega_{n \& x_n =G}(r)}{\sum_{n=1}^{5} \rho_n(r)\omega_n(r)}$$

$$= \frac{\rho_{1 \& x_1 =G}(r)\omega_{1 \& x_1 =G}(r) + \rho_{2 \& x_2 =G}(r)\omega_{2 \& x_2 =G}(r) + \rho_{3 \& x_3 =G}(r)\omega_{3 \& x_3 =G}(r)}{\rho_1(r)\omega_1(r) + \rho_2(r)\omega_2(r) + \rho_3(r)\omega_3(r) + \rho_4(r)\omega_4(r) + \rho_5(r)\omega_5(r)}$$

$$= \frac{(0)(0) + (0)(0) + (0)(0) + (0.0176)(0.2908) + (0)(0)}{(0.12)(0.0139) + (0.0754)(0.0361) + (0.0228)(0.0949) + (0.0176)(0.2908) + (0.0058)(1)}$$

$$= 0.2929$$

19.4 Software and Applications

The Hidden Markov Model Tookit (HTK) (http://htk.eng.cam.ac.uk/) supports hidden Markov models. Ye and her colleagues (Ye, 2008; Ye et al., 2002c, 2004b) describe the application of Markov chain models to cyber attack detection. Rabiner (1989) gives a review on applications of hidden Markov models to speech recognition.

Exercises

19.1 Given the Markov chain model in Example 19.1, determine the probability of observing a sequence of system states: *rmmrmrrmrrrrrmmm*.

19.2 A system has two states, misuse (*m*) and regular use (*r*), each of which can produce one of three events: *F, G,* and *H*. A hidden Markov model for the system has the initial state probabilities and state transition probabilities given from Example 19.1, and the state emission probabilities as follows:

$$P(F|m)=0.1 \quad P(G|m)=0.3 \quad P(H|m)=0.6$$

$$P(F|r)=0.5 \quad P(G|m)=0.2 \quad P(H|m)=0.3.$$

Use any path method to determine the probability of observing a sequence of five events: *GHFFH*.

19.3 Given the hidden Markov model in Exercise 19.2, use the best path method to determine the probability of observing a sequence of five events: *GHFFH*.

20
Wavelet Analysis

Many objects have a periodic behavior and thus show a unique characteristic in the frequency domain. For example, human sounds have a range of frequencies that are different from those of some animals. Objects in the space including the earth move at different frequencies. A new object in the space can be discovered by observing its unique movement frequency, which is different from those of known objects. Hence, the frequency characteristic of an object can be useful in identifying an object. Wavelet analysis represents time series data in the time–frequency domain using data characteristics over time in various frequencies, and thus allows us to uncover temporal data patterns at various frequencies. There are many forms of wavelets, e.g., Haar, Daubechies, and derivative of Gaussian (DoG). In this chapter, we use the Haar wavelet to explain how wavelet analysis works to transform time series data to data in the time–frequency domain. A list of software packages that support wavelet analysis is provided. Some applications of wavelet analysis are given with references.

20.1 Definition of Wavelet

A wavelet form is defined by two functions: the scaling function $\varphi(x)$ and the wavelet function $\psi(x)$. The scaling function of the Haar wavelet is a step function (Boggess and Narcowich, 2001; Vidakovic, 1999), as shown in Figure 20.1:

$$\varphi(x) = \begin{cases} 1 & \text{if } 0 \leq x < 1 \\ 0 & \text{otherwise} \end{cases}. \tag{20.1}$$

The wavelet function of the Haar wavelet is defined using the scaling function (Boggess and Narcowich, 2001; Vidakovic, 1999), as shown in Figure 20.1:

$$\psi(x) = \varphi(2x) - \varphi(2x-1) = \begin{cases} 1 & \text{if } 0 \leq x < \frac{1}{2} \\ -1 & \text{if } \frac{1}{2} \leq x < 1 \end{cases}. \tag{20.2}$$

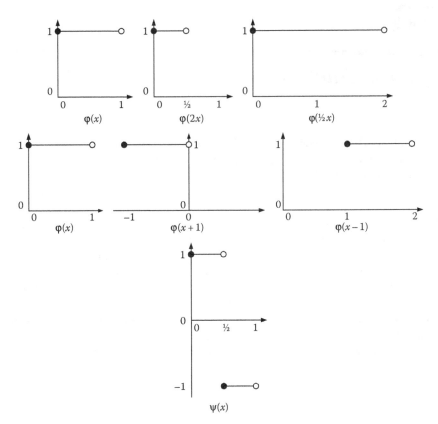

FIGURE 20.1
The scaling function and the wavelet function of the Haar wavelet and the dilation and shift effects.

Hence, the wavelet function of the Haar wavelet represents the change of the function value from 1 to −1 in [0, 1). The function φ(2x) in Formula 20.2 is a step function with the height of 1 for the range of x values in [0, ½), as shown in Figure 20.1. In general, the parameter a before x in φ(ax) produces a dilation effect on the range of x values, widening or contracting the x range by 1/a, as shown in Figure 20.1. The function φ(2x − 1) is also a step function with the height of 1 for the range of x values in [½, 1). In general, the parameter b in φ(x + b) produces a shift effect on the range of x values, moving the x range by b, as shown in Figure 20.1. Hence, φ(ax + b) defines a step function with the height of 1 for x values in the range of [−b/a, (1 − b)/a), as shown next, given a > 0:

$$0 \le ax + b < 1$$

$$\frac{-b}{a} \le x < \frac{1-b}{a}.$$

20.2 Wavelet Transform of Time Series Data

Given time series data with the function as shown in Figure 20.2a and a sample of eight data points 0, 2, 0, 2, 6, 8, 6, 8 taken from this function at the time locations 0, $\frac{1}{8}$, respectively, at the time interval of $\frac{1}{8}, \frac{2}{8}, \frac{3}{8}, \frac{4}{8}, \frac{5}{8}, \frac{6}{8}, \frac{7}{8}$, or at the frequency of 8, as shown in Figure 20.2b:

$$a_i, \quad i = 0, 1, \ldots, 2^k - 1, \quad k = 3 \text{ or}$$

$$a_0 = 0, a_1 = 2, a_2 = 0, a_3 = 2, a_4 = 6, a_5 = 8, a_6 = 6, a_7 = 8,$$

the function can be approximated using the sample of the data points and the scaling function of the Haar wavelet as follows:

$$f(x) = \sum_{i=0}^{2^k-1} a_i \varphi(2^k x - i) \qquad (20.3)$$

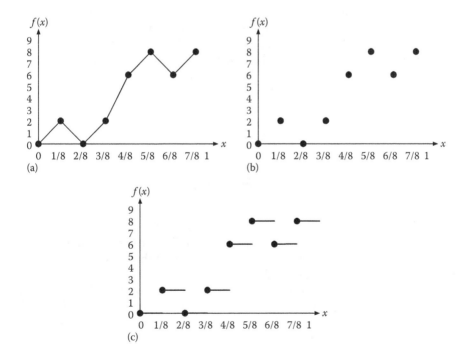

FIGURE 20.2
A sample of time series data from (a) a function, (b) a sample of data points taken from a function, and (c) an approximation of the function using the scaling function of Haar wavelet.

$$f(x) = a_0\varphi(2^3 x - 0) + a_1\varphi(2^3 x - 1) + a_2\varphi(2^3 x - 2) + a_3\varphi(2^3 x - 3) + a_4\varphi(2^3 x - 4)$$
$$+ a_5\varphi(2^3 x - 5) + a_6\varphi(2^3 x - 6) + a_7\varphi(2^3 x - 7)$$

$$f(x) = 0\varphi(2^3 x) + 2\varphi(2^3 x - 1) + 0\varphi(2^3 x - 2) + 2\varphi(2^3 x - 3) + 6\varphi(2^3 x - 4)$$
$$+ 8\varphi(2^3 x - 5) + 6\varphi(2^3 x - 6) + 8\varphi(2^3 x - 7)$$

In Formula 20.3, $a_i\varphi(2^k x - i)$ defines a step function with the height of a_i for x values in the range of $[i/2^k, (i+1)/2^k)$. Figure 20.2c shows the approximation of the function using the step functions at the height of the eight data points.

Considering the first two step functions in Formula 20.3, $\varphi(2^k x)$ and $\varphi(2^k x - 1)$, which have the value of 1 for the x values in $[0, 1/2^k)$ and $[1/2^k, 2/2^k)$, respectively, we have the following relationships:

$$\varphi(2^{k-1} x) = \varphi(2^k x) + \varphi(2^k x - 1) \tag{20.4}$$

$$\psi(2^{k-1} x) = \varphi(2^k x) - \varphi(2^k x - 1). \tag{20.5}$$

$\varphi(2^{k-1} x)$ in Equation 20.4 has the value of 1 for the x values in $[0, 1/2^{k-1})$, which covers $[0, 1/2^k)$ and $[1/2^k, 2/2^k)$ together. $\psi(2^{k-1} x)$ in Equation 20.5 also covers $[0, 1/2^k)$ and $[1/2^k, 2/2^k)$ together but has the value of 1 for the x values in $[0, 1/2^{k-1})$ and -1 for the x values in $[1/2^k, 2/2^k)$. An equivalent form of Equations 20.4 and 20.5 is obtained by adding Equations 20.4 and 20.5 and by subtracting Equation 20.5 from Equation 20.4:

$$\varphi(2^k x) = \frac{1}{2}\left[\varphi(2^{k-1} x) + \psi(2^{k-1} x)\right] \tag{20.6}$$

$$\varphi(2^k x - 1) = \frac{1}{2}\left[\varphi(2^{k-1} x) - \psi(2^{k-1} x)\right]. \tag{20.7}$$

At the left-hand side of Equations 20.6 and 20.7, we look at the data points at the time interval of $1/2^k$ or the frequency of 2^k. At the right-hand side of Equations 20.4 and 20.5, we look at the data points at the larger time interval of $1/2^{k-1}$ or a lower frequency of 2^{k-1}.

In general, considering the two step functions in Formula 20.3, $\varphi(2^k x - i)$ and $\varphi(2^k x - i - 1)$, which have the value of 1 for the x values in $[i/2^k, (i+1)/2^k)$ and $[(i+1)/2^k, (i+2)/2^k)$, respectively, we have the following relationships:

$$\varphi\left(2^{k-1} x - \frac{i}{2}\right) = \varphi(2^k x - i) + \varphi(2^k x - i - 1) \tag{20.8}$$

Wavelet Analysis

$$\psi\left(2^{k-1}x - \frac{i}{2}\right) = \varphi(2^k x - i) - \varphi(2^k x - i - 1). \tag{20.9}$$

$\varphi(2^{k-1}x - i/2)$ in Equation 20.8 has the value of 1 for the x values in $[i/2^k, (i+2)/2^k)$ or $[i/2^k, i/2^k + 1/2^{k-1})$ with the time interval of $1/2^{k-1}$. $\psi(2^{k-1}x - i/2)$ in Equation 20.9 has the value of 1 for the x values in $[i/2^k, (i+1)/2^k)$ and -1 for the x values in $[(i+1)/2^k, (i+2)/2^k]$. An equivalent form of Equations 20.8 and 20.9 is

$$\varphi(2^k x - i) = \frac{1}{2}\left[\varphi\left(2^{k-1}x - \frac{i}{2}\right) + \psi\left(2^{k-1}x - \frac{i}{2}\right)\right] \tag{20.10}$$

$$\varphi(2^k x - i - 1) = \frac{1}{2}\left[\varphi\left(2^{k-1}x - \frac{i}{2}\right) - \psi\left(2^{k-1}x - \frac{i}{2}\right)\right] \tag{20.11}$$

At the left-hand side of Equations 20.10 and 20.11, we look at the data points at the time interval of $1/2^k$ or the frequency of 2^k. At the right-hand side of Equations 20.10 and 20.11, we look at the data points at the larger time interval of $1/2^{k-1}$ or a lower frequency of 2^{k-1}.

Equations 20.10 and 20.11 allow us to perform the wavelet transform of times series data or their function representation in Formula 20.3 into data at various frequencies as illustrated through Example 20.1.

Example 20.1

Perform the Haar wavelet transform of time series data 0, 2, 0, 2, 6, 8, 6, 8. First, we represent the time series data using the scaling function of the Haar wavelet:

$$f(x) = \sum_{i=0}^{2^k - 1} a_i \varphi(2^k x - i)$$

$$f(x) = 0\varphi(2^3 x) + 2\varphi(2^3 x - 1)$$

$$+ 0\varphi(2^3 x - 2) + 2\varphi(2^3 x - 3)$$

$$+ 6\varphi(2^3 x - 4) + 8\varphi(2^3 x - 5)$$

$$+ 6\varphi(2^3 x - 6) + 8\varphi(2^3 x - 7).$$

Then, we use Equations 20.10 and 20.11 to transform the aforementioned function. When performing the wavelet transform of the aforementioned function, we use $i = 0$ and $i + 1 = 1$ for the first pair of the scaling functions at the right-hand side of the aforementioned function,

$i = 2$ and $i + 1 = 3$ for the second pair, $i = 4$ and $i + 1 = 5$ for the third pair, and $i = 6$ and $i + 1 = 7$ for the fourth pair:

$$f(x) = 0 \times \frac{1}{2}\left[\varphi\left(2^2 x - \frac{0}{2}\right) + \psi\left(2^2 x - \frac{0}{2}\right)\right] + 2 \times \frac{1}{2}\left[\varphi\left(2^{k-1} x - \frac{0}{2}\right) - \psi\left(2^{k-1} x - \frac{0}{2}\right)\right]$$

$$+ 0 \times \frac{1}{2}\left[\varphi\left(2^2 x - \frac{2}{2}\right) + \psi\left(2^2 x - \frac{2}{2}\right)\right] + 2 \times \frac{1}{2}\left[\varphi\left(2^{k-1} x - \frac{2}{2}\right) - \psi\left(2^{k-1} x - \frac{2}{2}\right)\right]$$

$$+ 6 \times \frac{1}{2}\left[\varphi\left(2^2 x - \frac{4}{2}\right) + \psi\left(2^2 x - \frac{4}{2}\right)\right] + 8 \times \frac{1}{2}\left[\varphi\left(2^{k-1} x - \frac{4}{2}\right) - \psi\left(2^{k-1} x - \frac{4}{2}\right)\right]$$

$$+ 6 \times \frac{1}{2}\left[\varphi\left(2^2 x - \frac{6}{2}\right) + \psi\left(2^2 x - \frac{6}{2}\right)\right] + 8 \times \frac{1}{2}\left[\varphi\left(2^{k-1} x - \frac{6}{2}\right) - \psi\left(2^{k-1} x - \frac{6}{2}\right)\right]$$

$$f(x) = 0 \times \frac{1}{2}\left[\varphi(2^2 x) + \psi(2^2 x)\right] + 2 \times \frac{1}{2}\left[\varphi(2^2 x) - \psi(2^2 x)\right]$$

$$+ 0 \times \frac{1}{2}\left[\varphi(2^2 x - 1) + \psi(2^2 x - 1)\right] + 2 \times \frac{1}{2}\left[\varphi(2^2 x - 1) - \psi(2^2 x - 1)\right]$$

$$+ 6 \times \frac{1}{2}\left[\varphi(2^2 x - 2) + \psi(2^2 x - 2)\right] + 8 \times \frac{1}{2}\left[\varphi(2^2 x - 2) - \psi(2^2 x - 2)\right]$$

$$+ 6 \times \frac{1}{2}\left[\varphi(2^2 x - 3) + \psi(2^2 x - 3)\right] + 8 \times \frac{1}{2}\left[\varphi(2^2 x - 3) - \psi(2^2 x - 3)\right]$$

$$f(x) = \left(0 \times \frac{1}{2} + 2 \times \frac{1}{2}\right)\varphi(2^2 x) + \left(0 \times \frac{1}{2} - 2 \times \frac{1}{2}\right)\psi(2^2 x)$$

$$+ \left(0 \times \frac{1}{2} + 2 \times \frac{1}{2}\right)\varphi(2^2 x - 1) + \left(0 \times \frac{1}{2} - 2 \times \frac{1}{2}\right)\psi(2^2 x - 1)$$

$$+ \left(6 \times \frac{1}{2} + 8 \times \frac{1}{2}\right)\varphi(2^2 x - 2) + \left(6 \times \frac{1}{2} - 8 \times \frac{1}{2}\right)\psi(2^2 x - 2)$$

$$+ \left(6 \times \frac{1}{2} + 8 \times \frac{1}{2}\right)\varphi(2^2 x - 3) + \left(6 \times \frac{1}{2} - 8 \times \frac{1}{2}\right)\psi(2^2 x - 3)$$

$$f(x) = \varphi(2^2 x) - \psi(2^2 x)$$
$$+ \varphi(2^2 x - 1) - \psi(2^2 x - 1)$$
$$+ 7\varphi(2^2 x - 2) - 1\psi(2^2 x - 2)$$
$$+ 7\varphi(2^2 x - 3) - 1\psi(2^2 x - 3)$$

$$f(x) = \varphi(2^2 x) + \varphi(2^2 x - 1) + 7\varphi(2^2 x - 2) + 7\varphi(2^2 x - 3)$$
$$- \psi(2^2 x) - \psi(2^2 x - 1) - \psi(2^2 x - 2) - \psi(2^2 x - 3).$$

Wavelet Analysis

We use Equations 20.10 and 20.11 to transform the first line of the aforementioned function:

$$f(x) = \frac{1}{2}\left[\varphi(2^1 x) + \psi(2^1 x)\right] + \frac{1}{2}\left[\varphi(2^1 x) - \psi(2^1 x)\right]$$

$$+ 7 \times \frac{1}{2}\left[\varphi(2^1 x - 1) + \psi(2^1 x - 1)\right] + 7 \times \frac{1}{2}\left[\varphi(2^1 x - 1) - \psi(2^1 x - 1)\right]$$

$$-\psi(2^2 x) - \psi(2^2 x - 1) - \psi(2^2 x - 2) - \psi(2^2 x - 3)$$

$$f(x) = \left(\frac{1}{2} + \frac{1}{2}\right)\varphi(2x) + \left(\frac{1}{2} - \frac{1}{2}\right)\psi(2x) + \left(\frac{7}{2} + \frac{7}{2}\right)\varphi(2x - 1) + \left(\frac{7}{2} - \frac{7}{2}\right)\psi(2x - 1)$$

$$-\psi(2^2 x) - \psi(2^2 x - 1) - \psi(2^2 x - 2) - \psi(2^2 x - 3)$$

$$f(x) = \varphi(2x) + 7\varphi(2x - 1)$$
$$+ 0\psi(2x) + 0\psi(2x - 1)$$
$$-\psi(2^2 x) - \psi(2^2 x - 1) - \psi(2^2 x - 2) - \psi(2^2 x - 3).$$

Again, we use Equations 20.10 and 20.11 to transform the first line of the aforementioned function:

$$f(x) = \frac{1}{2}\left[\varphi(2^{1-1} x) + \varphi(2^{1-1} x)\right] + 7 \times \frac{1}{2}\left[\varphi(2^{1-1} x) - \psi(2^{1-1} x)\right]$$

$$+ 0\psi(2x) + 0\psi(2x - 1) - \psi(2^2 x) - \psi(2^2 x - 1)$$

$$-\psi(2^2 x - 2) - \psi(2^2 x - 3)$$

$$f(x) = \left(\frac{1}{2} + \frac{7}{2}\right)\varphi(x) + \left(\frac{1}{2} - \frac{7}{2}\right)\psi(x)$$

$$+ 0\psi(2x) + 0\psi(2x - 1) - \psi(2^2 x) - \psi(2^2 x - 1)$$

$$-\psi(2^2 x - 2) - \psi(2^2 x - 3)$$

$$f(x) = 4\varphi(x) - 3\psi(x) + 0\psi(2x) + 0\psi(2x - 1)$$
$$-\psi(2^2 x) - \psi(2^2 x - 1) - \psi(2^2 x - 2) - \psi(2^2 x - 3). \quad (20.12)$$

The function in Equation 20.12 gives the final result of the Haar wavelet transform. The function has eight terms, as the original data sample has eight data points. The first term, $4\varphi(x)$, represents a step function at the height of 4 for x in [0, 1) and gives the average of the original data points, 0, 2, 0, 2, 6, 8, 6, 8. The second term, $-3\psi(x)$, has the wavelet function $\psi(x)$, which represents a step change of the function value from 1 to -1 or the step change of -2 as the x values go from the first half of the range [0, ½) to the second half of the range [½, 1). Hence, the second term, $-3\psi(x)$, reveals that the original time series data have the step change of $(-3) \times (-2) = 6$ from the first half set of four data points to the second half set of four data points as the average of the first four data points is 1 and the average of the last four data points is 7. The third term, $0\psi(2x)$, represents that the original time series data have no step change from the first and second data points to the third and four data points as the average of the first and second data points is 1 and the average of the third and fourth data points is 1. The fourth term, $0\psi(2x-1)$, represents that the original time series data have no step change from the fifth and sixth data points to the seventh and eighth data points as the average of the fifth and sixth data points is 7 and the average of the seventh and eighth data points is 7. The fifth, sixth, seventh and eighth terms of the function in Equation 20.12, $-\psi(2^2x)$, $-\psi(2^2x-1)$, $-\psi(2^2x-2)$ and $-\psi(2^2x-3)$, reveal that the original time series data have the step change of $(-1) \times (-2) = 2$ from the first data point of 0 to the second data point of 2, the step change of $(-1) \times (-2) = 2$ from the third data point of 0 to the fourth data point of 2, the step change of $(-1) \times (-2) = 2$ from the fifth data point of 6 to the sixth data point of 8, and the step change of $(-1) \times (-2) = 2$ from the seventh data point of 6 to the eighth data point of 8. Hence, the Haar wavelet transform of eight data points in the original time series data produces eight terms with the coefficient of the scaling function $\varphi(x)$ revealing the average of the original data, the coefficient of the wavelet function $\psi(x)$ revealing the step change in the original data at the lowest frequency from the first half set of four data points to the second half set of four data points, the coefficients of the wavelet functions $\psi(2x)$ and $\psi(2x-1)$ revealing the step changes in the original data at the higher frequency of every two data points, and the coefficients of the wavelet functions $\psi(2^2x)$, $\psi(2^2x-1)$, $\psi(2^2x-2)$ and $\psi(2^2x-3)$ revealing the step changes in the original data at the highest frequency of every data point.

Hence, the Haar wavelet transform of times series data allows us to transform time series data to the data in the time–frequency domain and observe the characteristics of the wavelet data pattern (e.g., a step change for the Haar wavelet) in the time–frequency domain. For example, the wavelet transform of the time series data 0, 2, 0, 2, 6, 8, 6, 8 in Equation 20.12 reveals that the data have the average of 4, a step increase of 6 at four data points (at the lowest frequency of step change), no step change at every two data points (at the medium frequency of step change), and a step increase of 2 at every data point (at the highest frequency of step change). In addition to the Haar wavelet that captures the data pattern of a step change, there are many other wavelet forms, for example, the Paul wavelet, the DoG wavelet, the Daubechies wavelet, and Morlet wavelet as shown in Figure 20.3, which capture other types of data patterns. Many wavelet forms are developed so that an appropriate wavelet form can be selected to give a close match to the

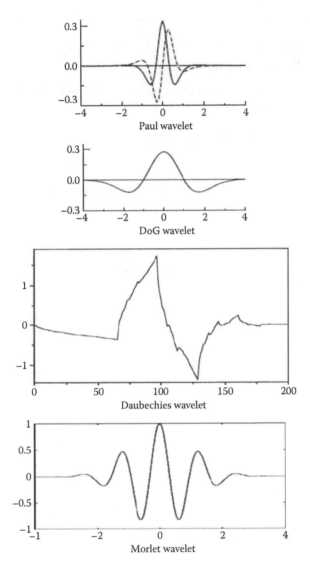

FIGURE 20.3
Graphic illustration of the Paul wavelet, the DoG wavelet, the Daubechies wavelet, and the Morlet wavelet. (Ye, N., *Secure Computer and Network Systems: Modeling, Analysis and Design*, 2008, Figure 11.2, p. 200. Copyright Wiley-VCH Verlag GmbH & Co. KGaA. Reproduced with permission).

data pattern of time series data. For example, the Daubechies wavelet (Daubechies, 1990) may be used to perform the wavelet transform of time series data that shows a data pattern of linear increase or linear decrease. The Paul and DoG wavelets may be used for time series data that show wave-like data patterns.

20.3 Reconstruction of Time Series Data from Wavelet Coefficients

Equations 20.8 and 20.9, which are repeated next, can be used to reconstruct the time series data from the wavelet coefficients:

$$\varphi\left(2^{k-1}x - \frac{i}{2}\right) = \varphi(2^k x - i) + \varphi(2^k x - i - 1)$$

$$\psi\left(2^{k-1}x - \frac{i}{2}\right) = \varphi(2^k x - i) - \varphi(2^k x - i - 1).$$

Example 20.2

Reconstruct time series data from the wavelet coefficients in Equation 20.12, which is repeated next:

$f(x) = 4\varphi(x)$
$\quad - 3\psi(x)$
$\quad + 0\psi(2x) + 0\psi(2x - 1)$
$\quad - \psi(2^2 x) - \psi(2^2 x - 1) - \psi(2^2 x - 2) - \psi(2^2 x - 3)$

$f(x) = 4 \times [\varphi(2^1 x) + \varphi(2^1 x - 1)]$
$\quad - 3 \times [\varphi(2^1 x) - \varphi(2^1 x - 1)]$
$\quad + 0 \times [\varphi(2^2 x) - \varphi(2^2 x - 1)] + 0 \times [\varphi(2^2 x - 2) - \varphi(2^2 x - 3)]$
$\quad - [\varphi(2^3 x) - \varphi(2^3 x - 1)] - [\varphi(2^3 x - 2) - \varphi(2^3 x - 3)] - [\varphi(2^3 x - 4) - \varphi(2^3 x - 5)]$
$\quad - [\varphi(2^3 x - 6) - \varphi(2^3 x - 7)]$

$f(x) = \varphi(2x) + 7\varphi(2x - 1)$
$\quad - \varphi(2^3 x) + \varphi(2^3 x - 1) - \varphi(2^3 x - 2) + \varphi(2^3 x - 3) - \varphi(2^3 x - 4)$
$\quad + \varphi(2^3 x - 5) - \varphi(2^3 x - 6) + \varphi(2^3 x - 7)$

$f(x) = [\varphi(2^2 x) + \varphi(2^2 x - 1)] + 7 \times [\varphi(2^2 x - 2) + \varphi(2^2 x - 3)]$
$\quad - \varphi(2^3 x) + \varphi(2^3 x - 1) - \varphi(2^3 x - 2) + \varphi(2^3 x - 3) - \varphi(2^3 x - 4) + \varphi(2^3 x - 5)$
$\quad - \varphi(2^3 x - 6) + \varphi(2^3 x - 7)$

$$f(x) = \varphi(2^2 x) + \varphi(2^2 x - 1) + 7\varphi(2^2 x - 2) + 7\varphi(2^2 x - 3)$$
$$- \varphi(2^3 x) + \varphi(2^3 x - 1) - \varphi(2^3 x - 2) + \varphi(2^3 x - 3) - \varphi(2^3 x - 4) + \varphi(2^3 x - 5)$$
$$- \varphi(2^3 x - 6) + \varphi(2^3 x - 7)$$

$$f(x) = \left[\varphi(2^3 x) + \varphi(2^3 x - 1)\right] + \left[\varphi(2^3 x - 2) + \varphi(2^3 x - 3)\right]$$
$$+ 7 \times \left[\varphi(2^3 x - 4) + \varphi(2^3 x - 5)\right] + 7 \times \left[\varphi(2^3 x - 6) + \varphi(2^3 x - 7)\right]$$
$$- \varphi(2^3 x) + \varphi(2^3 x - 1) - \varphi(2^3 x - 2) + \varphi(2^3 x - 3) - \varphi(2^3 x - 4)$$
$$+ \varphi(2^3 x - 5) - \varphi(2^3 x - 6) + \varphi(2^3 x - 7)$$

$$f(x) = 0\varphi(2^3 x) + 2\varphi(2^3 x - 1)$$
$$+ 0\varphi(2^3 x - 2) + 2\varphi(2^3 x - 3)$$
$$+ 6\varphi(2^3 x - 4) + 8\varphi(2^3 x - 5)$$
$$+ 6\varphi(2^3 x - 6) + 8\varphi(2^3 x - 7).$$

Taking the coefficients of the scaling functions at the right-hand side of the last equation gives us the original sample of time series data, 0, 2, 0, 2, 6, 8, 6, 8.

20.4 Software and Applications

Wavelet analysis is supported in software packages including Statistica (www.statistica.com) and MATLAB® (www.matworks.com). As discussed in Section 20.2, the wavelet transform can be applied to uncover characteristics of certain data patterns in the time–frequency domain. For example, by examining the time location and frequency of the Haar wavelet coefficient with the largest magnitude, the biggest rise of the New York Stock Exchange Index for the 6-year period of 1981–1987 was detected to occur from the first 3 years to the next 3 years (Boggess and Narcowich, 2001). The application of the Haar, Paul, DoG, Daubechies, and Morlet wavelet to computer and network data can be found in Ye (2008, Chapter 11).

The wavelet transform is also useful for many other types of applications, including noise reduction and filtering, data compression, and edge detection (Boggess and Narcowich, 2001). Noise reduction and filtering are usually done by setting zero to the wavelet coefficients in a certain frequency range, which is considered to characterize noise in a given environment (e.g., the highest frequency for white noise or a given range of frequencies for machine-generated noise in an airplane cockpit if the pilot's voice is the signal of interest). Those wavelet coefficients along with other unchanged wavelet coefficients are then used to reconstruct the signal with noise removed. Data compression is usually done by retaining the wavelet coefficients with the large magnitudes or the wavelet coefficients at certain frequencies that are considered to represent the signal. Those wavelet coefficients and other wavelet coefficients with the value of zero are used to reconstruct the signal data. If the signal data are transmitted from one place to another place and both places know the given frequencies that contain the signal, only a small set of wavelet coefficients in the given frequencies need to be transmitted to achieve data compression. Edge detection is to look for the largest wavelet coefficients and use their time locations and frequencies to detect the largest change(s) or discontinuities in data (e.g., a sharp edge between a light shade to a dark shade in an image to detect an object such as a person in a hallway).

Exercises

20.1 Perform the Haar wavelet transform of time series data 2.5, 0.5, 4.5, 2.5, −1, 1, 2, 6 and explain the meaning of each coefficient in the result of the Haar wavelet transform.

20.2 The Haar wavelet transform of given time series data produces the following wavelet coefficients:

$$f(x) = 2.25\varphi(x)$$
$$+0.25\psi(x)$$
$$-1\psi(2x) - 2\psi(2x-1)$$
$$+\psi(2^2 x) + \psi(2^2 x - 1) - \psi(2^2 x - 2) - 2\psi(2^2 x - 3).$$

Reconstruct the original time series data using these coefficients.

20.3 After setting the zero value to the coefficients whose absolute value is smaller than 1.5 in the Haar wavelet transform from Exercise 20.2, we have the following wavelet coefficients:

$$f(x) = 2.25\varphi(x)$$
$$+0\psi(x)$$
$$+0\psi(2x) - 2\psi(2x-1)$$
$$+0\psi(2^2 x) + 0\psi(2^2 x - 1) + 0\psi(2^2 x - 2) - 2\psi(2^2 x - 3).$$

Reconstruct the time series data using these coefficients.

References

Agrawal, R. and Srikant, R. 1994. Fast algorithms for mining association rules in large databases. In *Proceedings of the 20th International Conference on Very Large Data Bases*, Santiago, Chile, pp. 487–499.

Bishop, C. M. 2006. *Pattern Recognition and Machine Learning*. New York: Springer.

Boggess, A. and Narcowich, F. J. 2001. *The First Course in Wavelets with Fourier Analysis*. Upper Saddle River, NJ: Prentice Hall.

Box, G.E.P. and Jenkins, G. 1976. *Time Series Analysis: Forecasting and Control*. Oakland, CA: Holden-Day.

Breiman, L., Friedman, J. H., Olshen, R. A., and Stone, C. J. 1984. *Classification and Regression Trees*. Boca Raton, FL: CRC Press.

Bryc, W. 1995. *The Normal Distribution: Characterizations with Applications*. New York: Springer-Verlag.

Burges, C. J. C. 1998. A tutorial on support vector machines for pattern recognition. *Data Mining and Knowledge Discovery*, 2, 121–167.

Chou, Y.-M., Mason, R. L., and Young, J. C. 1999. Power comparisons for a Hotelling's T^2 statistic. *Communications of Statistical Simulation*, 28(4), 1031–1050.

Daubechies, I. 1990. The wavelet transform, time-frequency localization and signal analysis. *IEEE Transactions on Information Theory*, 36(5), 96–101.

Davis, G. A. 2003. Bayesian reconstruction of traffic accidents. *Law, Probability and Risk*, 2(2), 69–89.

Díez, F. J., Mira, J., Iturralde, E., and Zubillaga, S. 1997. DIAVAL, a Bayesian expert system for echocardiography. *Artificial Intelligence in Medicine*, 10, 59–73.

Emran, S. M. and Ye, N. 2002. Robustness of chi-square and Canberra techniques in detecting intrusions into information systems. *Quality and Reliability Engineering International*, 18(1), 19–28.

Ester, M., Kriegel, H.-P., Sander, J., and Xu, X. 1996. A density-based algorithm for discovering clusters in large spatial databases with noise. In E. Simoudis, J. Han, U. M. Fayyad (eds.) *Proceedings of the Second International Conference on Knowledge Discovery and Data Mining (KDD-96)*, Portland, OR, AAAI Press, pp. 226–231.

Everitt, B. S. 1979. A Monte Carlo investigation of the Robustness of Hotelling's one- and two-sample T^2 tests. *Journal of American Statistical Association*, 74(365), 48–51.

Frank, A. and Asuncion, A. 2010. UCI machine learning repository. http://archive.ics.uci.edu/ml. Irvine, CA: University of California, School of Information and Computer Science.

Hartigan, J. A. and Hartigan, P. M. 1985. The DIP test of unimodality. *The Annals of Statistics*, 13, 70–84.

Jiang, X. and Cooper, G. F. 2010. A Bayesian spatio-temporal method for disease outbreak detection. *Journal of American Medical Informatics Association*, 17(4), 462–471.

Johnson, R. A. and Wichern, D. W. 1998. *Applied Multivariate Statistical Analysis*. Upper Saddle River, NJ: Prentice Hall.

Kohonen, T. 1982. Self-organized formation of topologically correct feature maps. *Biological Cybernetics*, 43, 59–69.

Kruskal, J. B. 1964a. Multidimensional scaling by optimizing goodness of fit to a nonmetric hypothesis. *Psychometrika*, 29(1), 1–27.

Kruskal, J. B. 1964b. Non-metric multidimensional scaling: A numerical method. *Psychometrika*, 29(1), 115–129.

Li, X. and Ye, N. 2001. Decision tree classifiers for computer intrusion detection. *Journal of Parallel and Distributed Computing Practices*, 4(2), 179–190.

Li, X. and Ye, N. 2002. Grid- and dummy-cluster-based learning of normal and intrusive clusters for computer intrusion detection. *Quality and Reliability Engineering International*, 18(3), 231–242.

Li, X. and Ye, N. 2005. A supervised clustering algorithm for mining normal and intrusive activity patterns in computer intrusion detection. *Knowledge and Information Systems*, 8(4), 498–509.

Li, X. and Ye, N. 2006. A supervised clustering and classification algorithm for mining data with mixed variables. *IEEE Transactions on Systems, Man, and Cybernetics, Part A*, 36(2), 396–406.

Liu, Y. and Weisberg, R. H. 2005. Patterns of ocean current variability on the West Florida Shelf using the self-organizing map. *Journal of Geophysical Research*, 110, C06003, doi:10.1029/2004JC002786.

Luceno, A. 1999. Average run lengths and run length probability distributions for Cuscore charts to control normal mean. *Computational Statistics & Data Analysis*, 32(2), 177–196.

Mason, R. L., Champ, C. W., Tracy, N. D., Wierda, S. J., and Young, J. C. 1997a. Assessment of multivariate process control techniques. *Journal of Quality Technology*, 29(2), 140–143.

Mason, R. L., Tracy, N. D., and Young, J. C. 1995. Decomposition of T^2 for multivariate control chart interpretation. *Journal of Quality Technology*, 27(2), 99–108.

Mason, R. L., Tracy, N. D., and Young, J. C. 1997b. A practical approach for interpreting multivariate T^2 control chart signals. *Journal of Quality Technology*, 29(4), 396–406.

Mason, R. L. and Young, J. C. 1999. Improving the sensitivity of the T^2 statistic in multivariate process control. *Journal of Quality Technology*, 31(2), 155–164.

Montgomery, D. 2001. *Introduction to Statistical Quality Control*, 4th edn. New York: Wiley.

Montgomery, D. C. and Mastrangelo, C. M. 1991. Some statistical process control methods for autocorrelated data. *Journal of Quality Technologies*, 23(3), 179–193.

Neter, J., Kutner, M. H., Nachtsheim, C. J., and Wasserman, W. 1996. *Applied Linear Statistical Models*. Chicago, IL: Irwin.

Osuna, E., Freund, R., and Girosi, F. 1997. Training support vector machines: An application to face detection. In *Proceedings of the 1997 IEEE Computer Society Conference on Computer Vision and Pattern Recognition*, San Juan, Puerto Rico, pp. 130–136.

Pourret, O., Naim, P., and Marcot, B. 2008. *Bayesian Networks: A Practical Guide to Applications*. Chichester, U.K.: Wiley.

Quinlan, J. R. 1986. Induction of decision trees. *Machine Learning*, 1, 81–106.

Rabiner, L. R. 1989. A tutorial on hidden Markov models and selected applications in speech recognition. *Proceedings of the IEEE*, 77(2), 257–286.

Rumelhart, D. E., McClelland, J. L., and the PDP Research Group. 1986. *Parallel Distributed Processing: Explorations in the Microstructure of Cognition, Volume 1: Foundations*. Cambridge, MA: The MIT Press.

Russell, S., Binder, J., Koller, D., and Kanazawa, K. 1995. Local learning in probabilistic networks with hidden variables. In *Proceedings of the Fourteenth International Joint Conference on Artificial Intelligence*, Montreal, Quebec, Canada, pp. 1146–1162.

Ryan, T. P. 1989. *Statistical Methods for Quality Improvement*. New York: John Wiley & Sons.

Sung, K. and Poggio, T. 1998. Example-based learning for view-based human face detection. *IEEE Transactions on Pattern Analysis and Machine Intelligence*, 20(1), 39–51.

Tan, P.-N., Steinbach, M., and Kumar, V. 2006. *Introduction to Data Mining*. Boston, MA: Pearson.

Theodoridis, S. and Koutroumbas, K. 1999. *Pattern Recognition*. San Diego, CA: Academic Press.

Vapnik, V. N. 1989. *Statistical Learning Theory*. New York: John Wiley & Sons.

Vapnik, V. N. 2000. *The Nature of Statistical Learning Theory*. New York: Springer-Verlag.

Vidakovic, B. 1999. *Statistical Modeling by Wavelets*. New York: John Wiley & Sons.

Viterbi, A. J. 1967. Error bounds for convolutional codes and an asymptotically optimum decoding algorithm. *IEEE Transactions on Information Theory*, 13, 260–269.

Witten, I. H., Frank, E., and Hall, M. A. 2011. *Data Mining: Practical Machine Learning Tools and Techniques*. Burlington, MA: Morgan Kaufmann.

Yaffe, R. and McGee, M. 2000. *Introduction to Time Series Analysis and Forecasting*. San Diego, CA: Academic Press.

Ye, N. 1996. Self-adapting decision support for interactive fault diagnosis of manufacturing systems. *International Journal of Computer Integrated Manufacturing*, 9(5), 392–401.

Ye, N. 1997. Objective and consistent analysis of group differences in knowledge representation. *International Journal of Cognitive Ergonomics*, 1(2), 169–187.

Ye, N. 1998. The MDS-ANAVA technique for assessing knowledge representation differences between skill groups. *IEEE Transactions on Systems, Man and Cybernetics*, 28(5), 586–600.

Ye, N. 2003, ed. *The Handbook of Data Mining*. Mahwah, NJ: Lawrence Erlbaum Associates.

Ye, N. 2008. *Secure Computer and Network Systems: Modeling, Analysis and Design*. London, U.K.: John Wiley & Sons.

Ye, N., Borror, C., and Parmar, D. 2003. Scalable chi square distance versus conventional statistical distance for process monitoring with uncorrelated data variables. *Quality and Reliability Engineering International*, 19(6), 505–515.

Ye, N., Borror, C., and Zhang, Y. 2002a. EWMA techniques for computer intrusion detection through anomalous changes in event intensity. *Quality and Reliability Engineering International*, 18(6), 443–451.

Ye, N. and Chen, Q. 2001. An anomaly detection technique based on a chi-square statistic for detecting intrusions into information systems. *Quality and Reliability Engineering International*, 17(2), 105–112.

Ye, N. and Chen, Q. 2003. Computer intrusion detection through EWMA for autocorrelated and uncorrelated data. *IEEE Transactions on Reliability*, 52(1), 73–82.

Ye, N., Chen, Q., and Borror, C. 2004. EWMA forecast of normal system activity for computer intrusion detection. *IEEE Transactions on Reliability*, 53(4), 557–566.

Ye, N., Ehiabor, T., and Zhang, Y. 2002c. First-order versus high-order stochastic models for computer intrusion detection. *Quality and Reliability Engineering International*, 18(3), 243–250.

Ye, N., Emran, S. M., Chen, Q., and Vilbert, S. 2002b. Multivariate statistical analysis of audit trails for host-based intrusion detection. *IEEE Transactions on Computers*, 51(7), 810–820.

Ye, N. and Li, X. 2002. A scalable, incremental learning algorithm for classification problems. *Computers & Industrial Engineering Journal*, 43(4), 677–692.

Ye, N., Li, X., Chen, Q., Emran, S. M., and Xu, M. 2001. Probabilistic techniques for intrusion detection based on computer audit data. *IEEE Transactions on Systems, Man, and Cybernetics*, 31(4), 266–274.

Ye, N., Parmar, D., and Borror, C. M. 2006. A hybrid SPC method with the chi-square distance monitoring procedure for large-scale, complex process data. *Quality and Reliability Engineering International*, 22(4), 393–402.

Ye, N. and Salvendy, G. 1991. Cognitive engineering based knowledge representation in neural networks. *Behaviour & Information Technology*, 10(5), 403–418.

Ye, N. and Salvendy, G. 1994. Quantitative and qualitative differences between experts and novices in chunking computer software knowledge. *International Journal of Human-Computer Interaction*, 6(1), 105–118.

Ye, N., Zhang, Y., and Borror, C. M. 2004b. Robustness of the Markov-chain model for cyber-attack detection. *IEEE Transactions on Reliability*, 53(1), 116–123.

Ye, N. and Zhao, B. 1996. A hybrid intelligent system for fault diagnosis of advanced manufacturing system. *International Journal of Production Research*, 34(2), 555–576.

Ye, N. and Zhao, B. 1997. Automatic setting of article format through neural networks. *International Journal of Human-Computer Interaction*, 9(1), 81–100.

Ye, N., Zhao, B., and Salvendy, G. 1993. Neural-networks-aided fault diagnosis in supervisory control of advanced manufacturing systems. *International Journal of Advanced Manufacturing Technology*, 8, 200–209.

Young, F. W. and Hamer, R. M. 1987. *Multidimensional Scaling: History, Theory, and Applications*. Hillsdale, NJ: Lawrence Erlbaum Associates.

Index

A

Agglomerative hierarchical clustering, 141
Analysis of variance (ANOVA)
ANAVA, *see* Angular analysis of variance (ANAVA)
Angle between two vectors, 119, 120, 221, 222
Angular analysis of variance (ANAVA), 247
ANNs, *see* Artificial neural networks (ANNs)
Anomaly, xi, xii, 9, 14–15, 18, 251, 253, 257, 260, 261, 264, 265, 267, 269–274
Any path method, 291–297, 305
Apriori algorithm, 190, 192, 194–195
Artificial neural networks (ANNs)
architectures, 69–71, 86–88
back-propagation learning method, 80–86
determining connection weights, 71–80
processing units, 63–69
Association, xi, xii, 6, 9, 12–13, 18, 80, 85, 197, 198, 206, 210–211, 224, 225, 227
Association rule discovery, 189–194
Association rules
discovery, 189–194
and measures, 185–189
Attribute variable, xi, xii, 4–12, 16–18, 21, 22, 25, 26, 32–34, 36–40, 47, 49–51, 56–61, 88, 117, 118, 120–123, 136, 137, 147, 152, 154, 166, 170, 176, 177, 197–199, 217, 219, 271, 273
Attribute *vs.* target variables, 5–8
Autocorrelation, xiii, 17, 260, 261, 277–285
Autocorrelation function (ACF) coefficient, 278, 281–283, 285
Autoregressive and moving average (ARMA) model
ACF and PACF characteristics, 281–283
stationary series data, 279–281
transformations of nonstationary series data and, 283–285
Autoregressive, integrated, moving average (ARIMA) model, 277, 283–284
Autoregressive (AR) model, 279–282

B

Back-propagation learning, 63, 72, 79–86, 88, 89, 236
Baum–Welch reestimation method, 294–295
Bayes classifier, 31–35, 197
Bayesian network
probabilistic inference, 205–210
structure, 197–204
Bayes theorem, 31–35
Bellman's principle, 293
Best path method, 292–295, 305
Bias, 63, 67, 72–86, 88, 89, 91, 113, 123, 124

C

Categorical variable, 8–9, 12, 17, 18, 36, 37, 51, 56, 57, 91
Categorical *vs.* numeric variables, 8–9
Chi-square control charts, 272–274
Chi-square statistic, 273
Classification, xi, 9, 31, 37, 63, 91, 117, 185, 288
Classification and prediction patterns, 9–12
Cluster and association patterns, 12–13
Cluster distance, 141–142, 148–151, 153–157, 159–161, 163, 164, 166

323

Clustering
  density-based, xii, 13, 153–166
  hierarchical, xii, xiii, 13, 141–152, 164
  K-means, xii, 13, 153–166
Cluster linkage method
  average linkage, 141, 142, 146, 152
  centroid linkage, 141, 145, 146, 150, 151
  complete linkage, 141, 144, 146, 152
  single linkage, 141, 144, 146, 148, 149, 151, 152
Confidence, see Measure of association
Connection weight, 63, 66, 67, 71–80, 91, 167–170, 173, 236
Control limit
  lower, 253
  3-sigma, 253, 255, 265, 273
  upper, 253
Correlation coefficient, see Dissimilarity measure
Cosine similarity, see Dissimilarity measure
Counter-relationship, 270, 273–274
Covariance, 21, 119, 217–220, 223, 228, 229, 269–273
Criterion
  learning stop criterion, 85
Cumulative sum (CUSUM) control chart, see Univariate control chart
Cuscore control chart, see Univariate control chart
Cyber attack detection, 17, 122, 123, 136, 182, 184, 266, 274, 289, 305

D

Data compression, 317, 318
Data homogeneity, 40, 42–44, 46, 48, 49, 51, 55, 57, 58
Data mining, xi–xiv, 3–18, 31, 37, 61, 88, 91, 150, 210–211, 287
Data patterns
  classification and prediction patterns, 9–12
  cluster and association patterns, 12–13
  data reduction patterns, 13–14
  outlier and anomaly patterns, 14–15
  random fluctuation pattern, 179–181
  sequential and temporal patterns, 15–17

  spike pattern, 179–181
  steady change pattern, 179–181
  step change pattern, 179–181
Data record, xii, 3, 9–13, 17, 18, 34, 37–40, 42–44, 46, 49–51, 57–59, 66–69, 73, 75–78, 80–83, 85, 88, 89, 141–144, 147, 148, 185–190, 194, 195, 211, 217, 238
Data reduction patterns, xi, xii, 9, 13–14, 18
Data variables
  attribute vs. target variables, 5–8
  categorical vs. numeric variables, 8–9
DBSCAN, 165, 166
Decision boundary, 72–75, 77–79, 88, 93–94, 97
Decision threshold, 255, 257, 265, 266
Decision tree
  algorithm, 59–61
  binary, 37–51
  non-binary, 37, 51–56
Dendrogram, 149
Density-based clustering, see Clustering
Dependence of variables, 5–6
Determinant, 223
Deterministic trend, 278, 283
Dilation effect, 308
Dimension reduction, 13
Directed, acyclic graph, 198
Dissimilarity measure
  correlation coefficient, 117, 120
  Cosine similarity, 117, 120, 136, 166
  Euclidean distance, 117, 118, 120, 125, 136, 142, 154, 166, 234
  hamming distance, 117, 118
  Minkowski distance, 117, 118, 234
DNA sequence recognition, 293

E

Edge detection, 317, 318
Eigenvalue, 223–229
Eigenvector, 224–229
Elements of decision tree, 37–39
Emission probability, 291, 293, 294, 305
Empirical risk of classification, 92, 96
Error
  classification, 91, 92, 96, 113, 114
  output, 78, 80, 81, 85
  prediction, 87, 260, 261

Index

Error-correcting output coding method, 113
Euclidean distance, *see* Dissimilarity measure
Expectation maximization (EM) method, 294, 297, 299
Expected risk of classification, 91, 92
Exponentially weighted moving average (EWMA) control chart, 251, 252, 254, 257–261, 265, 267, 269, 272

F

False alarm rate, 253, 255, 265, 266
Function
    AND function, 66, 67, 74–76, 92, 101
    Gaussian radial basis, 111, 112
    hard limit, 64, 71, 72
    hyperbolic tangent, 65–67, 71
    kernel, 91, 109–114
    linear, 23, 65, 82, 89
    linearly separable, 73, 79
    neighborhood, 168, 170, 175, 176
    nonlinearly separable, 79
    NOT function, 79, 80, 88, 114
    OR function, 68, 70, 114
    polynomial, 111, 112
    sigmoid, 65, 71, 81, 82, 89
    sign, 64, 66–67, 71–73
    XOR function, 70–71, 79, 80, 82, 83, 89, 92, 105

G

Gauss–Newton method, 28
Generalization, 11–12, 17, 22, 28, 39, 40, 71, 75, 78, 87, 91, 92, 97, 208, 210, 234, 247, 253, 280, 308, 310
Gini-index, 40, 44–46, 48, 49, 57
Goodness-of-fit, 236, 241
Gradient decent search, 28, 80–81, 211

H

Hamming distance, *see* Dissimilarity measure
Handwritten character recognition, 293
Hidden layer, 70, 71, 79, 82
Hidden Markov model, xiii, 17, 287–305
Hidden units, 71, 79, 82, 84, 86–89, 114
Hierarchical clustering; *see also* Clustering
    agglomerative, 141
    nonmonotonic tree, 150–151
    procedure, 146–150
Histogram, 14, 177, 178, 180–182, 184
Hit rate, 253, 265, 266
Homogeneity, *see* Data homogeneity
Hotelling's T^2 statistic, 269–272

I

In-control process, 253, 262–264
Independence of variables, 5, 32, 197, 210–211, 257, 262, 273, 274
Individual difference scaling (INDSCALE), 247–248
Information entropy, 40–44, 46, 47, 49, 51–57, 60
Initial state probability, 288, 290–291, 294, 305
Input, 63, 66–73, 75–79, 81–83, 85, 89, 110, 111, 113, 123, 124, 167, 168, 247
Interval variable, *see* Variable
Inverse of a matrix, 25, 222, 272–273
Item set, 185–195

K

Karush–Kuhn–Tucker condition, 98, 107
Kernel function, *see* Function
K-means clustering, *see* Clustering
K-nearest neighbor classifier, xii, 12, 117–137
Knowledge organization, 247

L

Lagrange multiplier, 98, 99, 103, 106
Learning hidden Markov models, 294–304
Learning method, 71–73, 76–80, 89
Learning rate, 78, 81, 83, 85–86, 88, 89, 169, 170, 212, 237
Least squares method, 21, 23–28, 283
Length of a vector, 221

Lift, *see* Measure of association
Likelihood function, 24
Linear classifier, 93–108
Linearly separable problem, 91–96, 113
Linear regression models, *see* Regression model
Logistic regression model, *see* Regression model

M

MAP, *see* Maximum a posterior (MAP)
Marginalization, 203, 205, 206, 208, 210
Markov chain models, xiii, 17, 287–305
Matrix
    algebra, 220–227
    identity, 222, 223
    positive, definite, 227
    square, 25, 222–224
    symmetric, 222, 226, 227
Maximum a posterior (MAP)
    classification, 31, 32
    probability, 32
Maximum likelihood method, 21, 23–28, 283
Maximum likelihood (ML) probability, 32, 283
MDS, *see* Multidimensional scaling (MDS)
Mean, 21, 24, 85, 118, 123–125, 144, 153, 154, 179, 182, 201, 217–220, 251–255, 261–264, 269, 271, 273, 278–280, 283
Mean shift, 264, 270–271, 273
Measure of association
    confidence, 186, 189
    lift, 186, 189
    support, 186, 189
Minimum description length, 37, 39–40, 92
Minkowski distance, *see* Dissimilarity measure
Missing data, 57, 270
Mode
    dip test, 183, 184
Monotone regression algorithm, 235, 236, 239–241
Monotonic tree of hierarchical clustering, *see* Non-monotonic tree of hierarchical clustering

Moving average (MA) model, 280
Multidimensional scaling (MDS)
    algorithm, 233–246
    INDSCALE, 247–248
    number of dimensions, 246–247
Multilayer feedforward artificial neural network, 63. 80–86
Multivariate control chart
    chi-square control chart, 269, 272–274
    Hotelling's T^2 control chart, 269–274
    multivariate EWMA control chart, 269, 272
Multivariate EWMA control charts, 272
Multivariate statistics, 217–220

N

Naïve Bayes classifier, 31–35
Natural language processing, 293
Neighborhood function, *see* Function
Net sum, 63, 64, 82
Noise reduction and filtering, 317
Nominal variable, *see* Variable
Nonbinary decision tree, 51–56
Nonlinear classifier, 92, 113, 114
Nonlinearly separable problem, 91, 92, 105, 108, 114
Nonlinear regression models, 28
Non-monotonic tree of hierarchical clustering, 151
Nonstationarity, 277–279
Normal distribution, 24, 177, 179, 182, 183, 252, 253, 262, 263, 269, 270, 273, 278
Normalization, 118, 136, 137, 142, 152, 166, 176, 212
Numeric variable, *see* Variable

O

One-step ahead prediction model, 260, 261
Optimization, 91–92, 98, 100–104, 114
Ordinal variable, *see* Variable
Orthogonal vectors, 73, 74, 120, 222
Outlier, xi, xii, 9, 14–15, 18, 122, 251, 253, 278, 283
Outlier and anomaly patterns, 14–15
Out-of-control process, 253, 262, 263

Index

Output, 63, 64, 66–71, 73–81, 85, 89, 113, 167–169, 171, 173, 247
Output units, 63, 64, 66–69, 72, 73, 75, 76, 79–83
Overfitting, 87

P

Parameter estimation, 23–29
Partial autocorrelation function (PACF) coefficient, 278
PCA, see Principal component analysis (PCA)
Perceptron, 63, 71–80, 88, 89, 93, 111
Prediction, xi–xii, 9–12, 17, 21, 26, 29, 37, 59, 63–90, 235, 260–262, 283–284
Principal component, 111, 228–231, 274
Principal component analysis (PCA), xii, 14, 111, 217–231, 274
Probabilistic inference, 197, 205–210, 212
Probability
    conditional, 31, 199, 201, 203, 208, 210, 211
    joint, 201–203, 206, 208, 217–219, 263, 288, 289
    posterior, 32, 205, 208, 209, 213
    prior, 32, 201, 203, 208, 210, 211
Probability density function, 177, 183, 217, 218, 252, 263, 269
Probability distribution
    bimodal distribution, 179, 183
    multimodal distribution, 179, 183
    normal distribution, 24, 179, 182–184
    skewed distribution, 179, 182, 183
    uniform distribution, 179, 183
    univariate data, 177–182

Q

Quadratic programming, 91, 92, 96–105

R

Radius, 111, 112, 165, 166
Random walk, 278, 283
Ratio variable, see Variable
Receiver operating curve (ROC), 265–266
Reconstruction of time series data, 316–318

Recurrent ANN, 69, 71
Regression model
    linear, xii, 21–29
    logistic, 28
    nonlinear, xii, 21–29
Regression tree, xii, 12, 37–61
Residual, 260, 262, 283–284
R measure of data homogeneity, 57
ROC, see Receiver operating curve (ROC)

S

Scaling function, 307–309, 311, 314, 317
Seasonable cycle, 284
Self-organizing map, xii, 13, 167–176
Sequential and temporal patterns, 15–17
Sequential pattern, 9, 15–17
Shewhart control charts, 251–254, 261, 265
Shift effect, 308
Similarity measure, see Dissimilarity measure
Skewness, 182–184
Small data sets, 3–5
Soft margin support vector machine (SVM), 105, 108–110
Software
    Bayes Server, 213
    C4.5, 61
    CART, 61
    Hidden Markov Model Tookit (HTK), 305
    HUGIN, 210, 213
    IBM Intelligent Miner, 61
    Matlab, xiv, 35, 88, 101, 103, 114, 152, 165, 175, 284, 317
    Minitab, 267
    R, 183
    SAS
        SAS Enterprise Miner, 61
    SPSS
        SPSS AnswerTree, 61
    Statistica, 15, 28, 29, 152, 182, 183, 194, 211, 217, 230, 248, 251, 253, 267, 269, 270, 277, 278, 283, 317
    Weka, 35, 61, 88, 165, 175, 176, 194
Spectral decomposition of a matrix, 226
Speech recognition, 305

Split selection methods, 37, 40–44
SSE, *see* Sum of squared errors (SSE)
State transition probability, 287–291, 294, 305
Stationarity
   and nonstationarity, 278–279
   weak, 278, 279
Stress, 3, 5, 6, 8–11, 26, 29, 36, 63, 136, 234, 236, 239, 241, 246, 247, 256, 257, 259, 260
Structural risk minimization principle, 92, 93, 96
Sum of squared errors (SSE), 23, 29, 154
Supervised clustering, xii, 12, 117–137
Support, *see* Measure of association
Support vector, 98, 100, 103, 104, 108
Support vector machine, xii, xiii, 12, 91–115
Support vector machine (SVM)
   vs. ANN, 113–114
   geometric interpretation, 96–97
   linear classifier
      and linearly separable problem, 93–96
      and nonlinearly separable problem, 105–108
      quadratic programming problem, 98–105
   multi-class classification problems, 113
   nonlinear classifier and nonlinearly separable problem, 108–112
SVM, *see* Support vector machine (SVM)
Symmetry, 179, 182, 183, 222, 226, 227

T

Target variable, *see* Variable
Temporal pattern, xi, xiii, 9, 15, 17
Tensor product, 112
Test data, 16, 17
Time-frequency domain, 307, 309–311, 314, 317
Time lag, 278, 284
Time series analysis, xiii, 17, 277–285
Top-down construction, decision tree, 44–49

Training data, 17, 23, 32, 33, 37–39, 44, 49, 51, 56, 57, 73, 76, 78–80, 85–89, 91, 98, 101, 113–115, 120–125, 127–130, 136, 168, 198, 210, 211, 261, 288, 289, 294
Transfer function, 64–67, 70–73, 81, 82, 89
Transformation of a numeric variable to a categorical variable, 56–57
Transformation of nonstationary series data to stationary series data
   Box–Cox transformation, 284
   de-trending, 283
   differencing, 277, 283
   Log transformation, 284
   power transformation, 284

U

Univariate control chart
   attribute control chart, 251
   cumulative sum (CUSUM) control chart, 251, 254–257, 261, 264–267
   cuscore control chart, 251, 261–265, 267
   exponentially weighted moving average (EWMA) control chart, 251, 252, 254, 257–261, 265, 267
   multivariate control chart, xii, 15, 251, 263
   shewhart control chart, 251–254, 261, 265
   univariate control chart, xii, 15, 251–267, 270
   variable control chart, 251
   $\bar{x}$-control chart, 251–255, 265
Univariate data, xii, 13, 177–184

V

Variable
   attribute, xi, 5–12, 17, 18, 21, 22, 25, 26, 32–34, 36–40, 47, 49, 51, 56–61, 88, 117, 118, 120, 122, 123, 136, 137, 147, 152, 154, 166, 170, 176, 177, 197–199, 217, 219, 271, 273

categorical, 8–9, 11–12, 17, 18, 36, 37, 51, 56, 57
interval, 9
nominal, 8, 9
numeric, 8–9, 12, 17, 18, 37, 56–59, 184
ordinal, 8
ratio, 9
target, xi–xii, 5–12, 16, 17, 21, 26, 33, 34, 36–39, 50, 51, 57–59, 88, 117, 120, 121, 136, 197–199
Variance, 21, 24, 118, 182, 217, 219, 220, 228, 251, 270, 273, 278, 279, 283, 284
Variance-covariance matrix, 218, 219, 223, 228, 229, 269, 271–273
VC dimension, 92, 93, 96, 106
Viterbi algorithm, 293

W

Wavelet
  Daubechies wavelet, 307, 314, 315, 317
  definition, 307–308
  derivative of Gaussian (DoG) wavelet, 307
  function, 307–314
  Haar wavelet, 307–309, 311, 313, 314, 317, 318
  Morlet wavelet, 314, 315, 317
  Paul wavelet, 314, 315, 317
  time series data, 309–317
  transform, 307, 309–315, 317, 318
Weight, 63, 66, 67, 69, 71–86, 88, 89, 91, 113, 167–171, 173–175, 236, 247–248, 251, 257, 258, 261, 272